Implementing Multifactor Authentication

Protect your applications from cyberattacks with the help of MFA

Marco Fanti

BIRMINGHAM—MUMBAI

Implementing Multifactor Authentication

Group Product Manager: Mohd Riyan Khan

Publishing Product Manager: Prachi Sawant

Senior Editor: Runcil Rebello

Technical Editor: Arjun Varma

Copy Editor: Safis Editing

Project Coordinator: Ashwin Kharwa

Proofreader: Safis Editing

Indexer: Rekha Nair

Production Designer: Shankar Kalbhor

Marketing Coordinator: Agnes D'souza

First published: June 2023

Production reference: 1300523

Published by Packt Publishing Ltd.
Livery Place
35 Livery Street
Birmingham
B3 2PB, UK.

ISBN 978-1-80324-696-3

www.packtpub.com

To my extraordinary wife, who has always believed in me and supported my dreams.
To my two daughters, who are my greatest joy. To my inspiring mother, who instilled in me the conviction that everything is possible and equipped me with the wings to soar through any challenge. And to my father, who taught me the invaluable lesson of never ceasing to learn and embracing the endless wonders of knowledge.

This book is lovingly dedicated to each of you with immeasurable gratitude and affection.

I could not have written it without your passion and support.

Thank you for everything.

– Marco Fanti

Contributors

About the author

Marco Fanti's career skyrocketed from software engineering to cybersecurity as he discovered his passion for inventing innovative security tools. A prominent figure in the security community, he has collaborated with start-ups such as enCommerce and BehavioSec and giants such as Oracle and Accenture to create products that protect millions worldwide. A lifelong learner, Marco holds two MSc degrees (NYIT and NYU) and an MBA (UF), which enable him to craft bespoke solutions for clients by fusing the best features of various products. Originally from Brazil, Marco lives in Florida with his wife, perpetually exploring the cybersecurity frontier.

About the reviewers

Harvinder Nagpal has over 18 years of experience in workforce **Identity and Access Management (IAM)** within the broader cybersecurity realm. He has worked with customers of different sizes and various industry verticals to secure access to their on-premises and cloud-based applications, resources, and services on their cloud transformation journey. As a CISSP-certified cybersecurity practitioner, he has worked chiefly for product-based employers in different roles and responsibilities focused on product and customer adoption to improve customers' information security posture.

Ameya Khankar is a highly regarded and trusted business technology and cybersecurity professional focusing on the areas of technology risk, enterprise transformations, and digital governance. He advises large global enterprises in the US and globally as an expert on enterprise technology risks with a deep focus on strategies to strengthen their cybersecurity posture. He has advised a $4 billion organization in the past in defining their business transformation enterprise security strategy. Currently, he is advising a $9 billion organization to meet complex digital transformation and cybersecurity regulatory requirements.

Table of Contents

Part 2: Implementing Multifactor Authentication

3

4

5

Access Management with ForgeRock and Behavioral Biometrics   173

6

Federated SSO with PingFederate and 1Kosmos   221

7

MFA and the Cloud – Using MFA with Amazon Web Services   267

Part 3: Proven Implementation Strategies and Deploying Cutting-Edge Technologies

10

Appendix C

Installing Apache Tomcat Software   497

Preface

Unlock the full potential of **Multifactor Authentication** (**MFA**) with this dynamic, hands-on guide! Immerse yourself in a practical, engaging learning experience that will have you mastering multiple authentication products in no time. Then, prepare to enhance your applications' security and reduce the risk of cyber threats.

Embark on an adventure with Acme Software, a fictional company navigating the complex world of authentication products and mechanisms. Witness its pursuit of the perfect balance between security, risks, costs, and user experience. Learn from its journey and make empowered decisions to fortify your digital fortress.

This book offers step-by-step explanations, practical examples, and hands-on implementations of MFA concepts and technologies. Curious about **Identity and Access Management** (**IAM**)? We've got you covered! The book delves into IAM products with crystal-clear explanations that will help turn you into an expert.

Explore a diverse array of IAM products and enable secure **Single Sign-On** (**SSO**) for your enterprise and your customer-facing applications.

Witness as Acme Software explores the ideal products for its users, partners, and customers. To help you with your own learning experience, we provide instructions on obtaining free trial versions of the products used in the examples and how to build SaaS applications that use the security provided by the solutions demonstrated in each chapter!

By the end of this thrilling guide, you'll have the power to choose, deploy, and maintain an MFA solution that slashes the risk of successful malicious attacks during user authentication. So, join the ranks of cybersecurity champions and protect your digital realm with confidence!

Who this book is for

The target audience for this book includes the following:

- **IT professionals**: System administrators, network administrators, security engineers, and other IT staff responsible for implementing and maintaining secure authentication systems would benefit from a comprehensive understanding of MFA

- **Cybersecurity experts**: Professionals working in the cybersecurity field, including security consultants, researchers, and analysts, would find the book valuable to deepen their knowledge of MFA and stay current on best practices for protecting sensitive data and systems

- **Developers**: Software developers and engineers who build applications requiring secure authentication mechanisms can benefit from understanding MFA, its best practices, and how to integrate it into their applications effectively
- **Business decision-makers**: Executives, managers, and business owners responsible for the security of their organization's data and infrastructure can use the book to learn about MFA and make informed decisions regarding its implementation

Some chapters include examples of SaaS applications built using SDKs for the different authentication products. Although detailed instructions for building and deploying the applications are covered in the book, having a foundation in programming will make it easier for readers to grasp the content and apply the knowledge to their own projects.

What this book covers

Chapter 1, On the Internet, Nobody Knows You're a Dog

This inaugural chapter provides the fundamental groundwork for understanding the dynamic realm of digital identities and MFA. We begin with an overview of the concept of identity, both in its traditional and digital manifestations, and then delve into the nuances of two fundamental types of digital identity – workforce and customer identity. Critical in today's digitized world, these forms of identity offer unique challenges and opportunities that businesses must navigate effectively.

Next, we focus on the foundational pillar of digital security – authentication factors. These factors, which include something you know (such as a password), something you have (such as a token or a card), and something you are (such as a biometric characteristic that you have), make up the core of MFA.

The chapter continues with an introduction to the basic concepts and terminology related to digital identity and MFA. This vocabulary is relevant for the remaining chapters and for anyone aiming to understand and work in cybersecurity.

Finally, we delve into the concept of MFA in more detail, explaining its importance in contemporary cybersecurity strategies, how it operates, and why it has become the go-to solution for businesses and individuals seeking to enhance their digital security.

Chapter 2, When to Use Different Types of MFA

In *Chapter 2*, we dive deeper into the multifaceted nature of MFA. Recognizing that not all MFA solutions are created equal is critical, and we explore the contexts in which different types of MFA are most effectively utilized.

Given the rapidly evolving landscape of cybersecurity, the chapter also emphasizes the importance of staying up to date. We acknowledge that cyber criminals, or *bad actors*, always look for vulnerabilities and continually update their strategies. Thus, we present reliable sources of information for keeping pace with these changes.

Chapter 3, Preventing 99.9% of Attacks – MFA with Azure AD and Duo

Chapter 3 comprehensively explores **Azure Active Directory** (**Azure AD**) and how Acme Software can leverage it to improve its workforce's user management and security practices. As the cornerstone of Microsoft 365, Azure AD provides a robust, cloud-based IAM solution that caters to the company's needs, from centralizing user and group management to enforcing advanced security measures.

We commence by establishing the foundations of Azure AD, showcasing its essential benefits such as secure authentication, SSO capabilities, **conditional access** (**CA**), and MFA. The focus then shifts to the challenges associated with traditional password-based authentication. Drawing upon Microsoft's research, we delve into why passwords alone aren't sufficient for securing accounts and data, underscoring the necessity of MFA in the security equation.

From there, we guide readers through configuring Azure AD, presenting diverse authentication workflows tailored to different organizational roles' risk levels. Given the sophistication of attacks on passwords and MFA, we also demonstrate how to configure different authenticators, thereby enabling authorized individuals' access to sensitive company resources.

In recognition of the modern work-from-anywhere culture and the increasing prevalence of BYOD policies, we describe how Acme Software can employ Azure AD to ensure consistent security across applications accessed both internally and on public networks. We also introduce Duo, a **Two-Factor Authentication** (**2FA**) product from Duo Security.

Chapter 4, Implementing Workforce and Customer Authentication Using Okta

Chapter 4 takes a deep dive into Okta, a leading cloud-based identity management system offering two distinguished IAM products: Okta Workforce Identity and Okta Customer Identity. These products bring unique benefits to businesses, from comprehensive workforce management to secure customer interactions.

The first part of the chapter focuses on Okta Workforce Identity. This solution offers businesses an efficient way to manage and protect their workforce users, such as employees, contractors, and partners, from a single platform. First, we delve into its capabilities and discuss how its use allows businesses to maintain regulatory compliance while achieving their objectives. Next, we illustrate its implementation using a case study involving Acme's workforce applications, exploring the configuration and use of additional authenticators, with Duo as the authenticator of choice.

In the second part of the chapter, we switch focus to Okta Customer Identity. This tool enables businesses to securely manage end user identities and create frictionless application registration and login experiences. In addition, this solution provides businesses with the capacity to integrate authentication seamlessly into any cloud-based application. We delve into its features and demonstrate its use by exploring the development of MFA for customer-facing applications.

Chapter 5, Access Management with ForgeRock and Behavioral Biometrics

Chapter 5 takes us on a journey through the offerings of ForgeRock, another leading IAM solutions provider. In this chapter, we focus on how businesses such as Acme can effectively leverage ForgeRock's solutions to enhance the customer experience for external users while securing and enabling an agile workforce.

We start the chapter by taking readers through the experience of using ForgeRock.

Our next stop is the exploration of authentication trees, a noteworthy feature in ForgeRock's suite of solutions. Authentication trees offer a flexible and customizable approach to authentication that allows businesses to design their unique user journey, enhancing security and user experience.

Lastly, we delve into the innovative world of behavioral biometrics, a technology that brings a new level of security by studying the user's behavior during the login process. This cutting-edge technology enables businesses to increase security and reduce friction during the authentication process, providing a seamless blend of security and user convenience.

Chapter 6, Federated SSO with PingFederate and 1Kosmos

Chapter 6 is dedicated to a comprehensive exploration of PingFederate, a versatile solution for user authentication and SSO. In this chapter, we also introduce 1Kosmos, a provider of passwordless MFA that offers an improved, frictionless, and secure experience for workforce users.

We start the chapter by providing an overview of PingFederate and its ability to facilitate federated SSO. This allows users to access multiple applications with single login credentials, significantly enhancing user convenience without compromising security.

The chapter then pivots to introduce the concept of passwordless authentication – a technology that seeks to eliminate passwords as a point of vulnerability in the security architecture. We delve into how this innovative approach can enhance user experiences while maintaining high security standards.

Finally, we introduce 1Kosmos and its unique contribution to passwordless MFA with verified identities. 1Kosmos not only removes the password from the equation but also verifies the identity of users through robust biometric checks, adding another layer of security.

Chapter 7, MFA and the Cloud – Using MFA with Amazon Web Services

Chapter 7 introduces how businesses such as Acme can leverage **Amazon Web Services** (**AWS**) for their IAM needs. Given the trend of companies increasingly adopting cloud platforms to develop and deploy their products and services, understanding AWS's IAM services is crucial for workforce and customer enablement.

We introduce AWS IAM, explain its features and capabilities, and demonstrate how it can help businesses manage and secure access to their AWS resources effectively.

Next, we shift focus to workforce users. We discuss how AWS can be utilized to manage and protect workforce identities, ensuring secure access to necessary resources while maintaining the ease of operation for users.

Finally, we discuss Amazon Cognito, an AWS service that enables easy and secure user sign-up and sign-in. We cover how Cognito can be leveraged to authorize Acme's customers and end users, providing a seamless and secure user experience.

Chapter 8, Google Cloud Platform and MFA

Chapter 8 concludes our exploration of the *big three* cloud platform service providers—AWS, Microsoft Azure, and **Google Cloud Platform** (**GCP**)—each bringing unique strengths. In this chapter, we focus on GCP, rounding out our coverage of these dominant players in the cloud computing market.

We've previously delved into AWS and Azure, highlighting their unique offerings and applicability to businesses such as Acme. Now, we turn our attention to GCP, which prides itself on its machine learning and data analytics capabilities among its cloud services.

This chapter discusses Google Cloud Identity, examining its features and capabilities and how it fits into the overall landscape of GCP's cloud services. We also touch on the Google Cloud Identity Platform, GCP's robust IAM solution, which enables businesses to manage their user identities seamlessly across their applications.

Chapter 9, MFA without Commercial Products – Doing It All Yourself with Keycloak

Chapter 9 introduces readers to Keycloak, an open source IAM solution. As Acme Software seeks to explore options beyond traditional commercial products for its expanding IAM infrastructure, Keycloak offers a viable, cost-effective, and flexible alternative. This chapter aims to help Acme and readers understand Keycloak's potential to streamline authentication and authorization processes for the workforce and customers.

We begin by defining the Keycloak server, explaining its role in IAM, and elucidating its core features.

The chapter then explores the functionalities of Keycloak's administration console, providing insights into the flexibility and control it offers. We delve into using Keycloak for SSO, a feature that enhances user convenience and security.

Keycloak's MFA capabilities are also investigated, underscoring the software's commitment to robust security. By comparing Keycloak to other commercial products, readers will gain a comprehensive perspective on its relative strengths and areas of consideration.

Chapter 10, Implementing MFA in the Real World

Chapter 10 steers the reader toward a deeper understanding of cybersecurity from a business perspective. We explore the business implications of cybersecurity, its role in safeguarding organizational assets, and the associated legal and ethical responsibilities of an organization's leadership.

Firstly, we delve into the business side of cybersecurity, discussing the importance of authentication and the broader impact of cybersecurity on business functions. Next, we articulate how cybersecurity, far from being a mere technical concern, is intrinsically tied to a business's viability and reputation.

Subsequently, we provide insights on how to bolster cybersecurity within organizations. This section delves into proactive measures that businesses can adopt to stay ahead of emerging cybersecurity threats.

Finally, we offer practical strategies for implementing MFA in real-world settings. By highlighting the best practices and potential pitfalls, this chapter provides a roadmap for businesses to effectively leverage MFA to enhance their cybersecurity posture.

Chapter 11, The Future of (Multifactor) Authentication

Chapter 11, our final chapter, takes you on an expedition into the future, exploring how the emergence of Web 3.0 will reshape the landscape of digital identity and authentication. As we stand on the precipice of this digital revolution, we investigate the transformation that will ensue, emphasizing the implications for security, privacy, and user experience.

First, we introduce the concept of the Web 3.0 ecosystem, explaining its decentralization philosophy and how it will influence the nature of digital identity. Then, we discuss how **Personally Identifiable Information** (**PII**) will become more significant and unique in human and machine interactions in this new world.

We then delve into product trends, analyzing emerging technologies, such as verifiable credentials and innovative authentication mechanisms powered by blockchain and smart contracts.

Our exploration continues with the future of MFA, addressing topics such as passkey management, continuous authentication, and the potential of passkeys as a phishing-resistant MFA offering.

Chapter 11 culminates by pondering *what lies ahead*, leaving readers with a sense of anticipation and a broader understanding of the exciting possibilities that Web 3.0 brings to digital identity and authentication. This final chapter provides a peek into the future and equips readers with the knowledge required to adapt to and embrace the transformative wave of Web 3.0.

To get the most out of this book

This book assumes that readers are already familiar with at least one of the cloud platform service providers discussed – AWS, Microsoft Azure, and GCP. Therefore, we've made a few assumptions about your existing knowledge and experience:

- **Basic understanding of cloud computing**: You should understand the fundamentals of cloud computing, including **Software-as-a-Service** (**SaaS**) concepts

- **Familiarity with one or more cloud platforms**: Experience with AWS, Azure, or GCP is helpful, as many concepts in this book build upon the services and architecture of these platforms

Products covered in the book	Requirements
Azure AD	Azure AD Premium P1
Okta Workforce Identity	An Okta Standard account
Okta Customer Identity	An Okta Developer or Enterprise account
ForgeRock – Access Manager	A ForgeRock software platform account
Ping Identity – PingFederate	A Ping Identity account
AWS	An AWS root account
Google Cloud Identity	Google Cloud Identity or Workspace account
Keycloak	The Keycloak server's latest version (version 21.1.1 was used in this book)
Duo	A Duo Essentials account
1Kosmos – BlockID	A BlockID account
BehavioSec	A BehavioSec account

Knowing how to utilize Java and Docker for deploying applications is a helpful skill while reading this book. Some products we'll discuss can be locally installed and run as standalone Java applications or as Docker containers:

- **Java Platform**: Many enterprise-level applications, including some IAM products we will explore, are built with Java due to its robustness, portability, and scalability. Understanding how Java applications are deployed will give you a solid grasp of these solutions' underlying structure and functioning.

- **Docker containers**: Docker is a popular platform that uses containerization to package an application and its dependencies into a single object. Docker can simplify the deployment process and eliminate the *"but it works on my machine"* problem, making it an excellent tool for deploying applications for testing and development. Understanding how to pull, run, and manage Docker containers can significantly simplify installing and running the software products discussed in this book.

- **Running servers**: Certain products, such as Keycloak, run on a server that can be initiated using Java or run inside a Docker container. Understanding how to start these servers using either of these methods is essential for installing and testing these products in a local environment.

- **Troubleshooting and customization**: Understanding Java and Docker deployment also aids in troubleshooting any issues that might arise during the installation or operation of the software. Furthermore, if the product is open source, you can customize it to suit your needs better, and understanding the deployment process will be crucial for this.

Download the example code files

We have code bundles from our rich catalog of books and videos available at `https://github.com/PacktPublishing/`. Check them out!

Download the color images

We also provide a PDF file that has color images of the screenshots and diagrams used in this book. You can download it here: `https://packt.link/b4FmL`

Conventions used

There are a number of text conventions used throughout this book.

`Code in text`: Indicates code words in text, database table names, folder names, filenames, file extensions, pathnames, dummy URLs, user input, and Twitter handles. Here is an example: "It should look like this: `https://samltoolkit.azurewebsites.net/SAML/Login/9999`."

A block of code is set as follows:

```
html, body, #map {
  height: 100%;
  margin: 0;
  padding: 0
}
```

When we wish to draw your attention to a particular part of a code block, the relevant lines or items are set in bold:

```
[default]
exten => s,1,Dial(Zap/1|30)
exten => s,2,Voicemail(u100)
exten => s,102,Voicemail(b100)
exten => i,1,Voicemail(s0)
```

Any command-line input or output is written as follows:

```
$ mkdir css
$ cd css
```

Bold: Indicates a new term, an important word, or words that you see onscreen. For instance, words in menus or dialog boxes appear in **bold**. Here is an example: "Wait for the download to finish and click **File is Ready! Click here to download** to save the file."

> **Tips or important notes**
> Appear like this.

Get in touch

Feedback from our readers is always welcome.

General feedback: If you have questions about any aspect of this book, email us at customercare@packtpub.com and mention the book title in the subject of your message.

Errata: Although we have taken every care to ensure the accuracy of our content, mistakes do happen. If you have found a mistake in this book, we would be grateful if you would report this to us. Please visit www.packtpub.com/support/errata and fill in the form.

Piracy: If you come across any illegal copies of our works in any form on the internet, we would be grateful if you would provide us with the location address or website name. Please contact us at copyright@packt.com with a link to the material.

If you are interested in becoming an author: If there is a topic that you have expertise in and you are interested in either writing or contributing to a book, please visit authors.packtpub.com.

Share Your Thoughts

Once you've read *Implementing Multifactor Authentication*, we'd love to hear your thoughts! Scan the QR code below to go straight to the Amazon review page for this book and share your feedback.

https://packt.link/r/1803246960

Your review is important to us and the tech community and will help us make sure we're delivering excellent quality content.

Download a free PDF copy of this book

Thanks for purchasing this book!

Do you like to read on the go but are unable to carry your print books everywhere?

Is your eBook purchase not compatible with the device of your choice?

Don't worry, now with every Packt book you get a DRM-free PDF version of that book at no cost.

Read anywhere, any place, on any device. Search, copy, and paste code from your favorite technical books directly into your application.

The perks don't stop there, you can get exclusive access to discounts, newsletters, and great free content in your inbox daily

Follow these simple steps to get the benefits:

1. Scan the QR code or visit the link below

https://packt.link/free-ebook/9781803246963

2. Submit your proof of purchase
3. That's it! We'll send your free PDF and other benefits to your email directly

Part 1: Introduction

As our customers', co-workers', and partners' lives become increasingly entwined with the digital domain, the questions of what authentication is and why it's crucial to utilize various methods to establish one's identity online have never been more critical. Furthermore, in a world where high-profile cyberattacks have become an all-too-common headline, the urgency to address the authentication issue has reached all organizational levels.

In *Part 1*, we lay the groundwork by offering a clear and engaging explanation of what authentication is and why it's an indispensable aspect of our digital existence.

As you journey through this compelling guide, you'll encounter various types of authentication. The book provides an insightful analysis of the strengths and weaknesses of each method, as well as recommendations on when to use—or avoid—specific approaches, ensuring you have the tools to make informed decisions about your online security.

This part has the following chapters:

- *Chapter 1, On the Internet, Nobody Knows You're a Dog*
- *Chapter 2, When to Use Different Types of MFA*

1

On the Internet, Nobody Knows You're a Dog

In the ever-evolving landscape of cybersecurity, ensuring that proper access is given for the right reasons at the right time for digital identities is no longer just an optional feature – it's an indispensable component of securing modern applications. Moreover, as digital transformation accelerates, organizations must proactively protect their sensitive data and functions against persistent cybercriminals, hackers, and even insider threats.

To bring this critical topic to life, we invite you to join us on an engaging journey with ACME Software. This fictitious start-up grapples with the complexities of securing access to its business-critical data and functions. As ACME Software grows and expands, its workforce identities (corporate employees, contingent workers, and partners) and customer identities demand increasingly sophisticated authentication mechanisms to keep their information safe and sound.

Throughout this book, we will look at ACME Software while exploring its options and navigating the intricate world of modern authentication mechanisms. As we follow the start-up's story, you will discover not only the essentials of **multifactor authentication** (**MFA**) but also its practical applications, benefits, and potential pitfalls. By delving into real-life examples and scenarios, we aim to make this subject more engaging, accessible, and relatable, transforming what might otherwise be a dry, technical topic into a captivating learning experience.

This book will cover the following themes:

- The importance of securing digital identities in today's interconnected world
- An introduction to MFA, its principles, and its various forms
- A detailed examination of ACME Software's authentication requirements and the challenges it faces as it grows
- A comprehensive exploration of various MFA solutions, as well as their strengths and weaknesses

- Real-world examples of implementing and managing MFA solutions at ACME Software, demonstrating how to optimize security while maintaining user convenience

- The future of authentication – emerging trends and technologies that will shape the next generation of identity and access management

As we follow ACME Software's journey, we aim to equip you with the knowledge and understanding necessary to make informed decisions about MFA for your organization, empowering you to protect your valuable digital assets in a world of ever-increasing cyber threats.

In this chapter, we are going to cover the following topics:

- Identity and digital identity

- Additional authentication and security controls

Identity and digital identity

Identity is a universal concept that accompanies us throughout our lives, regardless of our cultural or national background. Immediately after birth, newborns around the world are identified in various ways. In some cultures, babies might receive bands on their wrists or ankles, while others may have different traditional identification methods. These methods often include the baby's name, date of birth, and other crucial information that helps distinguish them from others.

Governments and communities across the globe maintain records of their citizens' identities in various forms, such as birth certificates, family registers, or national ID systems. These records typically contain vital information such as names, birthdates, places of birth, and parentage.

Individuals from diverse cultures and nations rely on these records to establish and verify their identities. Moreover, the importance of these documents transcends geographical boundaries since people need them for various purposes, such as education, civic participation, and international travel. For example, these records may be required for enrolling in school, registering to vote, or obtaining necessary documents such as passports or driver's licenses.

The documents used to identify a person may change, depending on the context. For example, I need documents establishing my identity and employment authorization to apply for a job. On the other hand, I may need a passport rather than a driver's license when traveling abroad. And to open a bank account, I may require proof of residence and identification information. Collectively, these artifacts provide what is known as **personally identifiable information** (**PII**).

Let's look at the process of opening a bank account before the internet. A customer had to drive to the bank, meet with a bank representative, and present the required documents to open an account. Only then would they be issued an account number and be allowed to make transactions via that account. After applying for and receiving an **automated teller machine** (**ATM**) or debit card in the mail, they could use it to access their account. Every time they wanted to perform a transaction, they would need to go to a branch and authenticate themselves to a teller that would verify that they were the person they claimed to be and that they were authorized to perform the transaction they wanted. With an ATM card, they no longer needed to show their picture ID to confirm who they were. Anybody with that person's ATM card could do everything they were authorized to do at the ATM. When someone withdraws cash with an ATM card or makes a purchase with a debit card, the card reader takes information about the account from the card and sends it, along with the amount of the transaction, to the bank. To verify that the card was not stolen, the card reader requests the card's **personal identification number (PIN)**; once the PIN is entered correctly, the bank approves the transaction and withdraws the funds from the account.

Identity is a multifaceted concept encompassing the unique characteristics that define who or what a person or thing is. The amalgamation of physical, emotional, cultural, and social attributes creates the intricate tapestry of our individuality. In both the physical and digital realms, identity plays a crucial role in remembering, recognizing, and interacting with subjects, be they people or objects.

In today's increasingly interconnected world, our identities extend beyond the tangible realm, forming an integral part of our digital presence. This **digital identity** is a virtual representation of our real-world selves, encompassing various elements, such as usernames, passwords, biometrics, and personal preferences. It enables us to navigate the vast expanse of the internet, engage in online transactions, and interact with digital services.

The process of authentication is vital in both physical and digital environments. By verifying the identity of a subject, we ensure that they are who they claim to be and grant them access to specific services or actions based on their **authorization**. This process is essential for maintaining security and trust and enabling the seamless functioning of our increasingly digital lives.

In digital transactions, the owner of a digital identity is often referred to as the **security principal** or simply the **principal**. This term highlights the significance of the individual or entity at the heart of the authentication inquiry. As we engage in various online activities, our digital identities are the foundation for creating trust and facilitating secure transactions.

Just like identity existed before the internet, **two-factor authentication (2FA)** and MFA existed as well. The PIN on an ATM or debit card is one example of MFA (and 2FA, which is a subset of MFA). To verify (authenticate) my identity, I need to present my ATM card (something I have) and enter my PIN (something I know). Similarly, showing my driver's license to the bank teller is another example of MFA. The driver's license is the first factor (again, something I have), while matching the picture on the ID to me is the second factor (something you are).

Establishing identities is also critical, if not more important, online. Even though a large number of countries have established some form of online digital ID (you can see a list at `https://www.worldprivacyforum.org/2021/10/national-ids-and-biometrics/`), it is still rare to encounter customer-facing applications that will accept those digital IDs outside of the country that issued the ID.

The New Yorker published a cartoon in July 1993 where a large dog was sitting in front of a computer, speaking to another dog on the floor to his side, saying, *On the internet, nobody knows you're a dog*. It can be viewed here: `https://i.kym-cdn.com/photos/images/original/000/427/569/bfa.jpg`. Here's Dalle-2's interpretation of it:

Figure 1.1 – Dalle-2's interpretation of "On the internet, nobody knows you're a dog"

The saying quickly became popular and has been used to describe the anonymous nature of life online. As more and more applications become available online, identifying users is essential for several reasons.

For privacy reasons, users that register at a site may not want or permit their information and activities to be seen by somebody else. Therefore, companies must verify the user when they return to the site and validate their identity.

Companies that sell services need to make sure that the user registering is legitimate and that they are authorized to use those credentials. As Microsoft's investigation of the security breach by the group LAPSUS$ shows (`https://www.microsoft.com/en-us/security/blog/2022/03/22/dev-0537-criminal-actor-targeting-organizations-for-data-exfiltration-and-destruction/`), cybercriminals usually buy credit card numbers and other information on criminal underground forums and will also use the Redline password stealer, Loki, and other password stealers that are bought on the dark web or available for a subscription fee. They will use that information to open new accounts and spend money they don't intend to pay for. Companies in the financial services industry may also have other regulations they need to follow to prevent money laundering, for example.

Especially after the COVID-19 pandemic started, companies began to hire employees without ever seeing them. Onboarding employees has completely changed. It is not always possible to verify an employee's identity by looking at their physical documents (birth certificate, social security number, driver's license, and so on) before or when they start working. Even though identity verification is not something that affects the authentication of that user, it affects what we are fundamentally discussing in this book. If you give valid credentials to a bad actor, all the security in the world will not prevent that user from doing what those credentials allow them to do.

The process of registration is a crucial step in creating and managing a digital identity. It involves collecting and verifying information about a subject (a person or an entity) and linking it to a unique identifier in the digital realm. This identifier can be a username, email address, or any other unique attribute that distinguishes the subject from others. The relationship between a subject and their digital identity is established during the registration process, and it sets the foundation for future authentication and authorization.

The first step in the registration process is to collect relevant information about the subject. Data collection may include personal details such as name, address, date of birth, contact information, and digital credentials such as a username and password. In some cases, biometric data or other unique attributes may also be collected.

After collecting the necessary information, the next step is to verify the authenticity of the data provided by the subject. For example, data verification may involve checking the validity of an email address, confirming a phone number via SMS, or comparing the provided biometric data to a pre-existing database. This verification process ensures that the subject is who they claim to be and helps maintain the integrity of the digital identity system.

Once the data has been verified, an individual account is created for the subject. This account serves as the digital representation of the subject and is linked to their unique identifier (for example, username or email address). In addition, the account may include additional information, such as preferences, interests, and other data to help personalize the subject's digital experience:

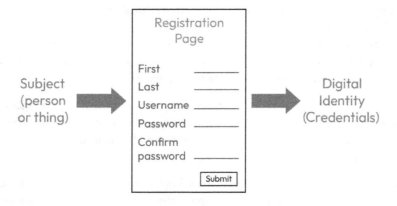

Figure 1.2 – Application registration

With the account created and linked to the subject's unique identifier, the subject can now use their digital identity to **authenticate** themselves when accessing online services.

The most common way of proving your identity online is by using a username and password:

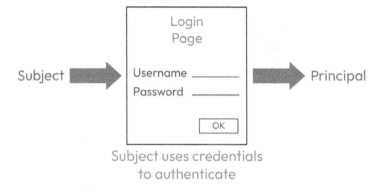

Figure 1.3 – Application authentication

As documents or other forms of identification are used to determine if a person is who they say they are, **authenticators** are used to assess the validity of claims from a subject engaged in a transaction online, confirming the digital identity of the subject.

In the physical world, governments and companies define the rules used to identify the users of their services or access to their systems. For example, a person must present a driver's license or another form of identification to travel to domestic destinations or withdraw money from their local bank. However, they need to show a passport to be able to travel internationally. In addition, government-issued identification may not be enough when going to a company's office, and badges may be required instead.

A digital identity is different. Even though it must be unique to the digital service it was created for, it does not uniquely identify the subject across all digital services.

Identity proofing, sometimes also referred to as **identity verification**, is required to validate that a subject is who they say they are. In a process similar to the one described earlier for the physical world, a person will present a driver's license or password, or other documents accepted by the identity-proofing service, and the identity-proofing service will provide **identity assurance** (the degree of certainty that the identity can be trusted to belong to the person).

Similarly, companies define their own rules to register for online (or virtual) identities and use them. In some cases, a username or email address is all that is required to create a new account. Others will need more information and, depending on the objective of the identity, validate the data used to create the new identity.

For internal users, the process is usually more complex. Legal or regulatory requirements may specify the information required for each user. The employer verifies that the worker is authorized to work in the country by validating some documents, for example.

Another difference may be **self-service**, where users can create their own accounts.

When self-service is not used, there are two ways of creating new identities. First, when companies are in their early stages, and the number of employees is small, they use manual processes to create accounts for their employees. Later, as the number of employees grows and the number of applications that those users have access to grows, an **identity management** platform or product usually performs automated identity creation and management.

Controlling access to systems, applications, and software and who is authorized to do what is called **access management**.

Workforce identity

Before they can offer services and applications to external customers, companies must start their identity work with everyone in the organization – employees, their contingent workforce, and business partners. **Workforce identity** software is used to manage identities for employees and the contingent workforce. Businesses may also use workforce identity to manage temporary or permanent identities for the contingent workforce and partners. **Identity federation** is the trust relationship between the company and an external (workforce) identity system to authenticate users. Identity systems usually work together with access management in what is called **identity and access management (IAM)** software.

The following are the typical requirements for workforce identity products:

- **Secure and frictionless experience**: Users need to be productive with their daily operations. The company must be able to use the product according to their required balance of secure and convenient access for workforce users.

- **Granular, centralized administration**: A workforce identity solution must provide sufficient capabilities to control the life cycle of the company's identities with a centralized administration giving full control to the identity infrastructure.

Customer identity

Businesses use **customer identity and access management (CIAM)** software to manage customer identities and offer a secure, seamless login experience for the company's applications. When building an internet-facing application, there are common features and standard requirements that companies usually ask for:

- **Self-service**: The first thing is self-service, account management, and many related features – starting with allowing users to sign up and sign in, managing their profile, changing their profile, changing their password, making account recovery, performing MFA, changing their authentication factors, and onboarding new devices. All of these things come under self-service account management. It would be best if you had a solution that allows you to do this for your customers and let your customers – the end users of your application – manage these profiles for themselves.

- **Scalability**: The second point is that it scales to tens of millions of users and has a large global coverage. This is different from workforce identity since usually, you have thousands or maybe tens of thousands of users. In the consumer space, you have tens of millions. On Azure, AWS, or Google Cloud, some companies have hundreds of millions of customers, and that number is always increasing. A system must allow millions of identities to be created for a large enterprise with a global presence in different countries and locations. The system must also be able to distribute these users or position them in a country closer to them; they may do this for data residency reasons. For example, users in Europe must have their data only in Europe.

- **Ease of use**: We usually want to attract as many users as possible in consumer identity. Ease of use is essential when onboarding customers in an online application. If the process is not user-friendly, it may discourage potential customers from completing the onboarding process and prevent them from using the application. The end users' onboarding and authentication journey must be as easy as possible while providing various options.

  Using social media accounts for onboarding can be convenient and efficient for users to create accounts and access online applications. In addition, this approach allows users to authenticate their identity and provide personal information while using their existing social media profiles rather than having to create a new account from scratch.

Again, this is different from workforce identity. The workforce is usually a captive audience that has to be created by an administrator and typically follows an HR process. Using the same process with external users will cause them to abandon the process. They will do business elsewhere. The journey to onboard end users has to be as seamless as possible.

One requirement that applies to customer or workforce IAM products is **single sign-on** (**SSO**). When **access management** (**AM**) products allow users to log in once for multiple applications, that is called SSO.

When there is a trusted relationship between separate organizations and companies that allow users to authenticate across domains, that is called **federated SSO**.

Different protocols are used for SSO. Some of them will be used in the practical implementation examples in this book, starting from *Chapter 3*:

- **SAML 2.0**: **Security Assertion Markup Language** (**SAML**) is an open standard created in 2005 to provide cross-domain SSO. In SAML, you have an **identity provider** (**IdP**), which is responsible for authenticating users and managing identities, a **relying party** (**RP**), which is a service requesting and receiving data from the IdP, and a **user agent** (**UA**), which is the user requesting the services. SAML is used by several SSO products (including Azure AD, as shown in *Chapter 3*) to authenticate users to online **Software-as-a-Service** (**SaaS**) applications such as Salesforce, Slack, and others.

- **OAuth 2.0**: OAuth allows users to share specific data with an application while keeping their credentials private. For example, a printing service can use OAuth to obtain permission from users to access their photos for printing. We are going to use OAuth for some examples in this book. The OAuth Playground website provides a detailed description of the steps involved in using OAuth, along with an example application that is free to use. OAuth Playground can be viewed at `https://www.oauth.com/playground/client-registration.html`:

Figure 1.4 – OAuth Playground client registration

After registering a new client on OAuth Playground, you can use the generated credentials to test the OAuth protocol:

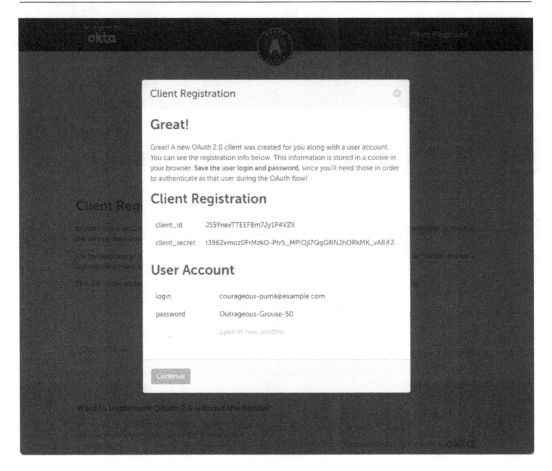

Figure 1.5 – OAuth Playground test credentials

To test these credentials, go to `https://www.oauth.com/playground/authorization-code.html` and enter the user account credentials that were generated in the previous step.

Now that the basic terminology is out of the way, let's dive into the main topic of this book: MFA.

Additional authentication and security controls

MFA is a method of verifying a user's identity by requiring them to present more than one piece of information. By combining multiple layers of security, MFA decreases the chances of compromised online access to an account.

What are authentication factors?

Authentication factors are different ways of proving identity. There are three different categories of authentication factors:

- **Something you know (knowledge)**: Passwords, PINs, answers to pre-selected security questions

- **Something you are (being or inheritance)**: Face recognition, fingerprint scan, voice recognition

- **Something you have (possession)**: SMS codes, one-time passwords, smart cards, ATM cards, mobile phones, key fobs:

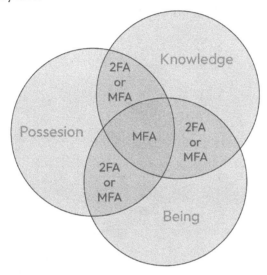

Figure 1.6 – Authentication factors

As can be seen in *Figure 1.7*, the three different authentication factors can be used individually, or combined, as part of the same authentication process. The process of combining two different factor types in the same authentication process is called 2FA or MFA. The process of combining three or more different categories of authentication factors used in the same authentication process is called MFA.

To be considered 2FA or MFA, the authentication factors should be from different categories.

Most websites use a username and password combination to verify users' identities. Some will attempt to increase security and require an answer to a security question as well. This is not MFA. Even though the user provided two factors to authenticate (password and answer to security questions), the second factor is also from the knowledge category. This is considered a two-step authentication process but a single factor.

Going back to our ATM example, MFA enhances security because it requires the hacker to obtain the two factors of authentication before being able to access your money. If your wallet is stolen or you lose your ATM card, the person that has your card cannot use it without knowing the pin as well. Similarly, if someone shoulder surfs (steals your PIN by spying over your shoulder as you use an ATM) and can use your PIN, they still don't have the ATM card needed to complete the transaction.

Most free email providers, such as Gmail, Outlook, iCloud, and Yahoo!, provide some form of 2FA:

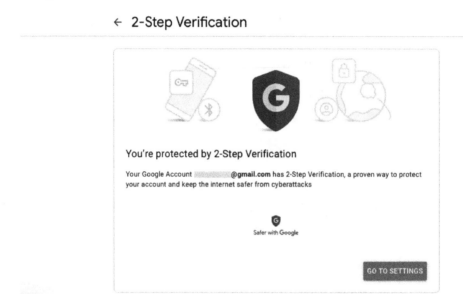

Figure 1.7 – Gmail 2-Step Verification confirmation

As we discuss MFA throughout this book, it is important to consider the needs of the organization and the types of users that are going to be using the systems. An authentication system needs to balance its security needs with the usability and risks of the application being accessed.

In certain industries and the government, special standards and regulations may also require (or prohibit) the use of different types of MFA systems.

`https://2fa.directory/us/` provides a list of websites for different industries and whether or not they support 2FA and is a good place to look to see what your competition is doing in this area.

Criminals can obtain user credentials in different ways. For example, they can buy user credentials on the dark web, try brute-force attacks, or use social engineering methods.

Another problem with passwords is that users reuse passwords across many different sites; they may share passwords with their colleagues. They may also write the passwords on post-it notes and attach them to their monitor at work or home.

All these issues make using passwords as the single method to identify users a significant security risk for companies.

If passwords are not enough, what else can organizations do? MFA, or at least 2FA, is the most common solution. Google, in their latest *Hacking Google* series, states "*Add 2FA to your account, and we do the rest regarding security.*" Microsoft says that 99.9% of identity attacks can be blocked by MFA (https://www.microsoft.com/en-us/security/blog/2019/08/20/one-simple-action-you-can-take-to-prevent-99-9-percent-of-account-attacks/).

On the other hand, MFA overuse may cause customers to choose to move to a friendlier site and do business with a different company or abandon a shopping cart or transaction completely. Therefore, the balance between usability and security has to be considered according to the risk involved with the transaction.

In some cases, the use of MFA is based on other signals that help the system decide when to ask for a second form of authentication – for example, detection that the user's IP address has traveled impossible distances, thus limiting the number of login attempts and increasing the time after each failure, and bot detection, among others.

Other tools may create a profile of the browser or mobile phone used by the users and ask for additional authentication if the phone changes or screen dimensions change, among other characteristics.

Behavioral biometrics can also be used to create a profile of the user and perform *continuous authentication* of the user based on their behaviors, not only when they log in:

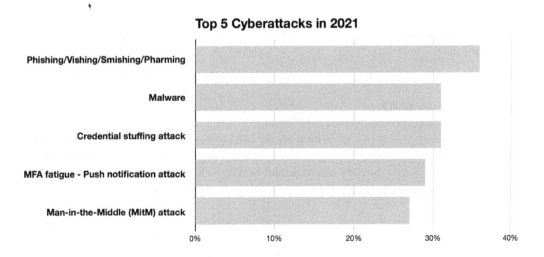

Figure 1.8 – Top five cyber attacks in 2022

According to a report by HYPR (`https://get.hypr.com/state-of-authentication-in-the-finance-industry-2022`), cyberattacks persistently targeted financial service institutions in 2022, as evidenced by the fact that 94% of those surveyed experienced some form of attacks within the last year. As shown in the preceding figure, the most common type of attack continues to be phishing, accounting for 36% of incidents. Other frequently occurring attacks included malware, credential stuffing, MFA fatigue attacks, and **Man-in-the-Middle (MitM)** attacks.

Phishing

Employees frequently fall for emails that promise bonuses, an urgent request from their CEO, or a request from the **Information Technology (IT)** department. Those emails ask users to click on a website or verify their credentials. Unfortunately, the whole company may be compromised when the employee clicks on the link or enters their credentials where they shouldn't.

Here are some other related attacks:

- When a hack is done via a phone call, this is known as **vishing**
- Similar to emails, SMS texts are sent to users in what is known as **smishing**
- When code to redirect the original browser request to a malicious website – without the knowledge or consent of the user – is installed on a server or personal computer, the attack is called **pharming**

Credential stuffing

Credential stuffing attacks occur when many username/password combinations are tried against a website. Bots usually perform this type of attack.

Malware

Malware, or malicious software, is a term that describes a malicious program or piece of code that is harmful to the user's computer.

Malware is normally used in conjunction with phishing to obtain the credentials from a user.

Account Take Over (ATO)

The reuse of credentials causes another typical attack. Most users commonly use the same email or username on many different apps. At the same time, passwords are also reused. If one account is compromised, bad actors can use the same credentials and try to log in to many other sites. **Account Take Over (ATO)** is usually the outcome of a successful credential stuffing attack.

MFA fatigue – push notification attack

A common way to prevent a credential stuffing attack is by using a second authentication step in addition to a username and password. For example, systems may require users to accept an app push notification or receive a phone call and press a key as a second factor. When an attack issues multiple MFA requests to the end user until the user accepts the authentication, this is called MFA fatigue. It is also known as a push notification attack.

Man-in-the-Middle attack

An MitM attack is a type of session hijacking attack. The attacker eavesdrops and interrupts an existing conversation by inserting themselves into the *middle* of the transfer.

The attacker pretends to be the other legitimate participant for both the user and the original web application, enabling them to intercept information and data from either side of the conversation. An MitM attack can be used for account takeover purposes or just for the duration of the session:

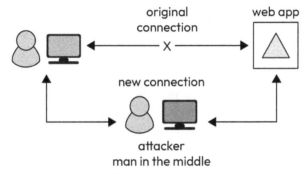

Figure 1.9 – MitM attack

In *Chapter 2*, we will discuss different types of authentication factors and what types can be used to prevent different types of attacks.

In addition to knowledge-based authentication factors, other commonly used authentication factors will be described next.

One-time password

A **one-time password** (**OTP**) is a mechanism for logging into an application or service using a unique password that can only be used once. OTP can be generated by security tokens or applications such as Google Authenticator or Microsoft Authenticator. SMS-based OTP is not recommended because of its vulnerabilities.

FIDO Alliance

The **Fast Identity Online** (**FIDO**) Alliance is an open industry association with a single goal: to create authentication standards to help reduce the world's reliance on passwords.

FIDO Universal 2nd Factor standard

Yubico and Google developed the **FIDO Universal 2nd Factor** (**FIDO U2F**) standard. After FIDO U2F was successfully tested with Google employees, the standard was contributed to the FIDO Alliance.

The WebAuthn specification

WebAuthn is a **World Wide Web Consortium** (**W3C**) specification that allows the creation and use of strong, public key-based credentials for authenticating users. It is designed to be a secure and convenient alternative to traditional username and password authentication methods and can be used to authenticate users on websites and other online platforms.

WebAuthn works with the FIDO **Client To Authenticator Protocol version 2** (**CTAP2**) to securely create and retrieve credentials on a security key. The two standards work together. Developers only use the WebAuthn specification; they don't have to worry about CTAP2. WebAuthn uses **public key infrastructure** (**PKI**) to create and manage the public keys that are used for authentication.

One of the main benefits of WebAuthn is that it allows users to authenticate using a variety of different devices, such as security keys, biometric sensors (such as fingerprint scanners or facial recognition cameras), and other types of hardware tokens. This makes it easier for users to authenticate securely and reduces the risk of password-based attacks such as phishing and brute-force attacks.

WebAuthn is supported by most modern web browsers and is becoming increasingly popular as a secure and convenient way to authenticate users on the web.

FIDO2

The FIDO2 specification includes World Wide Web Consortium's WebAuthn specification and FIDO Alliance's corresponding CTAP. The specifications are open and free for general use.

Passkeys

Passkeys are replacements for passwords based on FIDO Alliance and W3C standards. Passwords are replaced with strong credentials (cryptographic key pairs). In addition, passkeys are linked with the website or application they were created for, thus being safe from phishing. Passkeys are not a new thing, just a new name for WebAuthn/FIDO2 credentials, enabling a fully passwordless experience for the user. Even though passkeys are on a user's devices (something they have) and the relying party (the service provider that processes access to the applications) can ask for user verification, which is done by a biometric or PIN (something the user is or knows), some regulatory bodies still do not recognize passkeys as MFA.

This completes our introduction to MFA, authenticator factors, and the types of attacks companies face.

Summary

In this chapter, you learned why (digital) identity and authentication are fundamental parts of security. We also covered the basic concepts and terminology that will be used throughout this book. Finally, we introduced MFA.

In the next chapter, we are going to discuss the different types of authentication factors, how cybercriminals attempt to bypass them, and when to use or not to use different types of authentication factors.

When to Use Different Types of MFA

There is no magic bullet to solve all security needs. Hackers will actively look at ways to break it even if one existed, just as security companies create more and more solutions to improve the chances of better defending against threats. This chapter discusses when and when not to use different forms of **multifactor authentication** (**MFA**). We will also look at some websites for up-to-date information about MFA and new threats.

We are going to cover the following topics:

- Not all MFA is created equal – when to use different types of MFA
- Keeping up with bad actors – good sources for up-to-date information on MFA and related topics

Not all MFA is created equal – when to use different types of MFA

In May 2021, the President of the United States issued **Executive Order** (**EO**) 14028 to initiate a government-wide effort to improve cybersecurity. As part of this effort, the **Office of the Management and Budget** (**OMB**), part of the Executive Office of the President, issued a memorandum entitled *Moving the U.S. Government Toward Zero Trust Cybersecurity Principles*. The memo was sent on January 26, 2022, to all heads of executive departments and agencies of the government with specific cybersecurity standards and objectives that need to be in place by the end of the fiscal year 2024 (`https://www.whitehouse.gov/wp-content/uploads/2022/01/M-22-09.pdf`).

The initiative's goals are to "*ensure that baseline security practices are in place, migrate the Federal Government to a zero-trust architecture, and realize the security benefits of cloud-based infrastructure while mitigating associated risks.*"

The **Zero Trust model** is based on the principle that no actor, system, network, or service can be trusted. Therefore, we must verify anything and everything attempting to establish access. This initiative is a dramatic shift in how the government previously secured infrastructure, networks, and data and followed the principles recommended by all security experts.

In addition to relying on centralized, secure, enterprise-managed identity systems, the strategy emphasizes stronger enterprise identity and access controls, including MFA.

While the memorandum issued by the OMB was explicitly for government agencies, vendors, and contractors they work with, the guidance provided regarding MFA is one that I recommend all companies follow.

In addition to considering MFA "*a critical part of the Federal Government's security baseline,*" the memo also includes additional suggestions and requirements that we will discuss in the rest of this chapter:

"*Many approaches to multi-factor authentication will not protect against sophisticated phishing attacks. For agency staff, contractors, and partners, phishing-resistant MFA is required.*"

As discussed earlier, MFA will generally protect against some common methods of unauthorized account access that affect password-only systems. For example, one-time passwords, SMS, and magic links can be phished by bad actors and should be discontinued:

"*Agencies are encouraged to pursue greater use of passwordless multi-factor authentication as they modernize their authentication systems.*"

In authentication systems that include passwords as one of the factors in MFA, passwords can make it much easier to obtain access than when passwords are not one of the factors. We will discuss different systems that support passwordless MFA throughout this book and recommend using passwordless MFA whenever possible.

Why use MFA then?

Writing a book about MFA might be counterintuitive if bad actors can circumvent it. However, like most decisions in security, the answer is not always black and white. Instead, it depends on the risks that the user, the company, or the service provider are willing to accept and the inherent value of the protected information. By delving deeper into these aspects, we can understand the nuances of this issue and better appreciate the importance of MFA.

First and foremost, MFA provides an additional layer of security, making it more difficult for unauthorized users to access sensitive information. In addition, by requiring at least two independent authentication factors, MFA creates a more robust authentication process. Consequently, even if one factor is compromised, the likelihood of a successful breach is still reduced.

Now, it is true that MFA is not infallible and can be bypassed in some instances. However, the complexity of bypassing MFA often acts as a deterrent for potential attackers. The effort and resources required to circumvent MFA are significantly higher than traditional single-factor authentication methods. This is particularly important when considering the value of protected information; the more valuable the data, the more attractive it becomes as a target. Thus, MFA adds cost for potential attackers, who must weigh the investment against the potential payoff.

Furthermore, the risk tolerance of the parties involved also plays a role in the decision to use MFA. For users, companies, and service providers, the consequences of a security breach can be devastating – ranging from financial losses to reputational damage. The implementation of MFA serves as a demonstration of commitment to security, which may help mitigate these risks. Also, as we saw in the memo from the president, *"For agency staff, contractors, and partners, phishing-resistant MFA is required."*

Finally, it is essential to remember that security is constantly evolving. Security measures such as MFA must adapt and improve as new threats emerge to remain effective. In this ever-changing landscape, using MFA is not a guarantee of absolute security but rather a tool that helps minimize the likelihood of a breach.

Different types of MFA

Different types of authenticator factors can be used in **two-factor authentication** (**2FA**) and MFA. The cybersecurity and infrastructure agency classifies the MFA types into three tiers (https://www.cisa.gov/mfa):

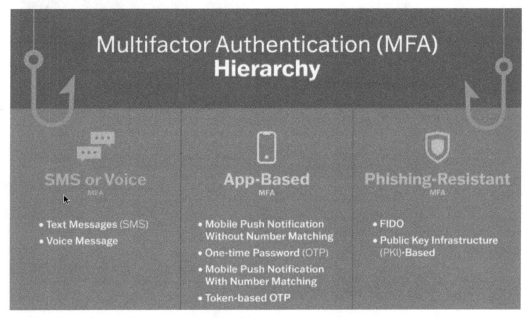

Figure 2.1 – The strengths of different types of authentication factors

The weakest types are text messages (SMSs) or voice messages. In addition to phishing, they are also susceptible to **SIM swap** attacks.

At the next level, app-based MFA is also divided into two types. Push notifications without number matching are weaker than the other types in this category – that is, **one-time password** (**OTP**), **token-based OTP** (**TOTP**), and push notifications with number matching.

Finally, the strongest level includes phishing-resistant MFA types such as FIDO and public-key infrastructure-based MFA. Recall that FIDO was discussed in *Chapter 1*. For more information, please go to `https://fidoalliance.org/what-is-fido/`.

SIM swap and why SMSs and voice messages are the weakest authenticator factor types to use

In February 2022, the FBI issued a public service announcement (`https://www.ic3.gov/Media/Y2022/PSA220208`) that included the following text:

"From January 2018 to December 2020, the FBI Internet Crime Complaint Center (IC3) received 320 complaints related to SIM swapping incidents with adjusted losses of approximately $12 million. In 2021, IC3 received 1,611 SIM swapping complaints with adjusted losses of more than $68 million."

Here is a screenshot from the FBI website:

Figure 2.2 – FBI's PSA on SIM swap schemes

A **SIM swap** is a type of social engineering attack in which a malicious actor uses social engineering or another method to convince a phone company to transfer the victim's phone number to a new SIM card that they control. This allows the attacker to intercept calls and text messages meant for the victim, potentially giving them access to sensitive information such as one-time codes used for recovering their accounts or MFA.

Let's see what happens if your phone number is transferred to a cybercriminal:

1. The hacker goes to the account login page and clicks on **Forgot password?**:

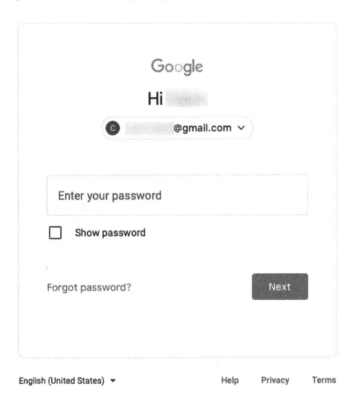

Figure 2.3 – Login page with Forgot password?

2. The hacker selects the phone number they now control as a recovery mechanism:

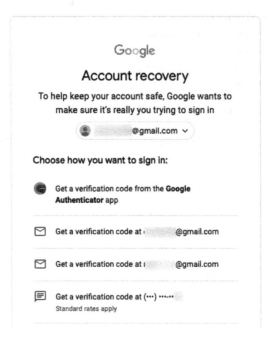

Figure 2.4 – Account recovery selection page

3.   The hacker enters the code:

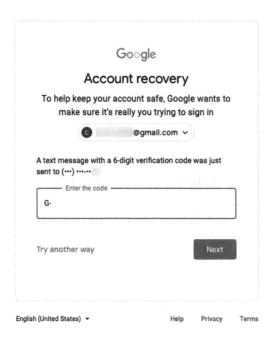

Figure 2.5 – Account recovery page

4. That's it – the account takeover is complete. The hacker can continue without changing the password or change the password and possibly complicate the process of the original owner recovering their account:

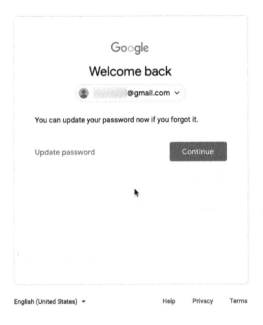

Figure 2.6 – The Welcome back page

SIM swap is one of the worst attacks because it is tough for the service provider to prevent. As seen in the preceding example, even though the account was protected by MFA, the cybercriminal was able to bypass the password and use only one factor to take over the account.

In a typical scenario, the security of services depends on the company providing them and the user using them. If the service provider allows the use of SMSs or voice messages as a recovery mechanism, the security of the service will also rely on the phone company, which neither the service provider nor the user can control. If SMSs or voice messages are only allowed as a second factor during a password-based login, and not for account recovery, the cybercriminal needs to obtain the user's ID and password, and also obtain control of the victim's phone using SIM swap.

What can the service provider do?

The service provider can mandate app-based or phishing-resistant authentication factors for account recovery and as a second factor of authentication. If that is not an option, it can at least recommend and educate users on the benefits of a stronger factor of authentication.

A service provider can also enhance the chances of a successful session or account takeover being detected by suggesting or mandating a second recovery mechanism:

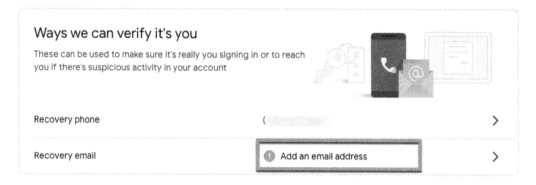

Figure 2.7 – Account recovery mechanism selection page

Users with an email address and a phone number as security mechanisms will be notified when one of the security mechanisms is used for account recovery:

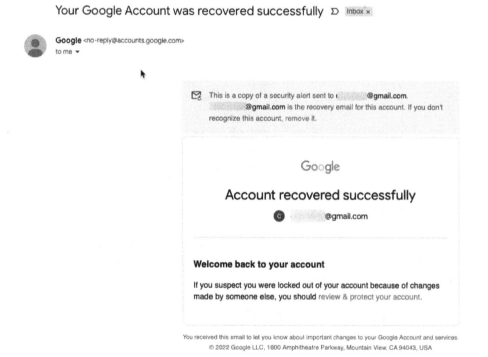

Figure 2.8 – Account recovered notification email

This is what the service provider can do to recover an account. Let's see in the next subsection what the user can do.

What can the user do?

To avoid being the victim of **SIM swap** fraud schemes, users should avoid using SMS messages and voice messages for 2FA as well as for recovering their accounts.

If quickly recovering the account is more important than avoiding the possibility of being the victim of a SIM swap, users should make sure that more than one method of recovery is enabled. This way, as we saw in the previous example, the user will be notified if someone recovers the account inappropriately, or if a user logs in to the account from an unknown computer.

MFA fatigue – also known as MFA push spam

As service providers and users become more security conscious and avoid SMS-based MFA, hackers increasingly use a technique that does not require SIM swap and can bypass authentication factors classified as more secure. This method is called **MFA fatigue**. MFA fatigue was used in confirmed cyberattacks on Cisco (`https://blog.talosintelligence.com/recent-cyber-attack/`), Microsoft (`https://www.microsoft.com/en-us/security/blog/2022/03/22/dev-0537-criminal-actor-targeting-organizations-for-data-exfiltration-and-destruction/`), and other companies. Several security companies, including Mandiant, have published reports about Russian actors using the technique (`https://www.mandiant.com/resources/blog/russian-targeting-gov-business`) to target the US government and private companies.

MFA fatigue is commonly initiated by compromising the user's identity to obtain initial access to the organization or the victim's account. This is typically done via the following methods:

- Deploying malware such as Redline Stealer or Loki Password Stealer
- Obtaining credentials and session tokens in criminal underground forums

- Recruiting current and former employees who have access to specific company networks, as depicted in *Figure 2.9*:

LAPSUS$ channel
We recruit employees/insider at the following!!!!

- Any company providing Telecommunications (Claro, Telefonica, ATT, and other similar)
- Large software/gaming corporations (Microsoft, Apple, EA, IBM, and other similar)
- Callcenter/BPM (Atento, Teleperformance, and other similar)
- Server hosts (OVH, Locaweb, and other similar)

TO NOTE: WE ARE NOT LOOKING FOR DATA, WE ARE LOOKING FOR THE EMPLOYEE TO PROVIDE US A VPN OR CITRIX TO THE NETWORK, or some anydesk

If you are not sure if you are needed then send a DM and we will respond!!!!
If you are not a employee here but have access such as VPN or VDI then we are still interested!!

You will be paid if you would like. Contact us to discuss that

@lapsusjobs   ↩ 624   👁 13.3K   📌 12:37 PM

Figure 2.9 – Recruitment message from the group LAPSUS$

When companies use mobile apps as a second factor of authentication, where a user sees a notification on their phone that they must approve, cybercriminals will attempt to cause the legitimate user to accept one of the repeated MFA prompts and let the cybercriminal in. In some cases, the attacker will also send a message to the victim pretending to be from the company and urging the user to accept the MFA push:

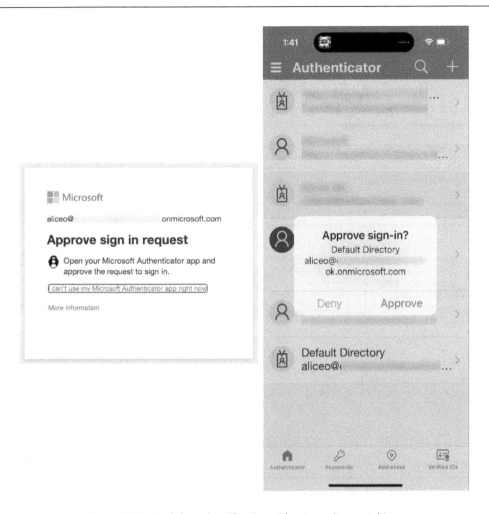

Figure 2.10 – Push-based notification without number matching

What can the service provider do?

Service providers can provide additional security against MFA fatigue by enabling **number matching**. If enabled, the user must enter a number in the authenticator that matches the number shown in the authentication sign-in. Microsoft considers number matching a critical security upgrade and will enforce number matching starting in 2023 (`https://learn.microsoft.com/en-us/azure/active-directory/authentication/how-to-mfa-number-match`). Duo, another product we will use as an authentication factor, starting from *Chapter 3*, also supports using a verification code to avoid MFA fatigue. This is called **Verified Duo Push**.

Another way that service providers can avoid scams based on MFA fatigue is by providing additional context for the push authentication, if enabled by the MFA application.

The following figures show number matching and other settings for Microsoft Authenticator. Starting February 27, 2023, Microsoft Authenticator will enforce number matching for all users tenant-wide, eliminating the need for this configuration. This is a crucial security enhancement over traditional second-factor push notifications:

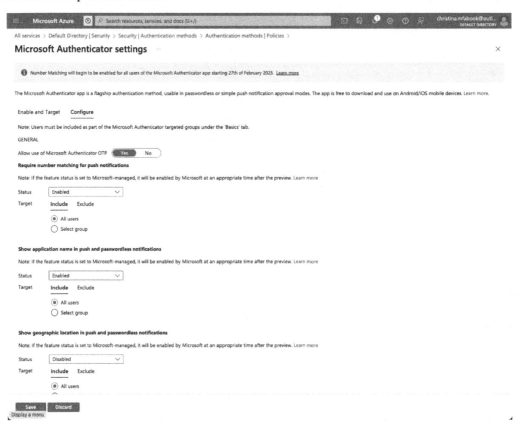

Figure 2.11 – Enabling number matching in Microsoft Authenticator

Number matching is a feature that requires the user to input numbers from the identity platform into their app to confirm the authentication request:

Figure 2.12 – MFA push with number matching in Microsoft Authenticator

As seen in the following figure, Microsoft allows the application name and the geographic location to also be shown during the authentication process:

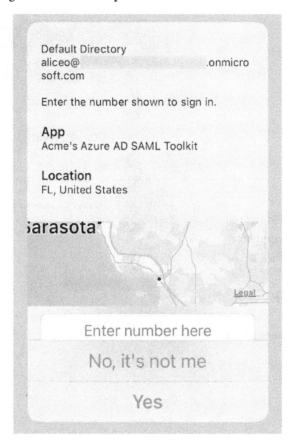

Figure 2.13 – MFA push with the application name and location in Microsoft Authenticator

Finally, we are going to look at the most secure authentication factor type: **phishing-resistant MFA**.

Phishing-resistant MFA

While writing this chapter, I received an email from my financial institution. Curiously, the picture in the email showed users protecting their passwords. However, passwords were not the focus of the document – phishing was.

The email described a typical financial institution phishing attack, where bad actors utilize mass email campaigns to a broad group of people designed to collect account credentials. In addition to mass email attacks, modern, targeted attacks are a more sophisticated way of infiltrating corporate security. They use social engineering to research information about users in select organizations and use real names to create an urgency for the target user to respond.

In either case, a hacker creates a fake email account or website with variations of legitimate email and website addresses, trying to make fake accounts appear authentic. They then send spear-phishing emails pretending to be a trusted sender (either the financial institution or, in the case of companies, somebody from the IT department – the boss or CEO of the target user).

When you download a file attached to an email or click on an email link, you can unsuspectingly install malware that can capture your credentials and access your accounts. In addition, you can be redirected to a fake website created to look exactly like your original website.

In *Figure 2.15*, you can see one example where cybercriminals created a copy of the popular encrypted email service Tutanota (`https://tutanota.com`) using a URL similar to the original, `https://tutanota.org`:

Figure 2.14 – Fake tutanota.org web page imitating tutanota.com

A typical phishing attack works like this:

1.   The user receives a phishing email or a WhatsApp or SMS message. Note the unknown from addresses and fake URL links in *Figure 2.16*:

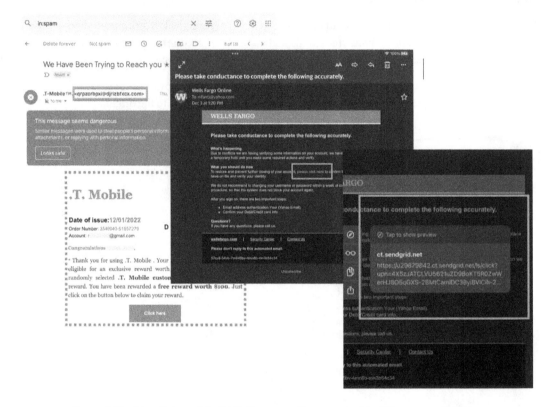

Figure 2.15 – Identifying a fake Wells Fargo email message with different email readers

2.   The user clicks on the link that points to a website unrelated to the original URL.

3.   The user doesn't notice that the URL is not what they were expecting it to be. In this case, it is `https://fakegoogle.com`:

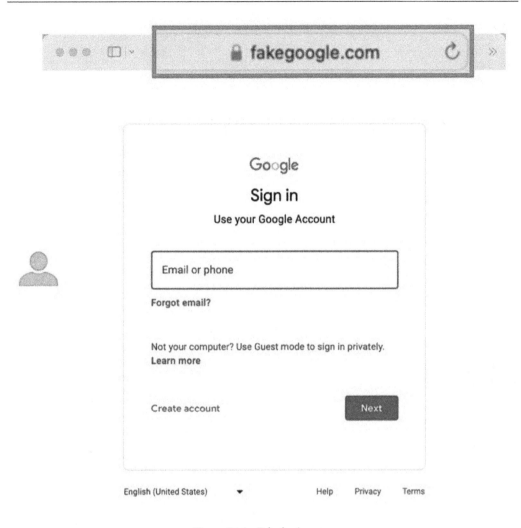

Figure 2.16 – Fake login page

4. The user enters their username. A **Man-in-the-Middle (MitM)** proxy forwards this username to the real website (`https://accounts.google.com`):

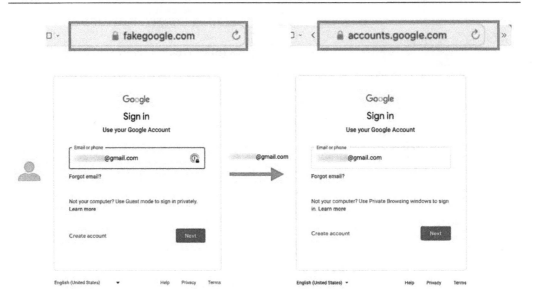

Figure 2.17 – Username used on the real login page

5. The user enters their password. Again, the MitM proxy forwards the password to the real website:

Figure 2.18 – Password used on the real login page

6. Finally, the user enters the OTP from the authenticator app. And again, the MitM proxy forwards the OTP to the real website:

Figure 2.19 – OTP copied to the real login page

7.   To avoid suspicion, the user is now redirected to the real website. The session ID cookies are then copied to the cybercriminal's browser, who can then impersonate the user on the real website:

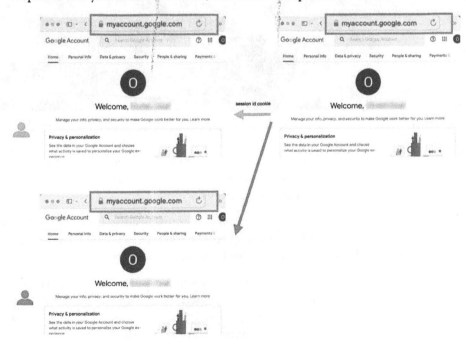

Figure 2.20 – Session ID copied to the hacker's page

Most phishing attacks can be prevented with user education. Unfortunately, cybercriminals only need to be successful once to be able to take over an account. At the same time, education may work for workforce users but not for external customers.

Another factor that makes phishing and MitM attacks easier for cybercriminals is the number of MitM toolkits, such as **evilginx2** and **Modlishka**, which come with videos and other guides that make it very easy for anyone to attempt phishing attacks.

As seen in *Figure 2.22*, the process described earlier (except for the phishing email or other social engineering scheme used to have the victim access the attacker's URL) is done by the tool. The session tokens that are captured can then be used by the criminals with very little effort:

Figure 2.21 – evilginx2 MitM toolkit

Being an online bad actor used to require specific skills and access to hard-to-find groups and resources. Finding valuable information on a compromised system was also time-sensitive and challenging, making it easier for defenders to detect malicious activity.

However, the rise of specialization in the online crime sector brought about **initial access brokers** (**IABs**), which specialize in gaining and maintaining access to compromised systems for others. This has led to the industry's professionalization, resulting in the emergence of polished marketplaces.

The most famous IAB is the **Genesis Marketplace**. The Genesis Marketplace, also referred to as the **Genesis Store** or **Genesis Market**, is an exclusive platform specializing in the sale of stolen credentials, cookies, and digital fingerprints collected from compromised systems. The marketplace not only offers the stolen data itself but also provides well-maintained tools to facilitate its use.

Since its establishment in 2017, Genesis has listed over 400,000 compromised systems, or **bots**, from over 200 countries. The attractiveness of Genesis lies not in the quantity of its data aggregation but in the quality of the stolen information that it offers and its commitment to maintaining the accuracy of this information.

A **bot**, in this context, refers to a compromised system that harvests credentials and includes automated malware that collects information. By utilizing persistent bots on victims' systems, Genesis can keep the stolen information up-to-date for its customers, making the data more valuable to attackers. Additionally, Genesis claims to have access to the compromised system, ensuring its fingerprints will be updated as they change.

In addition to its extensive collection of stolen data, Genesis offers a polished interface with effective data-correlation capabilities and well-maintained tools for its customers, including a robust search function. The marketplace also provides mainstream amenities such as an FAQ, user support, pricing in dollars (payable in Bitcoin), and professional copy editing, making it a convenient and user-friendly platform for malicious actors.

The increased sophistication of tools used by bad actors makes the use of phishing-resistant MFA even more critical.

Keeping up with bad actors – good sources for up-to-date information on MFA and related topics

The US government is an excellent source of information for security. There are multiple government websites we can go to for up-to-date security and (multifactor) authentication information. Similarly, the European Union, Australia, the UK, and Israel provide important and simplified guidance to their citizens that can be beneficial to anyone looking for more information on new security threats and how they affect MFA specifically.

Cybersecurity and Infrastructure Security Agency

Cybersecurity and Infrastructure Security Agency (CISA) (`https://www.cisa.gov/about-cisa`) is *"part of the Department of Homeland Security that leads the national effort to understand, manage, and reduce risk to our cyber and physical infrastructure. CISA connect their stakeholders in industry and government to each other and to resources, analyses, and tools to help them build their own cyber, communications, physical security, and resilience. This helps ensure a secure, resilient infrastructure for the American people."* In addition to CISA's **Zero Trust model** referenced in the memorandum mentioned earlier, CISA has many resources related to MFA and its benefits. For example, the MFA page (`https://www.cisa.gov/mfa`) includes links describing how to enable MFA for use with the government and consumers. Resources for consumers, for example, explain how to set up MFA for Microsoft accounts, Facebook, Gmail, and Apple ID.

Other valuable documents from CISA include the following:

- *Implementing Strong Authentication* (`https://www.cisa.gov/sites/default/files/publications/CISA_CEG_Implementing_Strong_Authentication_508_1.pdf`)

- *Implementing number matching in MFA applications* (`https://www.cisa.gov/sites/default/files/publications/fact-sheet-implement-number-matching-in-mfa-applications-508c`)

National Institute of Standards and Technology

National Institute of Standards and Technology (**NIST**) is a non-regulatory agency of the United States Department of Commerce responsible for promoting innovation and industrial competitiveness. NIST's activities are organized into laboratory programs focusing on physical science, engineering, applied technology, information security, and communication standards.

NIST publishes digital identity guidelines that the government and private companies widely use. For example, its famous *NIST Special Publication 800-63-3 Digital Identity Guidelines* (`https://pages.nist.gov/800-63-3/sp800-63-3.html`) describe identity guidelines in three significant areas:

- Enrollment and identity proofing (SP 800-63A)

- Authentication and life cycle management (SP 800-63B)

- Federation and assertions (SP 800-63C)

As part of authentication and life cycle management, it defines the **authentication assurance level** (**AAL**) that helps companies decide which types of MFA they require, depending on the level of assurance that the user being authenticated is who they claim to be. NIST also publishes a blog with cybersecurity insights (`https://www.nist.gov/blogs/cybersecurity-insights`) and several security-related special projects, such as *Multifactor Authentication for E-Commerce* (`https://www.nccoe.nist.gov/multifactor-authentication-e-commerce`).

National Security Agency

The **National Security Agency** (**NSA**) leads the US government in signals intelligence insights and cybersecurity services and products. The NSA's Cybersecurity Collaboration Center harnesses the power of industry partnerships to prevent and eradicate foreign cyber threats.

One of NSA's publications is the *Assessment of common multi-factor authentication solutions* (`https://media.defense.gov/2020/Sep/22/2002502665/-1/-1/0/CSI_MULTIFACTOR_AUTHENTICATION_SOLUTIONS_UOO17091520.PDF`). The document reviews commonly used MFA mechanisms against NIST standards.

Other US government websites include the **Government Services Administration** (creators of `Login.gov`, the public's one account and password for business with the government) and websites that focus on specific industries, such as the **National Credit Union Administration** (**NCUA**). One example of a publication from NCUA is *Guidance on Authentication in Internet Banking Environment* (`https://ncua.gov/regulation-supervision/letters-credit-unions-other-guidance/guidance-authentication-internet-banking-environment`).

The European Union is also a giant in cybersecurity and privacy topics. The website `https://digital-strategy.ec.europa.eu/en/policies/cybersecurity` in particular is a good example. It "*aims to build resilience to cyber threats and ensure citizens and businesses benefit from trustworthy digital technologies.*"

Other countries also provide resources related to MFA. The UK's **National Cyber Security Centre** (**NCSC**) is a good example. Its goal is "*Helping to make the UK the safest place to live and work online. Understands cyber security, and distills this knowledge into practical guidance.*" NCSC publishes guidance documents on passwords, authentication, phishing, and more. In addition, guidance documents are available for organizations (such as *Advice for organisations on implementing multi-factor authentication*: `https://www.ncsc.gov.uk/guidance/multi-factor-authentication-online-services` and customers to recognize common scams.

Similarly, in Australia, the Australian Cyber Security Council provides essential and simplified guidance to its citizens that can benefit anyone. An excellent place to start is `https://www.cyber.gov.au/`.

As listed in this section, the US government and other countries offer useful information on security and MFA through their websites. These resources provide up-to-date information on new security threats and how they impact the use of MFA by organizations and individual users.

Summary

In this chapter, you were introduced to how fraud schemes can impact different authentication factors and how to reduce the associated risks. You were also provided with a comprehensive overview of the various sources to follow to stay informed about the latest security threats and the measures that can be taken to counteract them.

With the information you have acquired in this chapter, you are now equipped to understand the importance of MFA in ensuring the security of an organization's workforce. In the next chapter, we will delve into a specific case study of how ACME Corporation has utilized Microsoft's **Azure Active Directory** (**AAD**) to implement MFA and improve the security of its workforce. This chapter will give you a hands-on understanding of how MFA can be integrated into an organization's existing infrastructure and how it can provide a more robust layer of security to protect against potential threats.

Part 2: Implementing Multifactor Authentication

In today's interconnected world, organizations of all sizes and across industries must prioritize authentication security to protect their valuable data, assets, and user identities. With cyber threats growing in scale and sophistication, implementing robust security measures is no longer optional—it's an absolute necessity. Therefore, a vital component of any organization's security arsenal is **Multifactor Authentication (MFA)**, which serves as an indispensable layer of protection against unauthorized access.

Part 2 delves into the critical subject of MFA, providing a comprehensive and accessible guide for deploying and configuring MFA on different **Identity and Access Management** (**IAM**) systems. This essential resource empowers organizations to fortify their digital defenses by detailing the intricacies of MFA, along with practical examples and step-by-step instructions for implementation. The importance of tailoring MFA solutions to both workforce and customer use cases is highlighted throughout this comprehensive guide. For workforce use cases, you'll learn how to balance robust security measures and seamless user experiences, ensuring employees can access critical systems and information efficiently and securely. Meanwhile, for customer use cases, the focus shifts to prioritizing ease of use and minimizing friction while maintaining solid protection against fraudulent activities and unauthorized access.

This part has the following chapters:

- *Chapter 3, Preventing 99.9% of Attacks – MFA with Azure AD and Duo*
- *Chapter 4, Implementing Workforce and Customer Authentication Using Okta*
- *Chapter 5, Access Management with ForgeRock and Behavioral Biometrics*
- *Chapter 6, Federated SSO with PingFederate and 1Kosmos*
- *Chapter 7, MFA and the Cloud – Using MFA with Amazon Web Services*
- *Chapter 8, Google Cloud Platform and MFA*
- *Chapter 9, MFA without Commercial Products – Doing It All Yourself with Keycloak*

Preventing 99.9% of Attacks – MFA with Azure AD and Duo

Acme Corporation, having an existing subscription to Microsoft 365, already benefits from access to **Azure Active Directory** (**Azure AD**). Azure AD is Microsoft's cloud-based **identity and access management** (**IAM**) solution. As the fundamental identity platform supporting Microsoft 365, Azure AD is engineered to administer user identities, authenticate users effectively, and provide access to applications and resources per designated permissions and policies.

This chapter will explore how Acme Corporation can use a premium version of Azure AD to streamline and enhance its workforce's user management and security processes, including Acme's employees, contractors, and partners. These benefits include centralized user and group management, secure authentication, **single sign-on** (**SSO**) capabilities, and the ability to enforce advanced security measures such as **conditional access** (**CA**) and **multifactor authentication** (**MFA**).

For many years, companies have discussed the problems associated with password-based authentication. Passwords, the primary method for securing accounts and data, are becoming increasingly vulnerable to attacks as technology advances. As far back as 2019, Microsoft published several articles and has shown examples where *your password doesn't matter* (`https://techcommunity.microsoft.com/t5/microsoft-entra-azure-ad-blog/your-pa-word-doesn-t-matter/ba-p/731984`) and also that MFA should be part of the solution (`https://www.microsoft.com/en-us/security/blog/2019/08/20/one-simple-action-you-can-take-to-prevent-99-9-percent-of-account-attacks/`).

We will understand how to configure Azure AD and provide different authentication workflows to match the level of risk associated with varying roles in the organization.

As we saw in *Chapter 2*, attacks on passwords and MFA have become increasingly sophisticated. Therefore, while configuring Azure AD, we will also learn how to configure different authenticators to ensure that only authorized individuals can access sensitive company data, applications, and systems. This chapter will show how companies can reduce the risk of unauthorized access, data breaches, and insider threats by implementing strong authentication measures for the workforce.

Modern workforce users with a work-from-anywhere working culture, combined with increased adoption of personal devices for enterprise users (**bring-your-own-device**, or **BYOD**) and mobile-first strategies means that access to Acme's applications is increasingly being made from home or through other access points outside of Acme's physical corporate network.

We will describe how Acme Software can utilize Azure AD to consistently secure all the company's applications accessed from the intranet and public networks. We will also look at Duo, a **two-factor authentication** (**2FA**) product from Duo Security (part of Cisco Secure).

In addition to enterprise applications, **Azure AD Business-to-Consumer** (**Azure AD B2C**) is a customer IAM solution that allows organizations to integrate social **identity providers** (**IdPs**), such as Google and Facebook, with their customer-facing applications. **Azure AD Business-to-Business** (**Azure AD B2B**) collaboration enables the secure sharing of resources and applications with external partners and organizations, using their existing IdPs.

In this chapter, we're going to cover the following main topics:

- Azure AD setup
- Enabling **Security Assertion Markup Language** (**SAML**)-based SSO for enterprise applications
- MFA using Azure AD
- CA policies
- What is Duo and why use it?

Technical requirements

This chapter requires a working Azure AD tenant with at least an Azure AD **Premium P1** or trial subscription. The account used for this chapter must belong to one of the following roles: Cloud Application Administrator, Global Administrator, or Application Administrator.

If your company does not have a Microsoft Azure subscription, a trial one can be created at `https://azure.microsoft.com/en-us/free/`. Another option (used in this book) is to create a Microsoft developer account at `https://developer.microsoft.com/en-us/microsoft-365/dev-program`.

For Duo integration, a Duo account with administrative privileges is needed. A free trial is available at `https://signup.duo.com/` if required.

If your company already has an Azure AD tenant, feel free to skip this section. Otherwise, let's see how to configure an Azure AD tenant with a few users to use for the rest of the chapter.

Azure AD setup

If you are going to use a Microsoft 365 developer account, the first step is to go to `https://developer.microsoft.com/en-us/microsoft-365/dev-program` and create a new account:

Figure 3.1 – Microsoft 365 Developer Program initial page

Setting up a developer account is the first step toward utilizing the various services provided by Microsoft for developers. However, to use Azure AD, it is necessary to have a Microsoft 365 E5 subscription.

Microsoft 365 is a cloud-based productivity and collaboration tool suite that includes Word, Excel, PowerPoint, and OneNote applications. It also provides cloud services such as Exchange Online, SharePoint Online, and OneDrive for Business. The E5 subscription is the most comprehensive offering, providing additional security and compliance features.

Once the developer account is set up, the next step is to obtain a Microsoft 365 E5 subscription. This subscription gives access to Azure AD, a cloud-based IAM service:

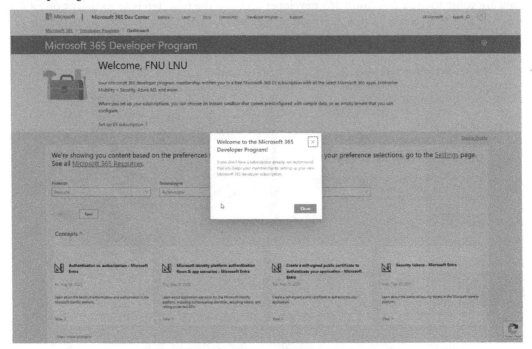

Figure 3.2 – Microsoft 365 Developer Program welcome page

With Azure AD, developers can integrate IAM into their applications and services, allowing users to sign in with their existing corporate credentials.

Two different dashboards can be used to administer Azure AD: the traditional Azure portal (`https://portal.azure.com`) or the new Entra admin center (`https://entra.microsoft.com/`).

`portal.azure.com` is the main portal for accessing and managing Azure cloud services. Azure is Microsoft's cloud computing platform that provides a wide range of services such as **virtual machines** (**VMs**), storage, networking, and analytics. The Azure portal provides a centralized location for managing and monitoring these services. We are only going to use the **Azure Active Directory** section of the portal in this chapter:

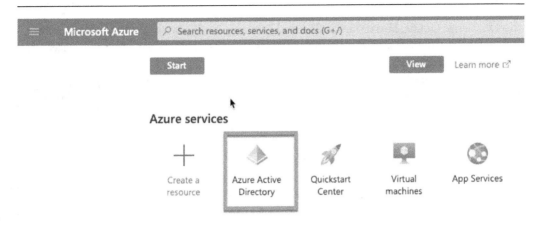

Figure 3.3 – Azure portal: services

You can see the **Azure Active Directory** overview page for the Microsoft Azure portal:

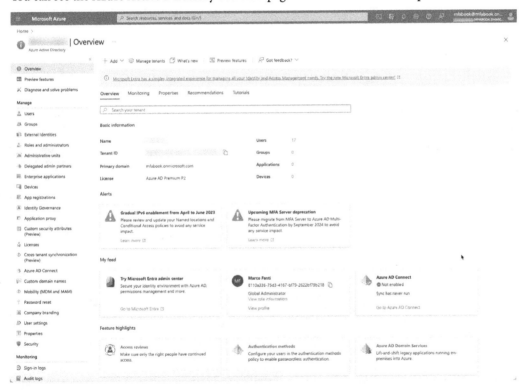

Figure 3.4 – Azure portal: Azure Active Directory

On the other hand, the Entra admin center, `entra.microsoft.com`, is the portal for managing Microsoft 365 services. Microsoft 365 is a suite of cloud-based productivity and collaboration tools that includes applications such as Word, Excel, PowerPoint, and OneNote, as well as cloud services such as Exchange Online, SharePoint Online, and OneDrive for Business. The Entra admin center portal provides a centralized location for managing these services, including user accounts, licenses, security settings, and more:

Figure 3.5 – Entra admin center

As with the Azure portal, We are only going to use the **Azure Active Directory** section of the Entra admin center in this chapter:

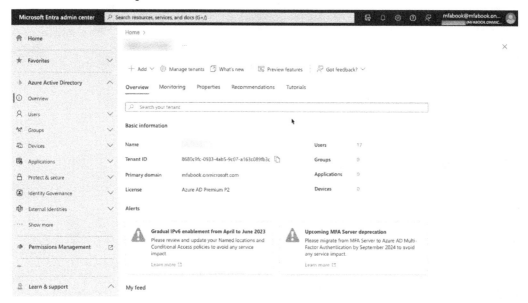

Figure 3.6 – Entra admin center: Azure Active Directory

If you are using a trial license, another benefit of using the Microsoft 365 Developer Program is access to a set of users and applications that can be preconfigured, as can be seen in the following screenshot. Moving forward in this chapter, it is important to note that the context and discussions will presume the existence of normal or regular users in addition to the system administrator:

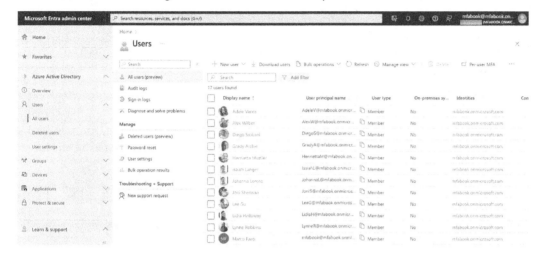

Figure 3.7 – Azure AD users

Azure AD is Microsoft's cloud-based IAM service, providing secure access to organizations' applications and resources. Azure AD enterprise applications are integrated with Azure AD for SSO, user authentication, and access management.

These applications can be **software-as-a-service** (**SaaS**) apps such as Office 365, Salesforce, or Dropbox, custom-developed applications, or third-party applications available in the Azure AD app gallery. When an application is integrated with Azure AD, users can authenticate and access it using administrative credentials.

Enabling SAML-based SSO for enterprise applications

As discussed in *Chapter 1*, SSO is very important for corporations. Azure AD provides secure authentication for users accessing Microsoft 365 services. It also enables SSO, allowing users to access multiple Microsoft 365 services and other cloud applications with a single set of credentials. Azure AD also enables integration with other third-party applications, allowing organizations to extend their SSO capabilities beyond Office 365 and manage access to other SaaS applications.

Azure AD offers multiple methods to achieve this integration, catering to various application types and requirements:

- **Azure AD app gallery**: The Azure AD app gallery provides preconfigured templates for thousands of popular enterprise applications, such as Salesforce, ServiceNow, and Workday. These templates simplify the integration process and enable SSO capabilities with minimal configuration.

- **Custom SAML or OpenID Connect (OIDC)-based applications:** For applications that support SAML or OIDC for authentication and authorization, you can create custom enterprise applications in Azure AD. This allows you to configure SSO and user provisioning with applications unavailable in the Azure AD app gallery.

- **Password-based SSO:** For applications that do not support SAML or OIDC, Azure AD offers password-based SSO. This method securely stores and manages user credentials for these applications, enabling users to access them without entering their usernames and passwords manually.

- **Microsoft Authentication Library (MSAL):** MSAL is a set of libraries and **software development kits (SDKs)** that simplifies integrating Azure AD authentication into custom-developed applications. MSAL supports various platforms, including .NET, JavaScript, Android, and iOS, enabling developers to implement modern authentication methods and secure access to APIs and resources.

The following screenshot shows some of the featured applications in the Microsoft Entra admin center:

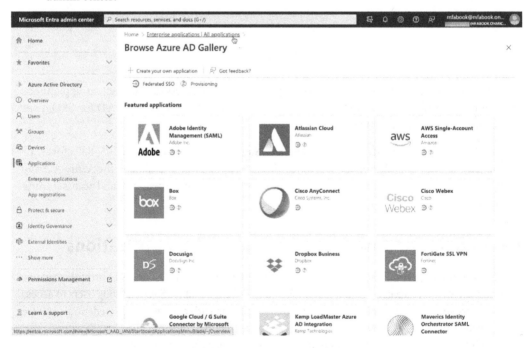

Figure 3.8 – Microsoft Entra admin center: Azure AD gallery

To start, Acme wanted to use SSO with its existing enterprise applications and decided to use applications already available in the Azure AD app gallery.

Adding an enterprise application

Instead of using an application used daily by its workforce, such as Salesforce, Acme started using Azure AD with a test application provided as part of Azure AD installation. The steps to do so are set out here:

1. Go to the **Azure Active Directory admin center** (`https://aad.portal.azure.com/`) and sign in using one of the following roles: Global Administrator, Cloud Application Administrator, or Application Administrator:

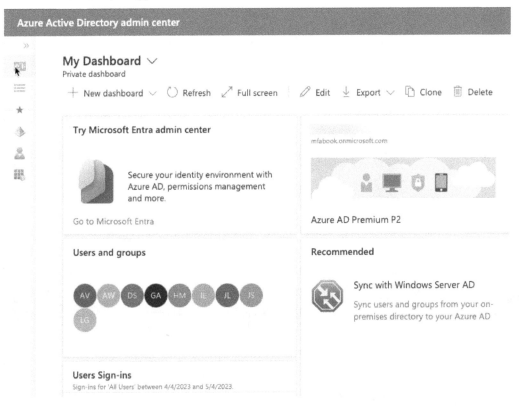

Figure 3.9 – Azure Active Directory admin center

2. To allow for more screen space, the text labels of the Azure AD menu blade are hidden by default. Select >> to show text labels:

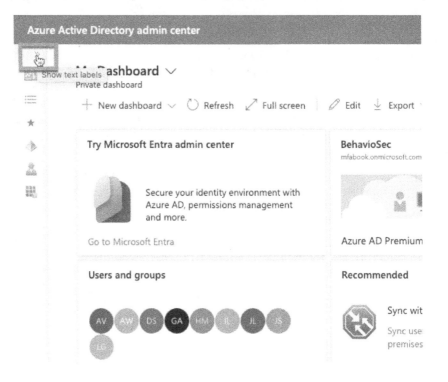

Figure 3.10 – Azure Active Directory admin center: Show text labels

3.   Select **All services**:

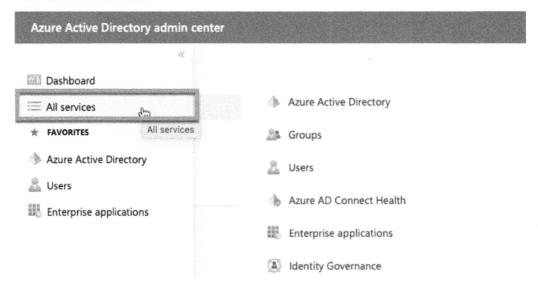

Figure 3.11 – Azure Active Directory admin center: All services

4. Select **Enterprise applications**:

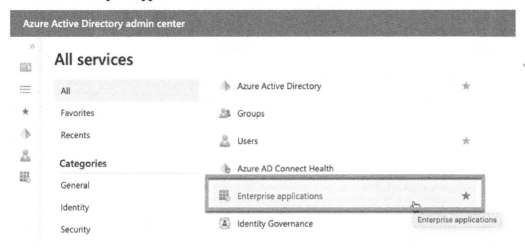

Figure 3.12 – Azure Active Directory admin center: Enterprise applications

5. An **All applications** pane opens and displays a list of the applications in your Azure AD tenant. In the **Enterprise applications** pane, select **New application**:

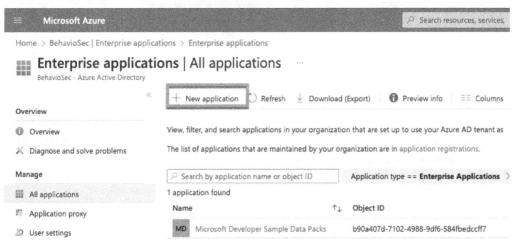

Figure 3.13 – The All applications pane

6. A **Browse Azure AD Gallery** pane opens and displays tiles for cloud platforms, on-premises applications, and featured applications. Search for and select the **Azure AD SAML Toolkit** application:

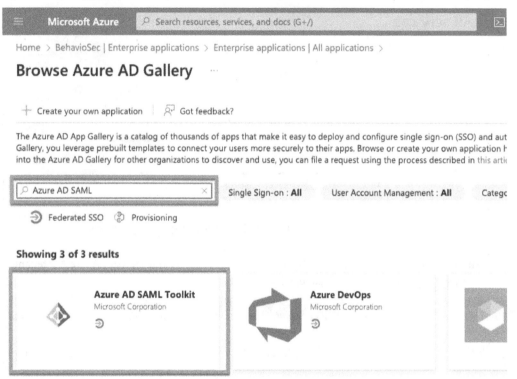

Figure 3.14 – Azure AD SAML Toolkit application

7. Enter a name that you want to use to recognize the instance of the application—for example, **Acme's Azure AD SAML Toolkit**. Click **Create**:

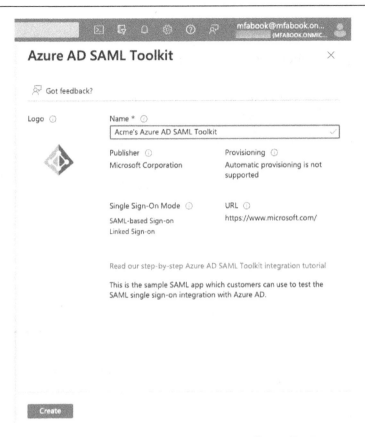

Figure 3.15 – The Acme's Azure AD SAML Toolkit application

Now that the **Acme's Azure AD SAML Toolkit** application has been created, let us assign a user account to it.

Assigning a user account to the Acme's Azure AD SAML Toolkit application

Acme does not have an IAM system that automatically assigns users to roles (and consequently to applications). To test our first enterprise application, we are going to create (if needed) and assign one or more users manually to the **Acme's Azure AD SAML Toolkit** application.

To add a user to an application, proceed as follows:

1. Select **Enterprise applications**, and then search for and select the **Acme's Azure AD SAML Toolkit** application.

2. In the left pane, select **Users and groups** or **Assign users and groups**:

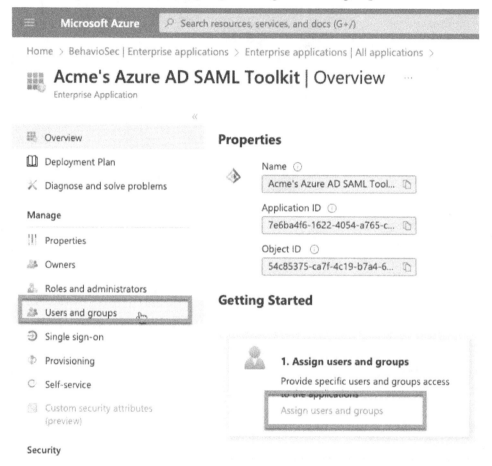

Figure 3.16 – Enterprise applications: Users and groups

3.   Select **Add user/group**:

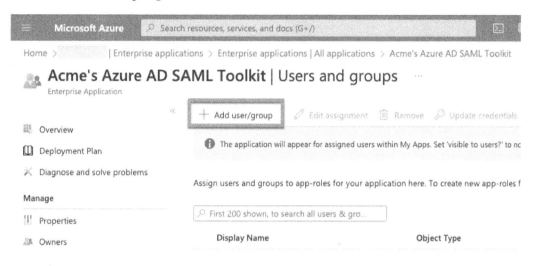

Figure 3.17 – Adding users to an application

4.   In the **Add Assignment** pane, select **None Selected** under **Users and groups**:

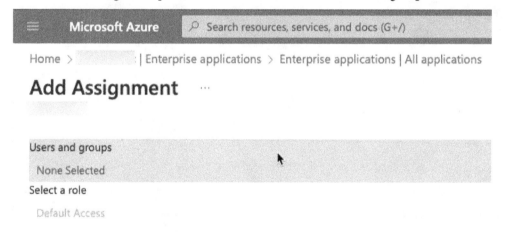

Figure 3.18 – Selecting users for an application

5.   Search for the user(s) that you want to assign to the application.

6.   Click **Select**:

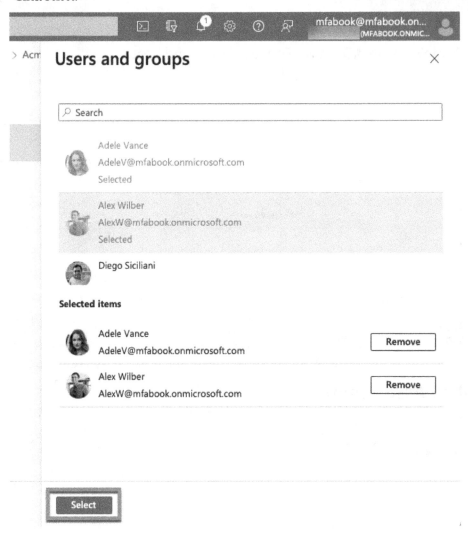

Figure 3.19 – Selecting users for an application (continued)

7. In the **Add Assignment** pane, click **Assign** at the bottom of the pane:

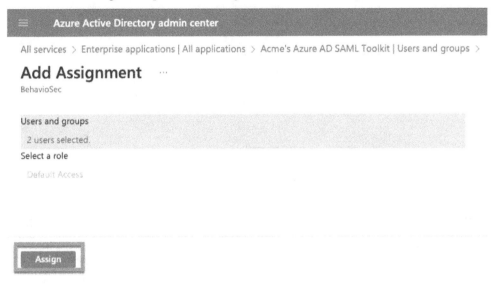

Figure 3.20 – Assigning users to an application

We have assigned user accounts to the **Acme's Azure AD SAML Toolkit** application. We are now ready to enable SSO for the application.

Enabling SAML-based SSO for the Acme's Azure AD SAML Toolkit application

Users can sign in to an application using their Azure AD credentials only after enabling SSO access to those applications.

To allow SSO access to an application, proceed as follows:

1. Select **Enterprise applications**, and then search for and select the **Acme's Azure AD SAML Toolkit** application.

2. In the left pane, select **Single sign-on**.

3. Select **SAML**:

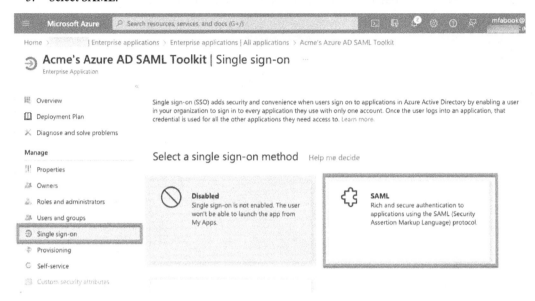

Figure 3.21 – Acme's Azure AD SAML Toolkit: Single sign-on pane

To begin the configuration of SSO in Azure AD, you must add identifier, sign-in, and reply URL values, and then download a certificate. Here's how you can do that:

1. Select **Edit** in the **Basic SAML Configuration** section on the **Set up Single Sign-On with SAML** pane:

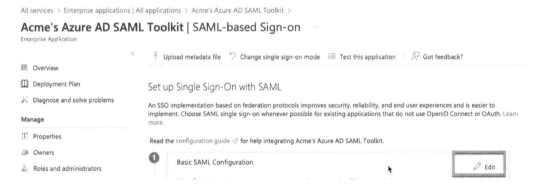

Figure 3.22 – The Set up Single Sign-On pane

2. For **Identifier (Entity ID)**, enter `https://samltoolkit.azurewebsites.net`.

3. For **Reply URL (Assertion Consumer Service URL)**, enter `https://samltoolkit.azurewebsites.net/SAML/Consume/`.

4. For **Sign on URL**, enter `https://samltoolkit.azurewebsites.net/SAML/Login/`.

5. Select **Save**:

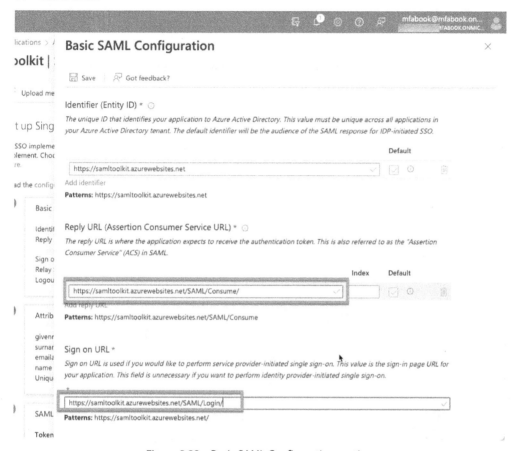

Figure 3.23 – Basic SAML Configuration section

The **Assertion Consumer Service (ACS)** URL is a critical component in the SAML authentication process and is commonly used for SSO between a **service provider (SP)** and an **identity provider (IDP)**.

The ACS URL is an endpoint on the SP's side where SAML assertions are sent by the IdP after the user has been successfully authenticated. In other words, it is the location where the SP receives and processes the SAML assertion containing the user's authentication and authorization information.

When a user attempts to access an SP application (such as a web application), the SP initiates the SSO process by sending a SAML authentication request to the IDP. The user is then authenticated by the IDP (for example, through a username and password or MFA). Once the user is authenticated, the IdP generates a SAML assertion containing the user's identity and access rights and sends it to the SP's ACS URL.

Important note

The values in **Reply URL (Assertion Consumer Service URL)** and **Sign on URL** are not the final ones. They are only placeholders to be replaced with the actual reply and sign-on URL created by the application.

6.   In the **Token signing certificate** section, click the **Download** link next to **Certificate (Raw)** to download the SAML signing certificate and save it to be used later:

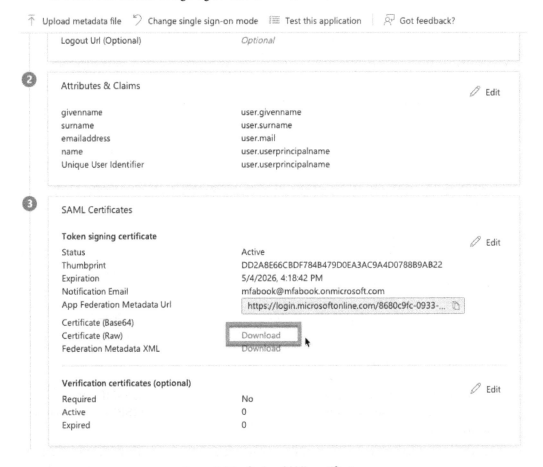

Figure 3.24 – Saving SAML certificate

We just completed the basic SAML-based sign-on configuration for the **Acme's Azure AD SAML Toolkit** application. We will now configure an instance of the application using the certificate we just saved, and we will then come back to edit the values on the basic configuration with the URLs of the instance we are creating.

Configuring SSO for the Acme's Azure AD SAML Toolkit application

Using SSO in the application requires you to register the user account with the application and add the SAML configuration values. Here's how to do that:

1. To start, go to `https://samltoolkit.azurewebsites.net` and click on **Register**:

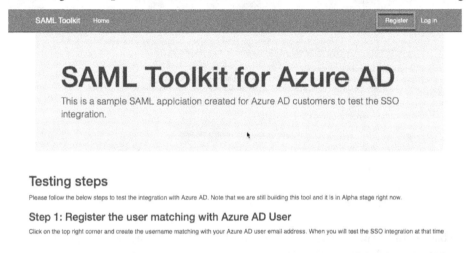

Figure 3.25 – Registering new SAML toolkit application

2. Using one of the email addresses assigned to the **Acme's Azure AD SAML Toolkit** application, create a new account. Click **Register**:

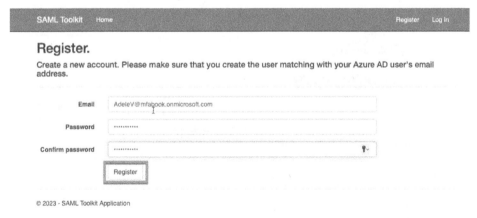

Figure 3.26 – Creating a matching user in the SAML toolkit application

3. Log in with the username and password created in the previous step. Click on the **SAML Configuration** option on the navigation bar:

Figure 3.27 – Logging in to the SAML toolkit application

4.   Select **Create**:

Figure 3.28 – After configuring, creating a new application

5.   Go back to the **Azure Active Directory admin center**. Save the values found in the **Azure AD Set up Acme's Azure AD SAML Toolkit** section of the **Acme's Azure AD SAML Toolkit** application's configuration:

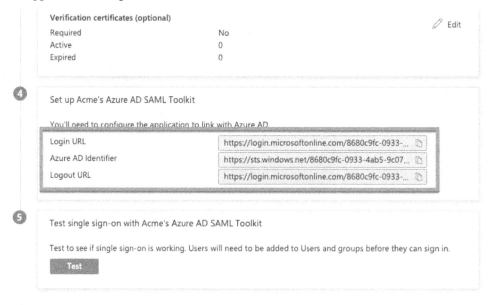

Figure 3.29 – Values for the new application

6. Fill in the SAML configuration. For **Login URL**, **Azure AD Identifier**, and **Logout URL**, enter the values saved in the previous step. Choose the file downloaded from the **Acme's Azure AD SAML Toolkit** application for the **Certificate RAW** field.

7. Click **Create**:

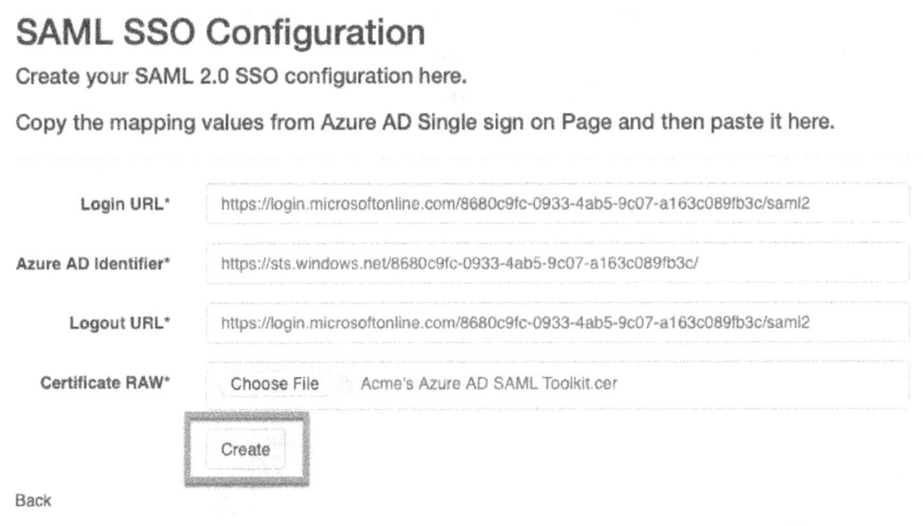

Figure 3.30 – After configuring, creating a new application

8. Now, copy the new **SP Initiated Login URL** and **Assertion Consumer Service (ACS) URL** values generated for your SAML SSO configuration, then go back to the basic configuration of the **Acme's Azure AD SAML Toolkit** application:

Figure 3.31 – Unique URLs for the new application

This completes the configuration of a new instance of the SAML toolkit application. We now go back to `https://aad.portal.azure.com/` to complete the SAML SSO configuration, as follows:

1. Click **Edit** in the **Basic SAML Configuration** section on the **Set up Single Sign-On** pane to update the SSO values for the **Acme's Azure AD SAML Toolkit** application.

2. For **Reply URL (Assertion Consumer Service URL)**, enter the **Assertion Consumer Service (ACS) URL** value that you previously recorded.

3. For **Sign on URL**, enter the **SP Initiated Login URL** value that you previously recorded.

4. Click **Save**:

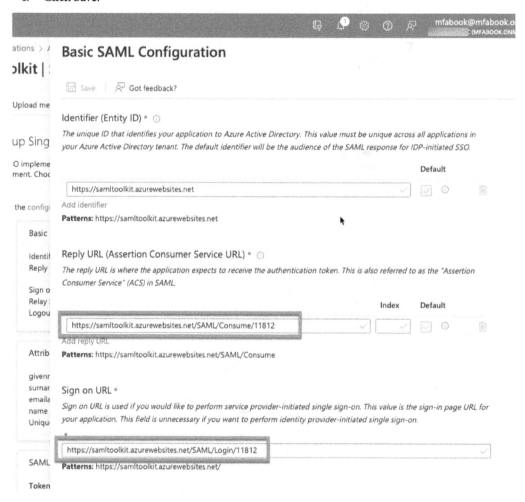

Figure 3.32 – Updating SSO values

We just completed the SAML SSO configuration for Acme's toolkit app, so let's test it to see if everything is working as expected.

Testing SSO in the Acme's Azure AD SAML Toolkit application

We are now ready to test signing on to the new app. To test SSO, do the following:

1. Save the **Sign on URL** value you just added to the **Basic SAML Configuration** section:

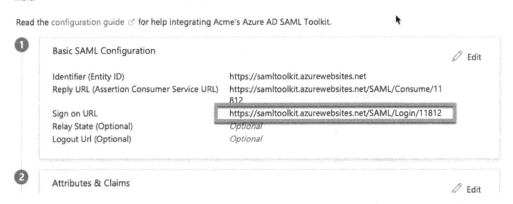

Figure 3.33 – Testing SSO

2. Open a new browser window in InPrivate or Incognito mode and browse to the URL you just saved. It should look like this: `https://samltoolkit.azurewebsites.net/SAML/Login/9999` (where `9999` is the number assigned for your own application when it was created).

3. Click **Log in** to sign in to the **Acme's Azure AD SAML Toolkit** application using the Azure AD credentials of the user account that you assigned to the application:

Figure 3.34 – Signing in to the Acme's Azure AD SAML Toolkit application

> **Important note**
> Enter the password assigned to the user in Azure AD, not the one you registered to the application. Remember—we are using Azure AD for SSO, so the user ID and password will be verified in Azure AD, not locally in the application.

On the **Sign in** page, enter the username and click **Next**, and then enter the password:

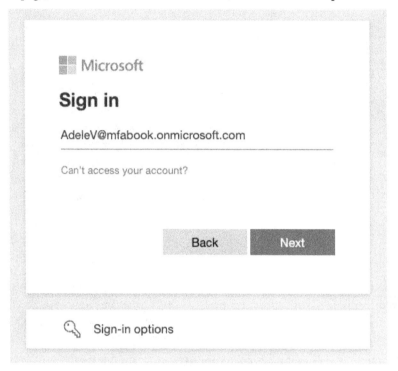

Figure 3.35 – Microsoft Sign in page

Depending on the version of Azure AD, you may be prompted to set up Microsoft Authenticator. Select **Ask later** to go directly to the application:

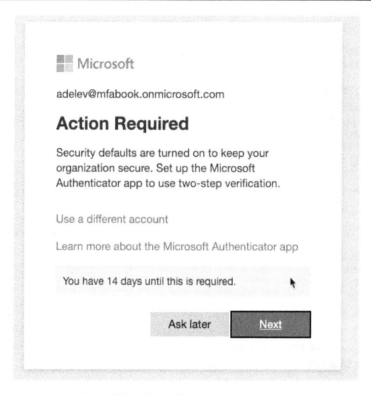

Figure 3.36 – Microsoft Action Required page

If everything was set up correctly, you should see the home page for the **SAML Toolkit for Azure AD** application:

Figure 3.37 – Successful SSO login to Acme's application

A lot of the work performed so far was to create an application for us to test with Azure AD's SSO. Adding SSO to cloud applications that you already have requires much less work than this first example (only the ones on the **Azure Active Directory admin center**).

Now that Acme has an application configured for SSO, Acme's management would like to start using MFA for authentication, instead of the username and password that we have used so far. Let's see how we can make SSO more secure with MFA.

MFA on Azure AD

So far, all employees at Acme have been using only passwords to authenticate to enterprise applications protected by Azure AD SSO.

Acme's management has read the following information: "*MFA can block over 99.9 percent of account compromise attacks*" (`https://www.microsoft.com/security/blog/2019/08/20/one-simple-action-you-can-take-to-prevent-99-9-percent-of-account-attacks/`).

Research shows that passwords are used in more than one account by the majority of users. As password leaks become more and more frequent, security experts recommend the use of at least two factors of authentication for most enterprise accounts (`https://techcommunity.microsoft.com/t5/azure-active-directory-identity/your-pa-word-doesn-t-matter/ba-p/731984`).

The US government also mandates the use of MFA by all of its agencies for authenticated access to federal systems by agency staff, contractors, and partners (`https://zerotrust.cyber.gov/federal-zero-trust-strategy/#identity`).

MFA works by requiring users to provide authentication factors to verify that users are really who they say they are. MFA requires a combination of two or more of the following authentication factors. Each factor must be from a different category:

- Something you know, such as a password, a passphrase, or a PIN
- Something you have in your possession, such as a smartphone, laptop, physical token, or smartcard
- Some form of biometric identification, such as a fingerprint, voice, or facial recognition

In addition to being used during login, MFA in Azure AD can also be used to secure a password reset attempt.

Also, as we are going to see with Duo, administrators can choose between different forms of secondary authentication for different users and applications.

Security defaults in Azure AD are one way to quickly enable Microsoft Authenticator for all users attempting to log in to Azure AD or SSO-enabled applications.

Acme's management decided not to use those defaults and to instead go for a more controlled approach to test MFA before using it for all employees and other users of the system.

To enable this controlled deployment approach, Acme chose CA policies. CA allows Acme to create and define policies for specific sign-in events and to request that additional actions be performed before a user is granted access to an application.

CA policies can be viewed as basic `if/then` statements that dictate the actions users must complete to access a resource. For example, if a payroll manager wants to access the payroll application, they must perform MFA.

Administrators primarily aim to achieve two objectives:

- Enable users to work efficiently at any time and from any location
- Safeguard the organization's resources

By implementing CA policies, administrators can effectively apply appropriate access controls when necessary, ensuring their organization's security. In addition, CA policies allow administrators to create and define policies based on your organization's infosec policies or as defined by industry and regulatory compliance standards.

Microsoft has enabled security defaults as an initial configuration for all tenants created after October 2019.

Security defaults in Azure AD include the following:

- MFA is required for all administrators
- All users must register for Azure AD MFA
- All users must perform MFA when necessary

To use CA, security defaults must be disabled. Before configuring CA, we must look at the steps required to disable Azure AD default security.

Disabling default security

To disable security defaults in your directory, follow these steps:

1. In **Azure Active Directory admin center**, select **Properties**:

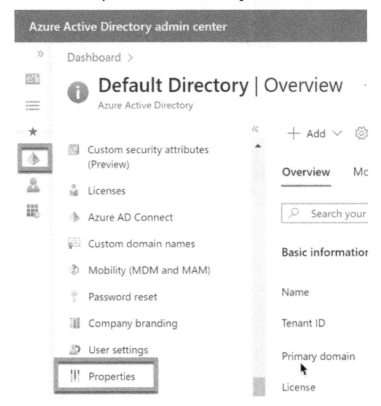

Figure 3.38 – Azure AD admin center: Properties

2. Click **Manage security defaults**:

Figure 3.39 – Azure AD security defaults

3. Set the **Enable Security defaults** toggle to **No**.

4. Choose **My organization is using Conditional Access** as the reason.

5. Click **Save**:

Figure 3.40 – Azure AD security defaults (continued)

If your tenant already has default security disabled, no changes are required.

Acme is using Microsoft's recommended way to enable and use Azure AD MFA: with CA policies. CA policies can be complex, and they can involve many different conditions and actions, so we are going to start with a simple scenario, using the SAML toolkit application we created earlier.

CA policies

To create Acme's first CA policy on Microsoft Azure AD, proceed as follows:

1. Search for and select **Azure Active Directory**. Then, select **Security** from the menu on the left-hand side:

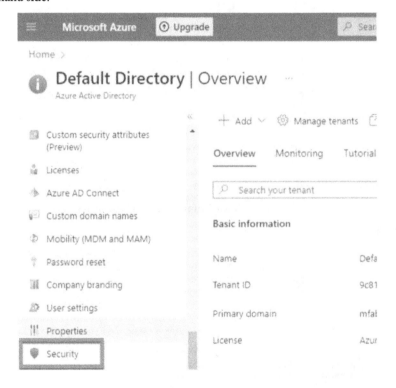

Figure 3.41 – Changing security CA

2. Click **Conditional Access**:

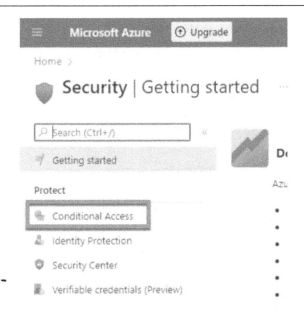

Figure 3.42 – Selecting Conditional Access

3.   Select **+ New policy**, and then select **Create new policy**:

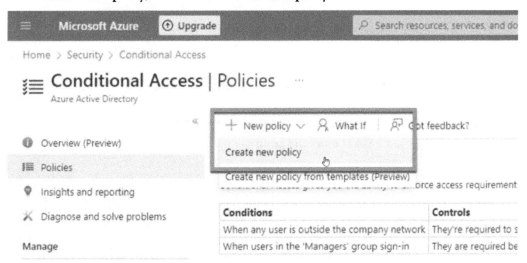

Figure 3.43 – Creating a new policy

4. Enter a name for the policy, such as **MFA Pilot**.

5. Under **Assignments**, select the current value under **Users or workload identities**:

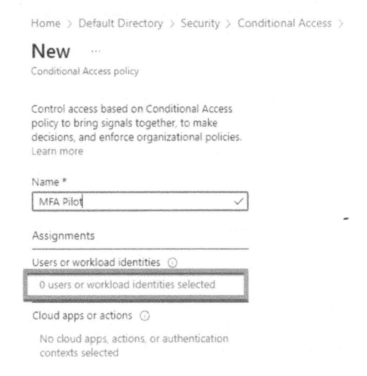

Figure 3.44 – Assignments

6. Under **What does this policy apply to?**, choose **Users and groups**.

7. Under **Include**, choose **Select users and groups**.

8. Select **Users and groups**:

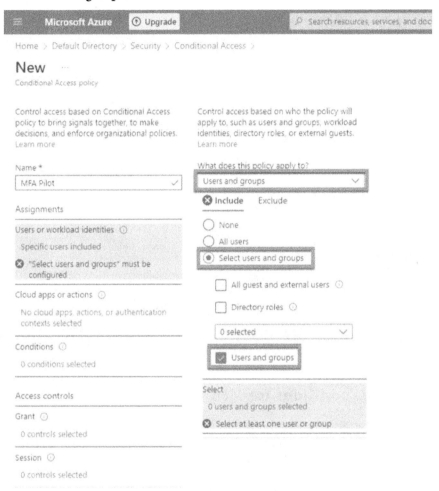

Figure 3.45 – Assigning users and groups

Since no users or groups are assigned yet, a list of users and groups opens automatically.

9. Browse for and select your Azure AD group, such as **MFA-Test**, then click **Select**:

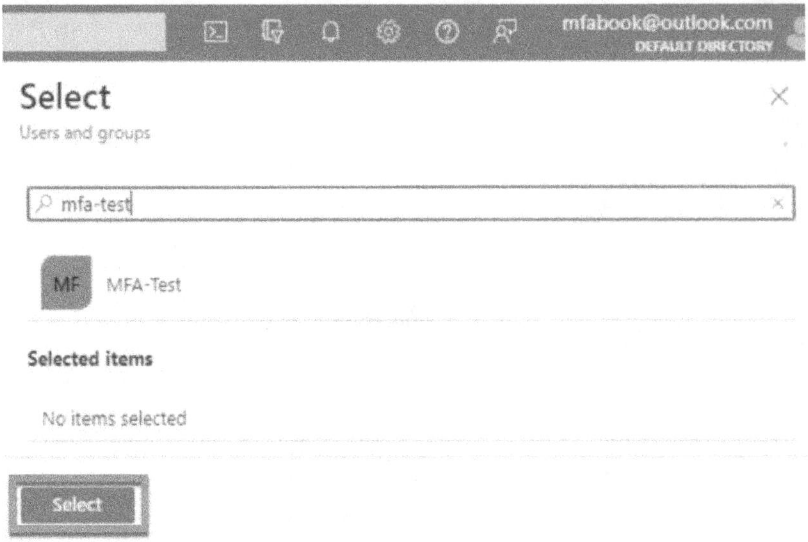

Figure 3.46 – Selecting groups

We started the creation of Acme's first CA policy and selected the group to apply the policy to. In the next section, we configure the conditions under which to apply the policy.

Configuring the conditions for MFA

So far, we have created an access policy that should be applied to a certain group of users. We now add the applications this policy applies to. When the users in the selected group try to access the applications where this policy applies, the policy will be triggered.

Configuring which enterprise applications require MFA

Configuring which enterprise applications this policy applies to allows an administrator to choose different conditional policies for different applications. Here are the steps to achieve this:

1. Select the current value under **Cloud apps or actions**.

2. Under **Select what this policy applies to**, verify that **Cloud apps** is selected.

3. Under **Include**, choose **Select apps**:

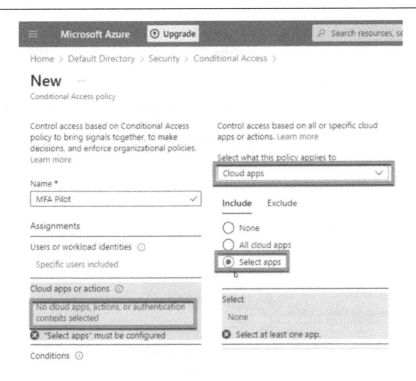

Figure 3.47 – Cloud apps or actions

4. Since no apps are yet selected, a list of apps opens automatically.

5. Browse the list of available cloud apps that can be used. Search and choose the **Acme's Azure AD SAML Toolkit** app. Then, click **Select**:

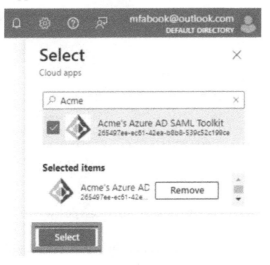

Figure 3.48 – Selecting cloud apps

The main reason we are creating this access policy is to require MFA when the selected group of users access the applications this access policy applies to. The next step is to configure MFA-only access.

Configuring MFA for access

Next, we configure access controls—the requirements for a user to be granted access—as follows:

1.   Under **Access controls**, select the current value under **Grant**.

2.   Select **Grant access**.

3.   Select **Require multi-factor authentication**.

4.   Click **Select**:

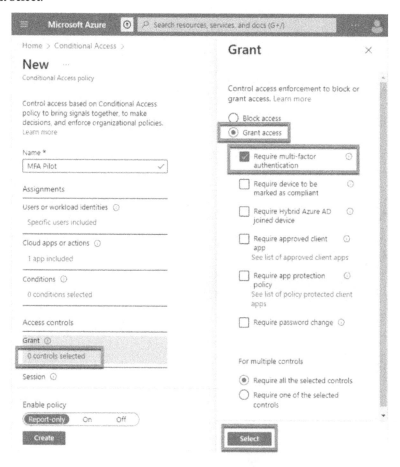

Figure 3.49 – Conditional Access policy

Policies can be created but don't come into effect until they are activated, which is our next and final step before being able to test this conditional policy.

Activating the CA policy

CA policies can be set to **Report-only** if you just want to see how the configuration would affect users without enforcing it. They can also be set to **Off** if you don't want to use the policy right now. In Acme's case, we are going to enable the CA policy to enforce Azure AD MFA for the users in the **MFA-Test** group, as follows:

1. Under **Enable policy**, select **On**:

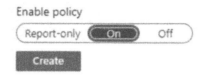

Figure 3.50 – Enabling a policy

2. To apply the CA policy, click **Create**.

Just as we tested SSO using username and password for the **Acme's Azure AD SAML Toolkit** application, we are now going to test the same application and verify that, given the correct users, the conditional policy will trigger MFA for the user signing in.

When implementing CA policies, it's crucial to use test groups to minimize potential disruptions to essential company work. Test groups allow administrators to evaluate the impact of new policies on a small subset of users before rolling them out to the entire organization. This approach ensures that any issues or unintended consequences can be identified and addressed without impacting the productivity of the whole workforce.

The process of using test groups for CA policies involves the following steps:

- **Create a test group**: Begin by identifying a small group of users to serve as the test group. These users should ideally represent a cross-section of organizational roles and responsibilities to ensure the policies are evaluated across different use cases.

- **Define the CA policy**: Develop the desired if/then statements that outline the required actions users must complete before accessing specific resources. For example, if a user wants to access sensitive financial data, they must be connected to the company's **virtual private network (VPN)** and complete MFA.

- **Apply the policy to the test group**: Assign the newly created CA policy to the group, ensuring that it affects only the selected users and not the entire organization.

- **Monitor and gather feedback**: Observe the test group's experience with the new policy, collect feedback, and identify any issues or areas for improvement. Assess whether the policy achieves the desired balance between security and productivity.

- **Adjust and refine the policy**: Based on the feedback and observations, make any necessary adjustments to optimize its effectiveness and minimize disruptions to users' workflow.

- **Gradual rollout**: Once the policy has been tested and refined, gradually roll it out to the rest of the organization. Monitor the implementation to ensure a smooth transition and address any issues arising during the rollout.

Administrators can methodically implement CA policies by starting with test groups while mitigating the risk of negatively impacting essential company operations. In addition, this approach allows a more controlled and effective integration of new security measures, ensuring both productivity and protection for the organization.

Testing Azure AD MFA

For the **MFA Pilot** CA policy, we choose **MFA-Test** as the group of users to which the policy applies. If we try to test signing on to Acme's SAML toolkit application using a user that is not part of the group, the user would still be able to sign on using a user ID and password only, bypassing Azure AD MFA. Here's what we need to do:

1. Open a new browser window in InPrivate or Incognito mode and browse to the login URL. Again, it should look like this: `https://samltoolkit.azurewebsites.net/SAML/Login/9999`.

2. Click **Log in**:

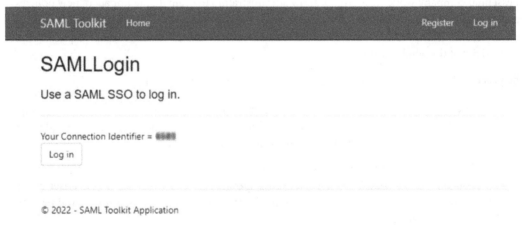

Figure 3.51 – Testing MFA login

Sign in with your non-administrator test user. Make sure the user is part of the **MFA-Test** group (or one of the group or groups selected for the CA policy).

3. You're required to register for and use Azure AD MFA. Click **Next** to begin the process:

Figure 3.52 – Starting MFA registration

4. You can choose to configure using the Microsoft Authenticator app or a different authenticator app (such as Authy or Google Authenticator).

5. Click **Next** to continue:

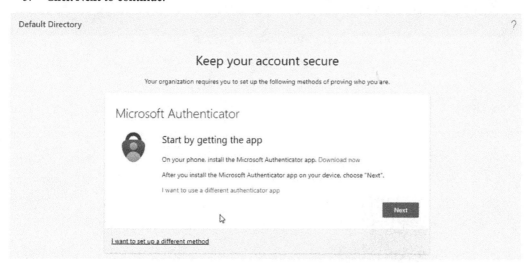

Figure 3.53 – Starting MFA app registration

6. Complete the instructions on the screen to configure the method of MFA that you've selected. Click **Next**:

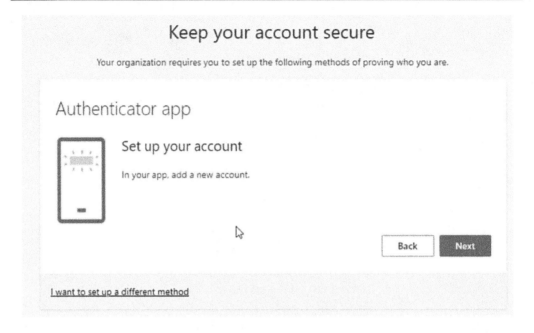

Figure 3.54 – Setting up the Authenticator app

7.   Scan the QR code. Click **Next**:

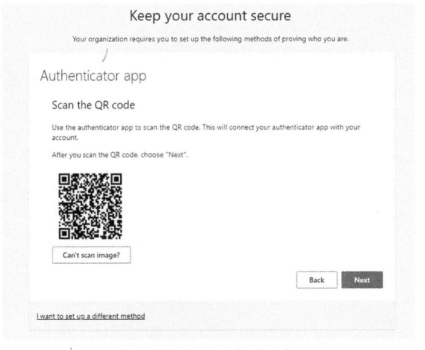

Figure 3.55 – Scanning the QR code

8. Enter the code from the Authenticator app to complete the setup:

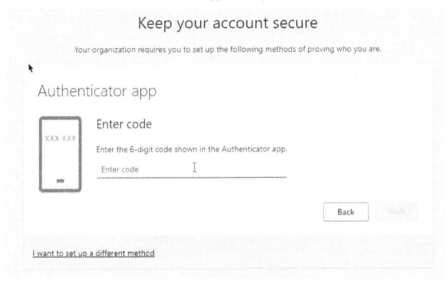

Figure 3.56 – Entering the code

9. Click **Next**. You should have successfully logged in to Acme's SAML toolkit application using Azure AD SSO and MFA.

After the authenticator app setup is complete, you won't be prompted for it again. Subsequent logins should look like the following, after the user ID and password:

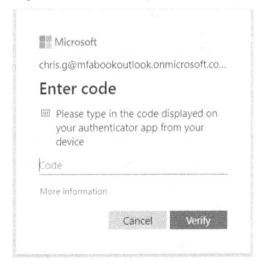

Figure 3.57 – MFA prompt after registration

Failure to enter the correct code will result in a new prompt:

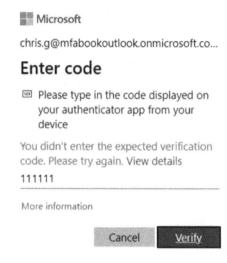

Figure 3.58 – Wrong authenticator app code entered

Now that Acme has started requiring users to perform MFA during sign-on, it also wants users to use MFA for password resets, enhancing the security of password resets as well.

Enabling combined security information registration in Azure AD

An important aspect of using passwords is being able to reset them when they expire or when users forget their passwords. Azure AD allows users to register authentication methods not only for Azure AD MFA but also for **self-service password reset (SSPR)**. Even though similar methods were used for Azure AD MFA and SSPR, users previously had to register for both features separately. To avoid confusion, Microsoft now offers combined registration. Users can register once and get the benefits of both Azure AD MFA and SSPR.

To enable combined registration, complete these steps:

1.   Log in to the Azure portal as a user administrator or global administrator.

2.   Go to **Azure Active Directory** | **User settings** | **Manage user feature settings**:

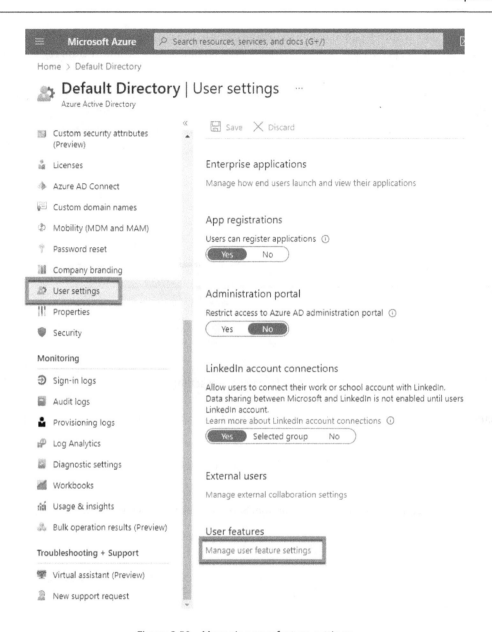

Figure 3.59 – Managing user feature settings

3.   In the **User features** panel, below the **Users can use the combined security information registration experience** text, click **All** to select a combined security information registration experience for all users.

New tenants have all users automatically enabled and have no choice to disable this option:

Figure 3.60 – Managing user settings

To enforce the rollout of Azure AD MFA, companies should also require MFA registration for all users. Here's how to do that:

1.   In the **Security** panel, select **Identity Protection** and then **MFA registration policy**.

2.   Select **All users**.

3.   Select **On** under the **Enforce policy** option. Save the policy by clicking **Save**:

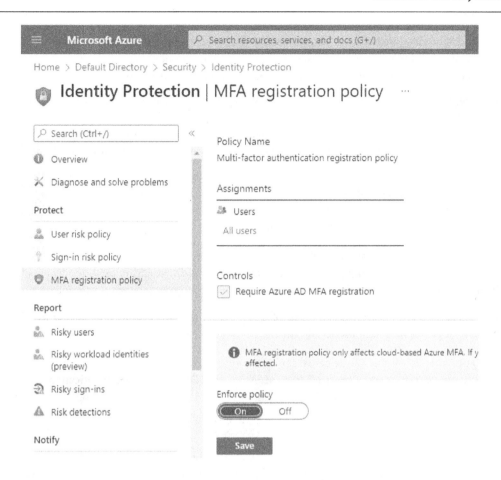

Figure 3.61 – MFA registration policy

In this section, we covered the creation of CA policies on Microsoft Azure AD. We shall look at Duo in the next section.

What is Duo and why use it?

Duo, the product we are going to introduce in this section, is an MFA solution that adds an extra layer of security to the user authentication process. Companies may choose to use Duo as a secondary factor of authentication for several reasons, such as the following:

- **Enhanced security**: Duo's MFA requires users to provide an additional authentication factor beyond their username and password. This added layer of protection helps protect against unauthorized access, even if an attacker compromises a user's credentials.

- **Ease of use**: Duo offers a user-friendly experience, with several options for secondary authentication factors such as push notifications, one-time passcodes, or biometric verification. Users can easily authenticate using smartphones, making the process quick and convenient.

- **Adaptive authentication**: Duo provides risk-based authentication, which evaluates the risk level of each login attempt based on factors such as the user's location, device, and network. This context-aware approach enables companies to apply stricter authentication requirements for high-risk scenarios while maintaining a seamless user experience for low-risk situations.

- **Device visibility and control**: Duo offers device insight and management features, helping organizations track and assess the security posture of devices accessing their systems. This information can be used to enforce access policies based on device security, ensuring that only trusted devices can access sensitive resources.

By using Duo as a secondary authentication factor, companies can strengthen their security posture, protect sensitive information, and provide a user-friendly authentication experience, all while maintaining compatibility with their existing systems and meeting compliance requirements.

Duo Security offers four different versions of its MFA solution to cater to the varying needs of organizations of different sizes and requirements. These versions are set out here:

- **Duo Free**: This version is designed for small organizations or those starting with MFA. It is free for up to 10 users. It provides basic MFA functionality, including authentication via Duo Push, SMS passcodes, and one-time passcodes generated by the Duo Mobile app or a hardware token. Duo Free also includes self-service user enrollment and basic administrative controls.

- **Duo Essentials**: As a more advanced version of Duo Free, Duo Essentials provides additional features and scalability. This version supports a variety of authentication methods, such as phone callbacks and biometric verification. Duo Essentials also offers SSO support, enabling users to access multiple applications with a single set of credentials, and includes more comprehensive reporting and user management options.

- **Duo Advantage**: Building on Duo Essentials, Duo Advantage adds extra capabilities, such as adaptive authentication with the help of the Duo Trust Monitor feature. It evaluates the risk level of each authentication attempt based on factors such as user location, device, and network, allowing for context-aware security policies. Duo Advantage also includes device visibility and control features that provide insight into the security posture of devices accessing the organization's resources. Additionally, this version supports more granular access policies based on user and device attributes.

- **Duo Premier**: As the most complete version of Duo's MFA solution, Duo Premier includes all the features of Duo Advantage and adds advanced capabilities designed for large enterprises and organizations with strict security requirements. It offers device trust and management, enabling organizations to enforce access policies based on device security and validate the trustworthiness of devices accessing their systems.

In order to minimize support issues related to Microsoft Authenticator, Acme chose to explore using Duo as an alternative. The paid versions of Duo offer dedicated help desk support from Duo Security, as well as a comprehensive and secure administration tool for managing mobile authentication and user administration tasks. Using Duo allows Acme to test Microsoft's capabilities for creating custom CA policies. Duo Security enables Acme to test enforcing specific forms of MFA, according to the application.

Integrating Duo and Microsoft Azure AD

To use Duo instead of Microsoft Authenticator, we are going to create a new CA policy. The policy is going to allow users to register Duo Authenticator during the login, if not previously registered. The policy will also enforce the use of Duo when required.

The steps required to enable Duo authentication for Microsoft Azure AD are set out here:

1. In `https://aad.portal.azure.com/`, select **Azure Active Directory** and then **Users**:

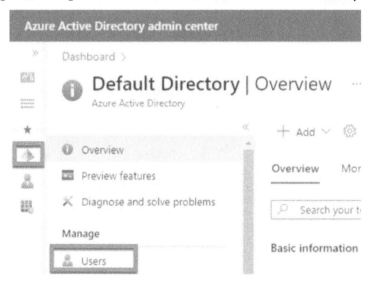

Figure 3.62 – Users pane

2. Select + **New user** to create a new user:

Figure 3.63 – Creating a new user

3. Select **Create user**:

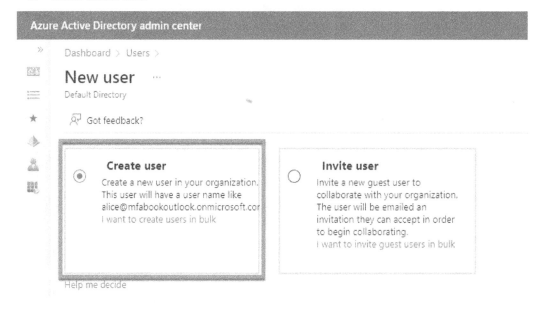

Figure 3.64 – Create user

4. Enter the user information. Click on **Roles** before saving the user:

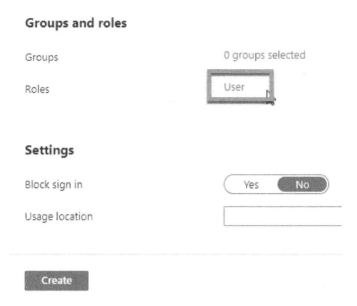

Figure 3.65 – Assigning roles

5.   Search for and select the **Global administrator** role. Click **Select**:

Figure 3.66 – Global administrator user

6.   Click **Create** to complete the creation of the user.

The next set of steps requires a Duo account:

1. Sign up for a Duo trial account (`https://signup.duo.com/https://signup.duo.com/`) if necessary:

Figure 3.67 – Signing up for a Duo account

2. Without exiting the Azure AD admin center, in a separate browser tab, log in to Duo (`https://admin.duosecurity.com/`). Select **Applications**.

3. Click **Protect an Application**:

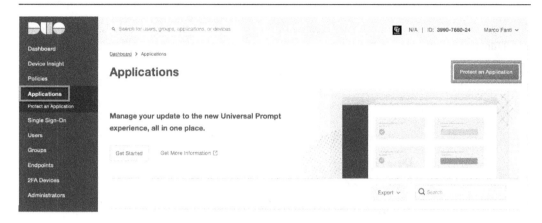

Figure 3.68 – Duo dashboard

4. Search for **Microsoft Azure Active Directory** in the applications list. Click **Protect**:

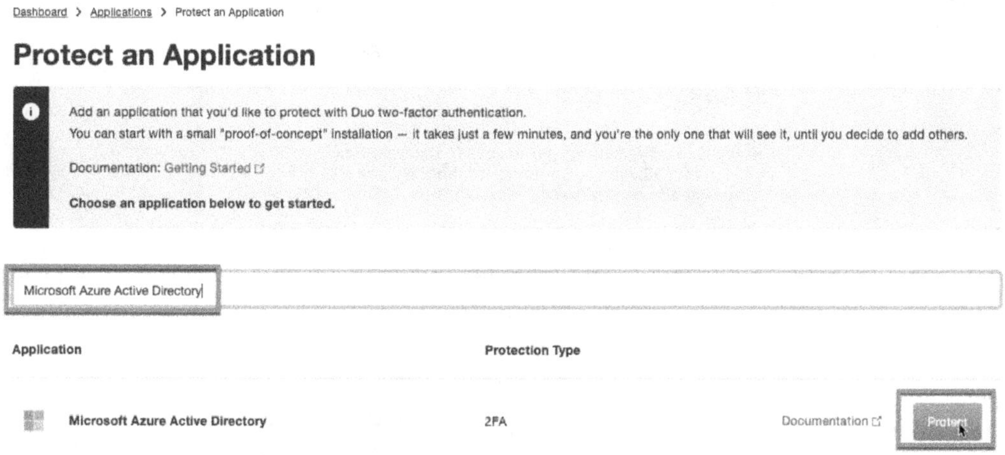

Dashboard > Applications > Protect an Application

Protect an Application

ℹ️ Add an application that you'd like to protect with Duo two-factor authentication.
    You can start with a small "proof-of-concept" installation — it takes just a few minutes, and you're the only one that will see it, until you decide to add others.

    Documentation: Getting Started ⬈

    Choose an application below to get started.

Microsoft Azure Active Directory

Application	Protection Type		
🔲 **Microsoft Azure Active Directory**	2FA	Documentation ⬈	Protect

Figure 3.69 – Protecting an application in Duo

5. Duo requires authorization to access your Azure AD tenant. Click the **Authorize** button:

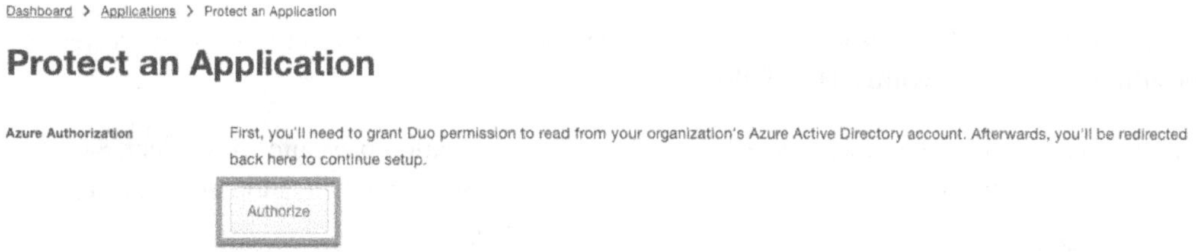

Dashboard > Applications > Protect an Application

Protect an Application

Azure Authorization   First, you'll need to grant Duo permission to read from your organization's Azure Active Directory account. Afterwards, you'll be redirected back here to continue setup.

    Authorize

Figure 3.70 – Authorizing application protection in Duo

6. Sign in with the Duo admin user created in the previous set of steps. Click **Accept** to allow Duo access to the Azure AD tenant:

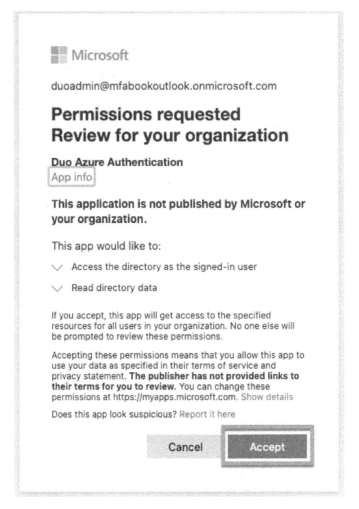

Figure 3.71 – Azure AD permissions request

After this step, **Global Administrator** role privileges can be removed from the account created for Duo administration in Azure AD, as follows:

1. Go back to the **Microsoft Azure Active Directory** application page and click **select**. Save the custom control JSON text in the **Details** section to complete the Duo authentication setup:

Successfully added Microsoft Azure Active Directory to protected applications. Add another.

Dashboard > Applications > Microsoft Azure Active Directory

Microsoft Azure Active Directory

Authentication Log | 🗑 Remove Application

See the Duo Azure Conditional Access Documentation ☐ to integrate Duo with Azure Conditional Access.

Details

This application is configured against Azure Active Directory Tenant ID: 9c81c6e0-c312-4a1a-a685-8cffc3fe78d0

Custom control

```
{
    "AppId": "21249e6b-bc2a-4dcd-9dd9-d6cbd7496814",
    "ClientId":
"YXBpLWVkZGU5N2I4LmR1b3NlY3VyaXR5LmNvbTpESTJYUzlVU0M4TDY1RU9DUEtK0Q==",
    "Controls": [
        {
            "ClaimsRequested": [
                {
                    "Type": "DuoMfa",
                    "Value": "MfaDone",
                    "Values": null
                }
            ],
            "Id": "RequireDuoMfa",
            "Name": "RequireDuoMfa"
        }
    ],
    "DiscoveryUrl": "https://us.azureauth.duosecurity.com/.well-
known/openid-configuration",
    "Name": "Duo Security"
}
```

Use this custom control to configure Azure Conditional Access.

Figure 3.72 – Duo JSON configuration

2. Click on the **Save Changes** button when done changing any of the details of the application (such as the application name). You may choose to keep the Duo admin panel open to copy the custom control JSON text into the Azure admin center portal later or choose to save the text separately.

We are now done with the creation of the application in the Duo administration application. We are going to create a custom control that will be used in CA policies when Duo is specifically required for MFA.

Using the Duo custom control

Microsoft Azure AD provides custom control functionality to allow third-party companies such as Duo Security to integrate their authentication apps with Azure AD. We are going to use custom controls to allow Duo to be used in a CA policy, as follows:

1. Log in to your Azure AD tenant in the Microsoft Azure portal as a global administrator (if you aren't already logged in).

2. Go to **Azure Active Directory** | **Security** | **Conditional Access**.

3. Click **Custom controls**. Then, click **New custom control**:

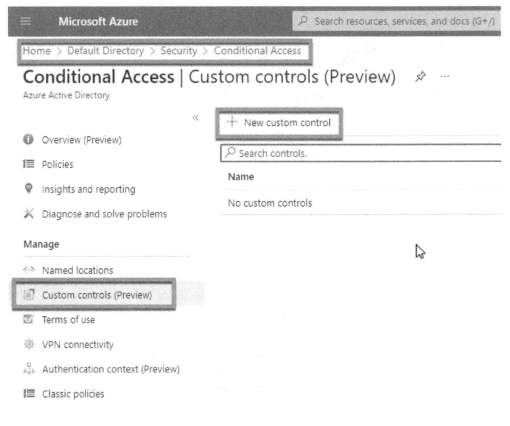

Figure 3.73 – CA

4. Paste in the *custom control* JSON text you copied earlier, removing the example custom control JSON text.

5. Click the **Create** button. Azure creates a new custom control, `RequireDuoMfa`:

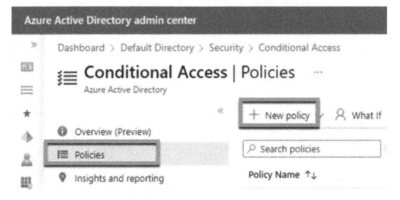

Figure 3.74 – New custom control for CA

The custom control we just created can now be used in CA policies, which is what we are doing next:

1. Without leaving the **Conditional Access** pane, click **Policies** and then click + **New Policy**:

Figure 3.75 – Creating a new policy

2. Enter a name for the new policy, such as **Require Duo MFA**:

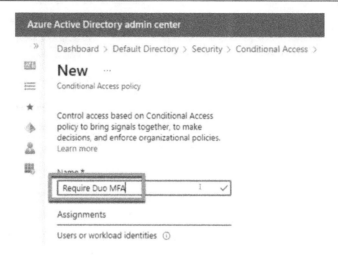

Figure 3.76 – New policy for Duo CA

Just as with our first CA policy, we can assign this new policy to selected users or Azure security groups. We can also assign this new policy to specific Azure cloud apps or to any of the other Azure conditions such as client platform or network.

3.   For this example, click on **0 users or workload identities selected** under **Assignments, User or workload identities**:

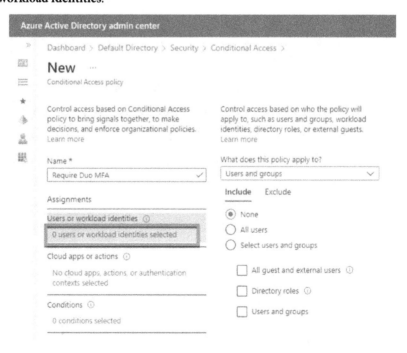

Figure 3.77 – Selecting users or workload identities

4. Click **Select users and groups** and then **Users and groups** in the **Include** pane:

Figure 3.78 – Selecting users and groups

5. Click on a user or security group to select it, then click the **Select** button in the **Select** pane. In this example, the new Duo policy assignment includes the **Duo Users** Azure AD group, so members of that group will require Duo 2FA when logging in to Azure AD:

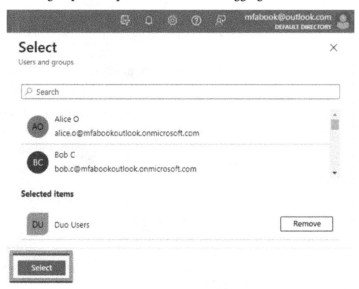

Figure 3.79 – Users and groups for CA

6. Next, click on **No cloud apps, actions, or authentication contexts selected** under **Cloud apps or actions**. On the **Include** tab, select the **Select apps** radio button:

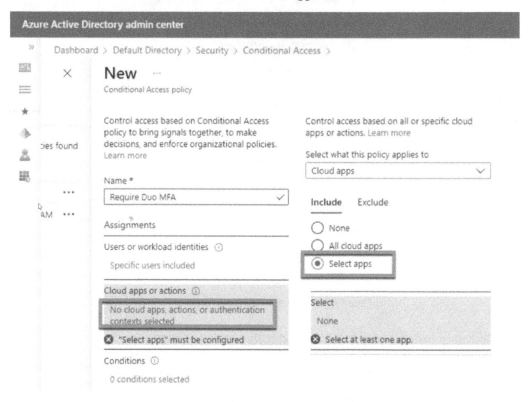

Figure 3.80 – Cloud apps for CA

7. Click **Select apps** and choose the Azure AD applications where you want Duo authentication before access.

    For this example, we are choosing **Office 365** and the **Acme's Azure AD SAML Toolkit** app. This means that the policy (and the Duo custom control in it) gets applied when the users assigned that policy access those two applications only. Other Azure and Office applications not specified by the policy assignment will not exercise the CA control at login and therefore will not require Duo:

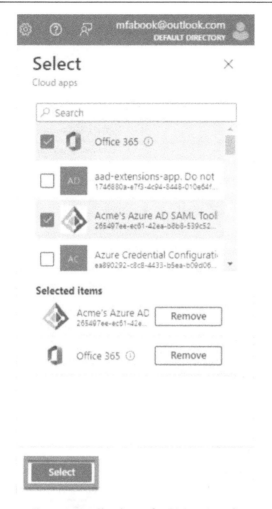

Figure 3.81 – Cloud apps for CA (continued)

> **Important note**
>
> Duo custom control should not be applied to CA policies requiring Duo for external guest logins. This is not supported.

8. Click **0 controls selected** under **Access controls, Grant**. To allow users access with Duo authentication, select the **Grant access** radio button, and check the box next to the **RequireDuoMfa** custom control you created in the previous steps. While you may choose to combine or require satisfying multiple controls before granting user access, this example simply adds the Duo authentication requirement to the new policy. Click **Select** when done:

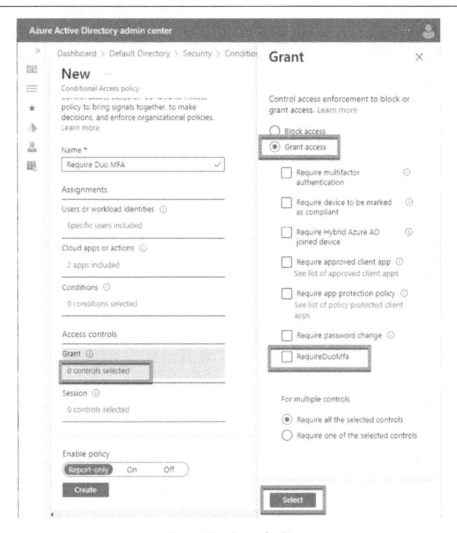

Figure 3.82 – Grants for CA

The final step to creating the new Duo policy is to enable it.

9. Click the **On** toggle switch underneath **Enable policy**, and then click **Create**. Azure creates and enables the new **Require Duo MFA** policy:

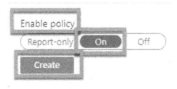

Figure 3.83 – Enabling and creating a new policy for CA

We have finished creating the CA policy that requires Duo when users from the Duo users group try to access Office 365 or the **Acme's Azure AD SAML Toolkit** application. And now, we can test the new configuration.

Testing Duo

To test Duo, sign in to one of the cloud apps configured in the CA policy. Don't forget to use a user that is part of that policy. Proceed as follows:

1. Open a new browser window in InPrivate or Incognito mode and browse to `https://office.com`.

2. Sign in with your test user. Make sure the user was selected (or is a member of a group selected) in the CA policy.

3. If this is the first time the user is using Duo, after the email and password, the user will be welcomed to the Duo setup. Click **Next**:

Figure 3.84 – Welcome to Duo Security

4. On the next screen, select **Duo Mobile**:

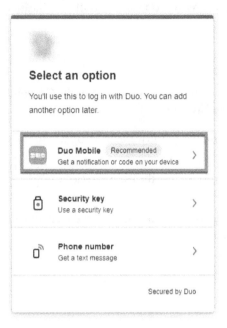

Figure 3.85 – Selection page

5. Next, enter the country code and your phone number. Click **Add phone number**:

Figure 3.86 – Adding a phone number

6.   Finally, a confirmation page that your phone number has been added will appear. Click **Continue**:

Figure 3.87 – Finishing registration of the new Duo app

And that's it. Office 365 is now protected with Duo for those users in the selected group:

Figure 3.88 – Duo-protected Office 365

This completes our test of using a custom control in a Microsoft Azure AD CA policy to enforce the use of Duo as a second factor of authentication for selected users and cloud-based apps.

The same functionality can be extended for more apps, users, and groups of users, allowing for very granular control of the authentication factor required to access all applications protected by Microsoft Azure AD.

Summary

In this chapter, we went over the process of enabling SSO for enterprise applications, using different Azure AD authentication factors for MFA with the use of Azure AD CA policies. Microsoft claims that the use of MFA protects against 99.9% of attacks. To improve on that claim, we also discussed the use of Duo as a second factor of authentication, and how to configure Duo for a more secure Azure AD MFA.

In the next chapter, we will look at Okta, another provider of IAM products, and their solution for workforce users as well as users external to the company.

Later in the book, we are going to look again at Microsoft's CA policies and how they can be used for continuous authentication.

4
Implementing Workforce and Customer Authentication Using Okta

In the previous chapter, we focused on using Azure AD and Duo Multi-Factor Authentication to safeguard the Acme workforce, including employees, contractors, and temporary users.

In this chapter, we will shift attention to **Okta**, a central cloud-based identity management vendor that offers two distinct **identity and access management (IAM)** products. The first product protects workforce users, equivalent to the configuration of Azure AD presented in *Chapter 3*. Okta's second product provides authentication for external customers. This product is designed to enable secure access to applications and services for customers not belonging to the organization's workforce (comparable to Azure AD B2C).

The first is **Okta Workforce Identity**. Okta Workforce Identity allows companies to manage employees, contractors, and partners effortlessly from a unified platform. The same platform also protects applications and user devices. Okta Workforce Identity gives companies complete control over identities, helping those companies comply with regulations and achieve their business goals.

Okta Customer Identity is the second solution, and it allows companies to create and manage end user identities. Companies can use Okta Customer Identity to provide registration and login portals for their applications without additional friction to their customers. By offloading customer identity management to Okta, companies can build authentication into any cloud-based application, creating secure, easy-to-use experiences for those customers.

In this chapter, we will describe how to implement Okta Workforce Identity to secure Acme's workforce applications and how to configure and use additional authenticators. We will use Duo as the authenticator to understand how Azure AD and Okta Workforce Identity compare. The second part of this chapter will examine the Okta Customer Identity product so that we can explore **multi-factor authentication (MFA)** for the customer-facing applications the company is starting to develop.

We are going to cover the following main topics:

- Getting started with Okta Workforce Identity
- Importing users into Okta Workforce Identity
- Okta Workforce Identity MFA
- Enabling additional authenticators in Okta Workforce Identity
- Okta Customer Identity solution implementation
- App-level MFA policies for Okta Customer Identity

Technical requirements

This chapter requires two Okta accounts – one for the Workforce Identity solution and another for the Customer Identity solution. Trial versions of both account types are available at `https://www.okta.com/free-trial-c/` (Workforce Identity) and `https://www.okta.com/free-trial/customer-identity-c/` (Customer Identity).

Workforce Identity with Okta

Our first encounter with Okta will be with the Workforce Identity solution. Similar to what we did with Microsoft Azure AD, the Okta Workforce Identity solution is going to be used for Acme's employees, contractors, and partners.

Creating a Workforce Identity account

This step is optional and only required if you don't have an Okta Workforce Identity account:

1. To create a Workforce Identity trial account, go to `https://www.okta.com/free-trial-c/`. Select **Workforce Identity Trial** if it's not selected already:

Figure 4.1 – Workforce Identity Trial

2. Enter the required information and click **Get Started**:

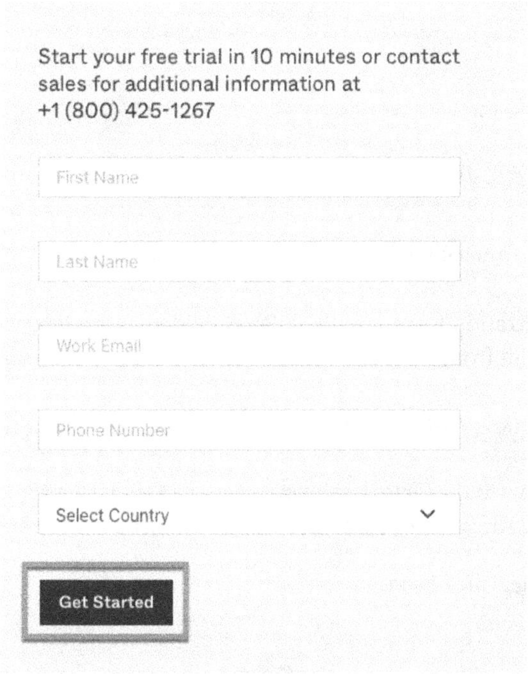

Start your free trial in 10 minutes or contact sales for additional information at +1 (800) 425-1267

First Name

Last Name

Work Email

Phone Number

Select Country

Get Started

Figure 4.2 – Workforce Identity registration page

This will initiate the process for a new 30-day trial account with Okta:

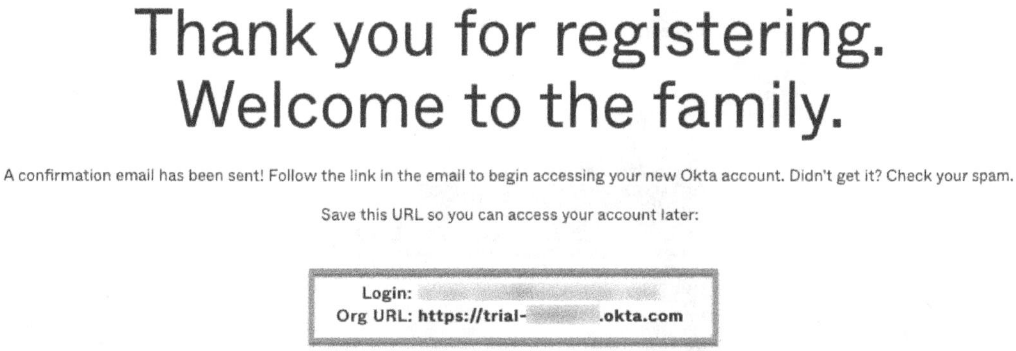

Thank you for registering. Welcome to the family.

A confirmation email has been sent! Follow the link in the email to begin accessing your new Okta account. Didn't get it? Check your spam.

Save this URL so you can access your account later:

Login:
Org URL: **https://trial-   .okta.com**

Figure 4.3 – Workforce Identity welcome page

3.   Copy and save the **Login** and **Org URL** details:

Save this URL so you can access your account later:

Figure 4.4 – Workforce Identity organization URL

After submitting the registration form, use the URL provided to log into your newly created org. Only after receiving an email from Okta and confirming the account will you be allowed to log in.

Signing into your Workforce Identity account for the first time

Open a new browser window to the Okta dashboard using the **Org URL** property shown in *Figure 4.4* (it will look something like `https://trial-9999999.okta.com`):

1.   Enter your username. Click **Next**:

Figure 4.5 – Dashboard login – Username

2. Enter your password. Click **Verify**:

Figure 4.6 – Dashboard login – Password

When logging in for the first time, Okta will require you to set up **two-factor authentication** (**2FA**). Two options are available. When possible, you should use a mobile app or something stronger for 2FA. We recommend using the Okta Verify app, especially for an administrator account.

3. Select **Okta Verify**. Click **Setup**:

Figure 4.7 – Setting up security methods

4.   Click **Set up later** to skip the optional phone security method setup:

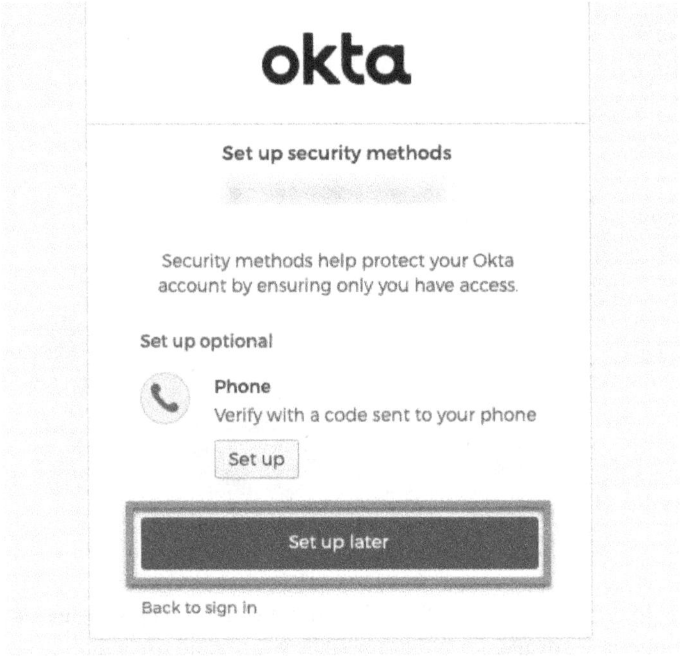

Figure 4.8 – Optional security method setup

At this point, you will see the Okta dashboard. We haven't set up any apps, so the dashboard will be empty. Since we logged in as an administrator, the administrator dashboard will also be available.

Configuring Okta

Before we can use the application dashboard, we need to configure a few things:

1.   To start configuring Okta, click **Admin**:

Figure 4.9 – Okta dashboard

The **Get started with Okta** screen will appear:

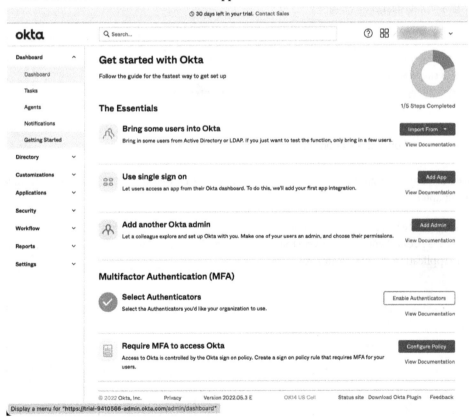

Figure 4.10 – Okta administrator dashboard

Initially, the getting start screen will display **1/5 Steps Completed**. For security purposes, the basic authenticator's configuration is predefined by Okta, which is the completed step. We will see how to change that configuration later.

The essentials

The remaining four steps are considered the minimum configuration required to use Okta for managing workforce users. We registered some users in Azure AD in the previous chapter. We can create similar users in Okta Workforce Identity. Alternatively, we can import those users if we want to use the same users we created previously.

Bringing some users into Okta

Okta allows users to be added manually or imported from Active Directory or LDAP systems. Users can also be imported from a **comma-separated values (CSV)** file:

Figure 4.11 – Importing users via the Okta Administrator dashboard

Since we have users in Azure AD, let's export those users so that we can use the same users from the previous chapter:

1. In a separate tab, go to the Azure Active Directory admin center (`https://aad.portal.azure.com/`) and sign in using the Global Administrator, Cloud Application Administrator, or Application Administrator role.

2. Select **Users | Bulk Operations | Download Users**:

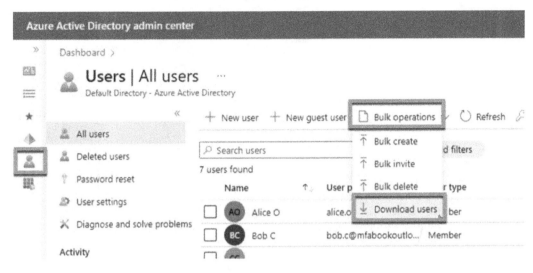

Figure 4.12 – Exporting users from Azure AD

3. After choosing a filename, click **Start**:

Figure 4.13 – Download users

4. Wait for the download to finish and click **File is Ready! Click here to download** to save the file. Click **X** to close the pane:

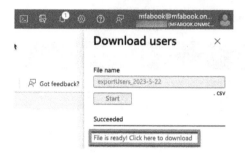

Figure 4.14 – Download succeeded

5.   Back in the Okta Admin tool, select **Import From** under the **Bring some users into Okta** step:

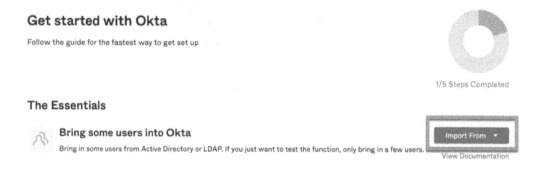

Figure 4.15 – Importing users from CSV

6.   Click on **this template** to save the Okta import users template:

Figure 4.16 – Saving the Okta import users CSV template

The Azure AD file contains **userPrincipalName**, **displayName**, **surname**, **mail**, **givenName**, **id**, and **userType** as fields, among others:

userPrincipalName	displayName	surname	mail	givenName	id	user
LeeG@mfabook.onmicrosoft.com	Lee Gu	Gu	LeeG@mfabook.onmicrosoft.com	Lee	0edaf3e7-5ec4-4482-b6c6-75ee3f82cd0d	Mem
IsaiahL@mfabook.onmicrosoft.com	Isaiah Langer	Langer	IsaiahL@mfabook.onmicrosoft.com	Isaiah	1f14ee1b-d7f1-448b-aaf3-3f3906d04537	Mem
AdeleV@mfabook.onmicrosoft.com	Adele Vance	Vance	AdeleV@mfabook.onmicrosoft.com	Adele	2b9d201b-a7af-4cb0-b4d0-7a136fa88271	Mem
NestorW@mfabook.onmicrosoft.com	Nestor Wilke	Wilke	NestorW@mfabook.onmicrosoft.com	Nestor	2c972533-2925-4f12-b37c-079e5d26241d	Mem
HenriettaM@mfabook.onmicrosoft.com	Henrietta Mueller	Mueller	HenriettaM@mfabook.onmicrosoft.com	Henrietta	3bac06fe-db74-4346-b4b8-2b8be0cfed1e	Mem
MeganB@mfabook.onmicrosoft.com	Megan Bowen	Bowen	MeganB@mfabook.onmicrosoft.com	Megan	3cb38adb-e0e6-4d31-b241-590ea1371ce9	Mem
GradyA@mfabook.onmicrosoft.com	Grady Archie	Archie	GradyA@mfabook.onmicrosoft.com	Grady	44a8ba59-8af4-4791-a4cb-3ae5e9aacc03	Mem
JohannaL@mfabook.onmicrosoft.com	Johanna Lorenz	Lorenz	JohannaL@mfabook.onmicrosoft.com	Johanna	66560cff-16f7-4933-a58b-356716d4fbe7	Mem
PradeepG@mfabook.onmicrosoft.com	Pradeep Gupta	Gupta	PradeepG@mfabook.onmicrosoft.com	Pradeep	684bebda-1771-4257-b0b1-21b49266d47b	Mem
mfabook@mfabook.onmicrosoft.com	Marco Fanti	Fanti	mfabook@mfabook.onmicrosoft.com	Marco	8110a336-75d3-4167-bf79-2622bf79b218	Mem
PattiF@mfabook.onmicrosoft.com	Patti Fernandez	Fernandez	PattiF@mfabook.onmicrosoft.com	Patti	937ddaed-8138-473f-93b5-ce7c380d8c3e	Mem
LynneR@mfabook.onmicrosoft.com	Lynne Robbins	Robbins	LynneR@mfabook.onmicrosoft.com	Lynne	9ac2a69f-f8bb-4aa5-9e53-c86b87449892	Mem
JoniS@mfabook.onmicrosoft.com	Joni Sherman	Sherman	JoniS@mfabook.onmicrosoft.com	Joni	a1ead197-5e20-4294-9535-bd19599d6783	Mem
DiegoS@mfabook.onmicrosoft.com	Diego Siciliani	Siciliani	DiegoS@mfabook.onmicrosoft.com	Diego	b75e2042-8d7e-4bd5-8ef1-8778c30d3a7b	Mem
AlexW@mfabook.onmicrosoft.com	Alex Wilber	Wilber	AlexW@mfabook.onmicrosoft.com	Alex	e4cfad2c-0f3e-4592-bdfb-3fd36daa01b6	Mem
LidiaH@mfabook.onmicrosoft.com	Lidia Holloway	Holloway	LidiaH@mfabook.onmicrosoft.com	Lidia	eccf185e-b19e-4071-b674-064e9b4a9b1d	Mem
MiriamG@mfabook.onmicrosoft.com	Miriam Graham	Graham	MiriamG@mfabook.onmicrosoft.com	Miriam	f44445e5-53d3-452a-835b-4290f5f97699	Mem

Figure 4.17 – Exported Azure AD CSV file

The Okta template contains the **login**, **firstName**, **lastName**, **email**, and **displayName** fields, which are going to be used in the import:

login	firstName	lastName	middleName	honorificPrefix	honorificSuffix	email	title	displayName
LeeG@mfabook.onmicrosoft.com	Lee	Gu				LeeG@mfabook.onmicrosoft.com		Lee Gu
IsaiahL@mfabook.onmicrosoft.com	Isaiah	Langer				IsaiahL@mfabook.onmicrosoft.com		Isaiah Langer
AdeleV@mfabook.onmicrosoft.com	Adele	Vance				AdeleV@mfabook.onmicrosoft.com		Adele Vance
NestorW@mfabook.onmicrosoft.com	Nestor	Wilke				NestorW@mfabook.onmicrosoft.com		Nestor Wilke
HenriettaM@mfabook.onmicrosoft.com	Henrietta	Mueller				HenriettaM@mfabook.onmicrosoft.com		Henrietta Mueller
MeganB@mfabook.onmicrosoft.com	Megan	Bowen				MeganB@mfabook.onmicrosoft.com		Megan Bowen
GradyA@mfabook.onmicrosoft.com	Grady	Archie				GradyA@mfabook.onmicrosoft.com		Grady Archie
JohannaL@mfabook.onmicrosoft.com	Johanna	Lorenz				JohannaL@mfabook.onmicrosoft.com		Johanna Lorenz
PradeepG@mfabook.onmicrosoft.com	Pradeep	Gupta				PradeepG@mfabook.onmicrosoft.com		Pradeep Gupta
mfabook@mfabook.onmicrosoft.com	Marco	Fanti				mfabook@mfabook.onmicrosoft.com		Marco Fanti
PattiF@mfabook.onmicrosoft.com	Patti	Fernandez				PattiF@mfabook.onmicrosoft.com		Patti Fernandez
LynneR@mfabook.onmicrosoft.com	Lynne	Robbins				LynneR@mfabook.onmicrosoft.com		Lynne Robbins
JoniS@mfabook.onmicrosoft.com	Joni	Sherman				JoniS@mfabook.onmicrosoft.com		Joni Sherman
DiegoS@mfabook.onmicrosoft.com	Diego	Siciliani				DiegoS@mfabook.onmicrosoft.com		Diego Siciliani
AlexW@mfabook.onmicrosoft.com	Alex	Wilber				AlexW@mfabook.onmicrosoft.com		Alex Wilber
LidiaH@mfabook.onmicrosoft.com	Lidia	Holloway				LidiaH@mfabook.onmicrosoft.com		Lidia Holloway
MiriamG@mfabook.onmicrosoft.com	Miriam	Graham				MiriamG@mfabook.onmicrosoft.com		Miriam Graham

Figure 4.18 – Updated Okta CSV file template

For the import, perform the following steps:

1.   For the desired users, do the following:

      A.   Copy **userPrincipalName** to **login** and **email** in the Okta template.

      B.   Copy **displayName** to **displayName** in the Okta template.

      C.   Copy **surname** to **lastName**.

      D.   Copy **givenName** to **firstName**.

2.   Click on **Upload CSV**:

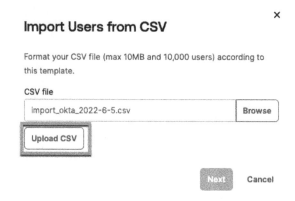

Figure 4.19 – Selecting the CSV file to import

3.   After a message stating **CSV file parsed successfully!** is displayed, click **Next**:

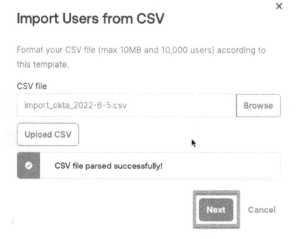

Figure 4.20 – Import Users from CSV

4. Select **Automatically activate new users** and click on **Import Users**:

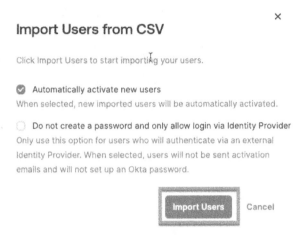

Figure 4.21 – Complete import

If the file is correct, the users should be imported into Okta, and the number of imported users will be displayed.

5. Click on **Done**:

Figure 4.22 – Number of users imported

Another way to confirm that the correct users were imported is by going to the **Directory | People** page in the **Admin** dashboard.

6.   Expand **Directory** and click on **People**:

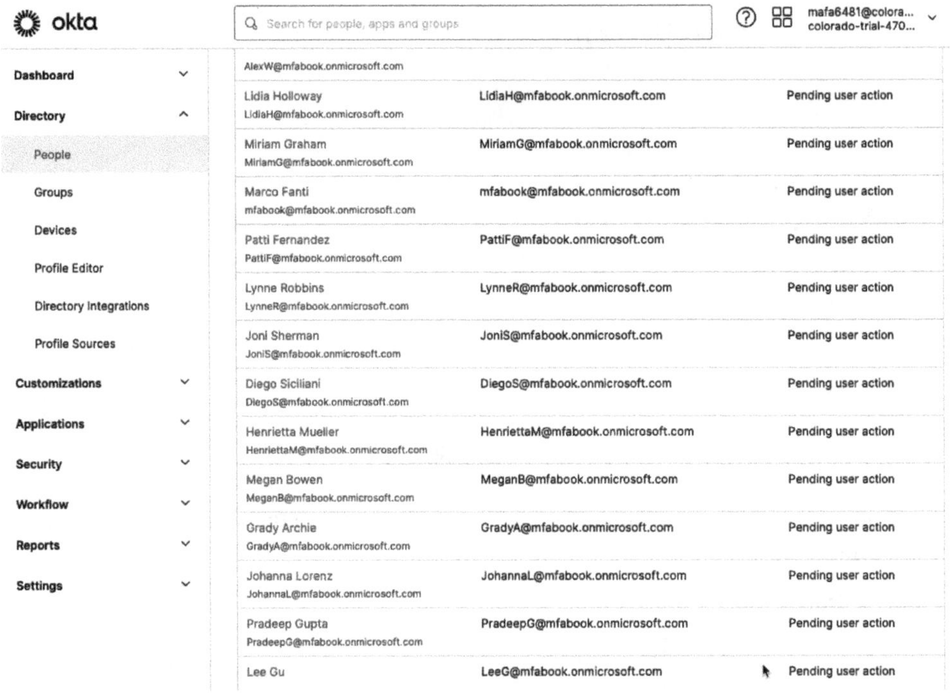

Figure 4.23 – Directory – People

The Okta app integration catalog is a centralized repository of pre-built integrations between Okta's Workforce Identity platform and various cloud applications and services. This integration catalog is used to simplify and streamline the process of connecting Okta's IAM platform with various third-party applications, making it easier for organizations to manage user identities, access controls, and security policies across multiple systems.

The Okta app integration catalog offers many integrations with popular cloud applications, including Salesforce, Workday, Box, Slack, and others. These integrations enable organizations to connect their existing applications quickly and easily with Okta's IAM platform, reducing the complexity and cost of managing multiple identity silos and authentication workflows.

As mentioned previously, the Okta app integration catalog features several popular integrations, including Salesforce, ServiceNow, Office 365, and Workday.

Salesforce integration enables organizations to manage user identities and access controls for Salesforce, one of the most widely used **customer relationship management** (**CRM**) platforms. By integrating Okta with Salesforce, organizations can centralize user identity and access management, simplify user authentication, and enhance security.

ServiceNow integration allows organizations to manage user identities and access controls for ServiceNow, a popular cloud-based platform that provides IT service management, operations management, and business management. By integrating Okta with ServiceNow, organizations can improve the user experience, streamline access management, and enforce strong security policies.

Office 365 integration enables organizations to manage user identities and access controls for Microsoft Office 365, a cloud-based suite of productivity tools that includes Word, Excel, and PowerPoint. By integrating Okta with Office 365, organizations can simplify user authentication, streamline access management, and enhance security.

Workday integration enables organizations to manage user identities and access controls for Workday, a leading cloud-based human resources management platform. By integrating Okta with Workday, organizations can centralize user identity and access management, streamline user authentication, and enforce strong security policies:

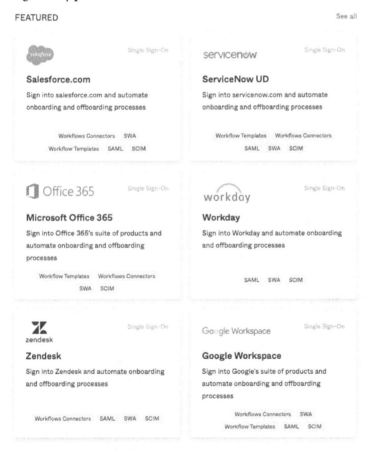

Figure 4.24 – Featured enterprise application integrations

Because Salesforce is one of the most widely used platforms and provides a 30-day trial version that supports SAML for **single sign-on** (**SSO**), we will use it as an example of SSO with enterprise applications for this chapter. Any enterprise application from Okta's App Integration Catalog can also be used.

Integrating enterprise applications to use SSO

Let's test SSO with enterprise applications:

1.   Expand **Dashboard**. Click on **Getting Started**:

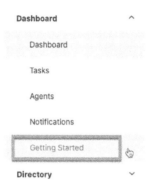

Figure 4.25 – Dashboard – Getting Started

2.   Click on **Add App**:

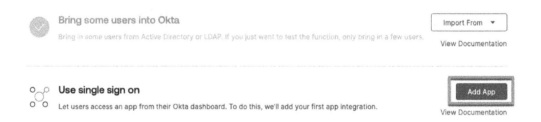

Figure 4.26 – Add App

3. In the **Browse App Integration Catalog** pane, select **Salesforce.com**:

Figure 4.27 – Browse App Integration Catalog

4. To add Salesforce.com to SSO, click **Add Integration**:

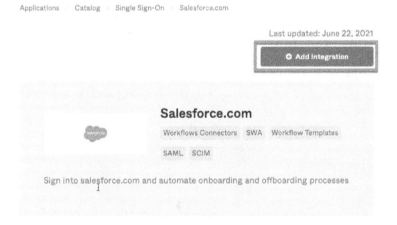

Figure 4.28 – Adding Salesforce.com to SSO

Before going to the next step, you should log in to the Salesforce.com setup dashboard and search for my domain:

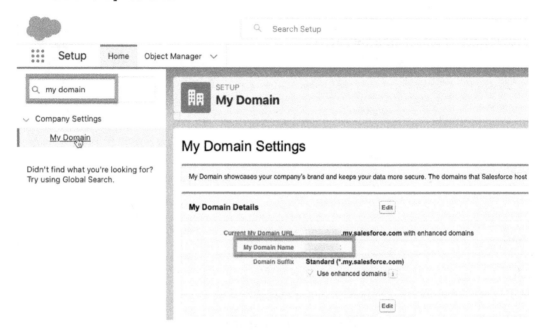

Figure 4.29 – Salesforce.com custom domain (My Domain) settings

5. Change the **Application Label** property if a more descriptive value is desired. This is the value that will appear on the users' dashboard. Select an appropriate **Instance Type** (**Sandbox** for a trial version of Salesforce.com). Also, enter a **Custom Domain** for your Salesforce.com environment. Then, click **Next**:

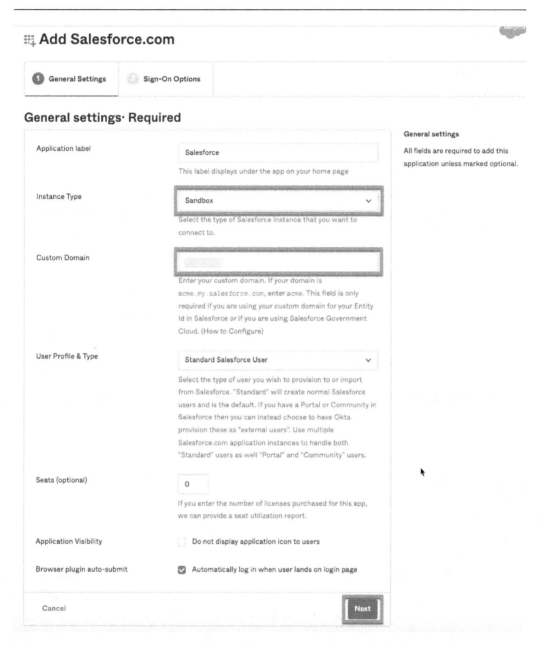

Figure 4.30 – Salesforce.com – General settings

6.   On the **Sign-On Options Required** page, select **SAML 2.0**:

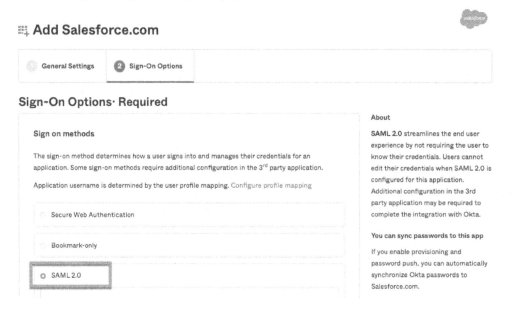

Figure 4.31 – Salesforce.com sign-on methods section

7.   Click **View Setup Instructions**. This will open a new browser tab containing instructions:

Figure 4.32 – Salesforce.com SAML 2.0 setup instructions

8.   Click on the URL provided in the setup instructions and follow *steps 1* through *7*:

Figure 4.33 – Salesforce.com SAML 2.0 setup steps

9. Copy the **Identity Provider Login URL** and **Custom Logout URL** properties from *step 6*, as depicted in the following screenshot:

Figure 4.34 – Salesforce instructions – step 6

10. In the **Advanced Sign-On Settings** section, enter your **Login URL (Identity Provider Login URL**, which you saved in the previous step) and **Logout URL**:

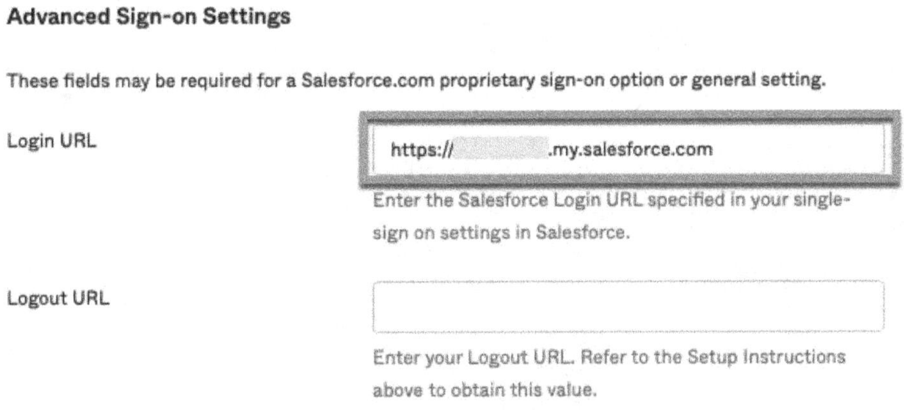

Figure 4.35 – Salesforce.com – Advanced Sign-on Settings

11. In the **Credential Details** section, change **Application username format** to **Email**. Click **Done**:

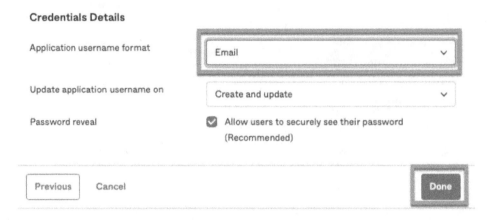

Figure 4.36 – Salesforce.com – Credential Details

Before Salesforce.com is available for users to use, they must be assigned to the application, either directly or by group. To assign users directly to Salesforce.com, follow these steps:

1. From the **Assignments** tab, click **Assign | Assign to People**:

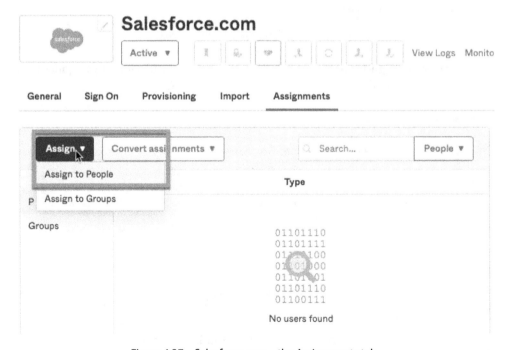

Figure 4.37 – Salesforce.com – the Assignments tab

2. Select the desired users from the list and click **Done**:

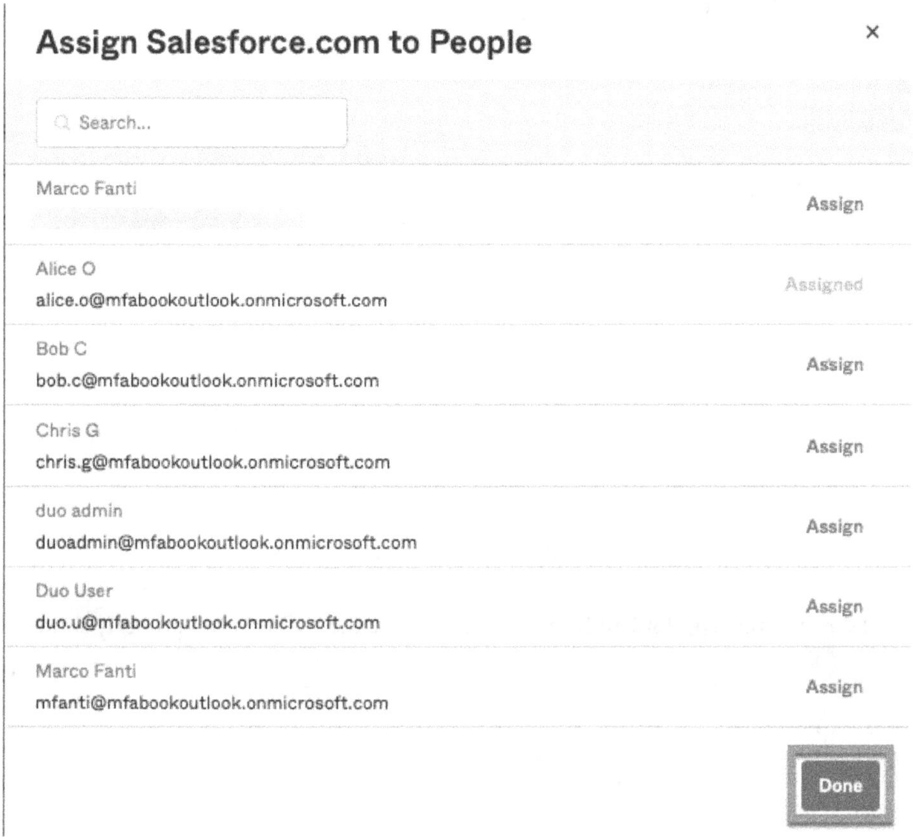

Assign Salesforce.com to People   ×

Q Search...	

Marco Fanti	Assign
Alice O alice.o@mfabookoutlook.onmicrosoft.com	Assigned
Bob C bob.c@mfabookoutlook.onmicrosoft.com	Assign
Chris G chris.g@mfabookoutlook.onmicrosoft.com	Assign
duo admin duoadmin@mfabookoutlook.onmicrosoft.com	Assign
Duo User duo.u@mfabookoutlook.onmicrosoft.com	Assign
Marco Fanti mfanti@mfabookoutlook.onmicrosoft.com	Assign

Done

Figure 4.38 – Salesforce.com assignments selection list

This statement signifies that you have completed the necessary steps to integrate Okta Workforce Identity SSO with Salesforce.com. The purpose of this integration is to streamline user access and enhance security between the two platforms.

By completing the configuration process, organizations can now benefit from the combined power of Okta and Salesforce.com, ensuring a seamless and secure user experience for their workforce.

Now, let's try logging into Salesforce.com using Okta:

1. Click on the **General** tab:

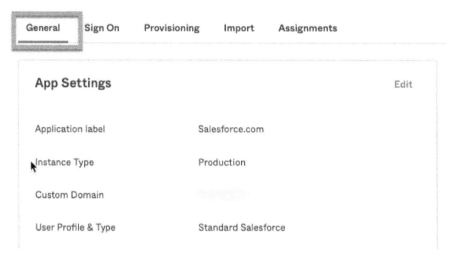

Figure 4.39 – Salesforce.com – the General tab

2. Scroll down to the **App Embed Link** section and save the **Embed Link** property shown:

Figure 4.40 – Office Portal's URL

3. Open a new browser window in InPrivate or incognito mode and browse to the **Embed Link** property you just saved. A sign-in page will appear. Enter your **Username** and click **Next**:

Figure 4.41 – Salesforce.com – signing in using Okta

4. Select **Push Notification (Okta Verify)** as the first factor of authentication:

Figure 4.42 – Salesforce.com first factor verification

5. Enter the Okta password for the user to complete the setup. Then, click **Verify**:

Figure 4.43 – Salesforce.com second factor verification

Once the sign-in process is complete, the Salesforce application will be available for the user:

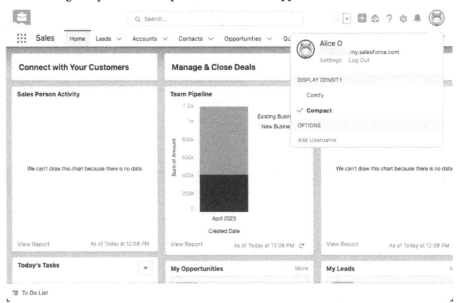

Figure 4.44 – Salesforce.com application dashboard

With that, we have tested Slasforce.com's SSO integration using Okta's Workforce Identity:

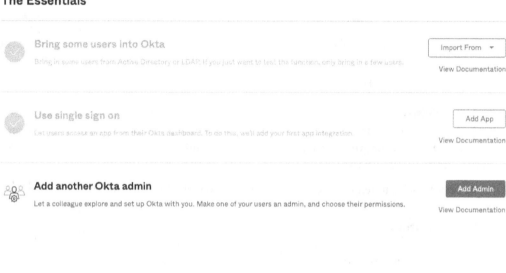

Get started with Okta

Follow the guide for the fastest way to get set up

3/5 Steps Completed

The Essentials

Bring some users into Okta

Bring in some users from Active Directory or LDAP. If you just want to test the function, only bring in a few users.

Import From ▾

View Documentation

Use single sign on

Let users access an app from their Okta dashboard. To do this, we'll add your first app integration.

Add App

View Documentation

Add another Okta admin

Let a colleague explore and set up Okta with you. Make one of your users an admin, and choose their permissions.

Add Admin

View Documentation

Multifactor Authentication (MFA)

Select Authenticators

Select the Authenticators you'd like your organization to use.

Enable Authenticators

View Documentation

Require MFA to access Okta

Access to Okta is controlled by the Okta sign on policy. Create a sign on policy rule that requires MFA for your users.

Configure Policy

View Documentation

Figure 4.45 – The Get started with Okta page

At this point, we are not going to add any other application, so we will go back to the getting started page and continue configuring Okta.

Adding another Okta administrator

Before continuing, let's add another Okta administrator. This will prevent Acme from being locked out of the account if the first admin user that's created is unavailable:

1. From the administrator dashboard, click on the **Add Admin** button:

Figure 4.46 – Adding a second administrator

The **Administrators** page of the Okta dashboard will open. On this page, you can configure administrative controls in your Okta organization using role assignments. Administrators can be individual users or groups that get this role. The **Administrators** page also allows you to create resource sets and roles.

2. (Optional) Click **x** to remove the **Welcome to the updated administrators page!** message:

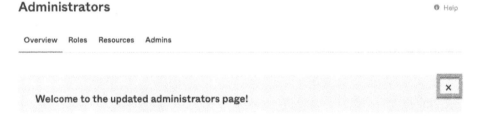

Figure 4.47 – Welcome message

3. Click on the **+Add administrator** button:

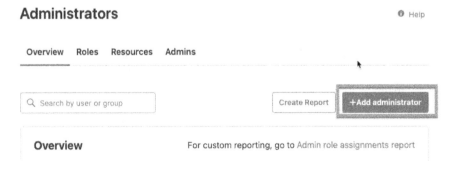

Figure 4.48 – Adding an administrator manually

4. Search for a user or group in the **Admin** search field:

Figure 4.49 – New administrator selected

5. Expand the **Role** field and select one of the existing roles. You can also create a new role using this field. For this example, select **Application Administrator**:

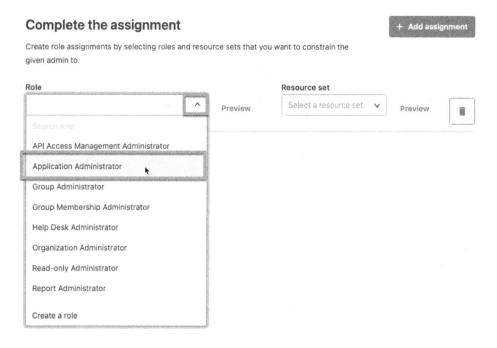

Figure 4.50 – Adding an assignment to an administrator

6. Because **Application Administrator** is a standard role (predefined in Okta), it cannot be constrained to only certain resources. Click on **Save Changes** to finish creating the new administrator. If more assignments are needed for the same administrator, click + **Add assignment** and repeat *step 5*:

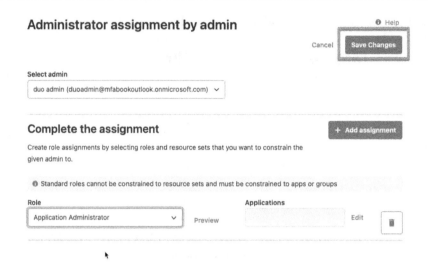

Figure 4.51 – Saving changes to a new administrator

At this point, we have completed 3 out of the 5 steps that Okta recommends. We have imported users into the solution, added an application to use SSO, and created a backup administrator account.

Okta provides a pre-configured set of authenticators, which can be enabled out of the box. As we saw previously, when we log into Salesforce.com using Okta, we are prompted to use Okta Verify as an authenticator, in addition to a password.

However, depending on the specific security requirements of an organization or application, it may be necessary to add new authenticators or modify the existing ones. Next, we will learn how to change the configuration or add new authenticators by adding Duo Security, just as we did with Azure AD in the previous chapter.

Configuring authenticators

Let's learn how to configure authenticators:

1. On the getting started page, click on the **Enable Authenticators** button:

Figure 4.52 – Selecting authenticators

The first page you will see is the **Setup** page for authenticators. By default, Okta Verify, Password, and Phone are available for authentication. Email can also be used as an authenticator for self-service account recovery. Self-service account recovery allows users to reset their passwords or unlock their accounts without the need to contact an administrator. As we did with Microsoft's Azure AD, let's add Duo as an authenticator for Okta.

2. Click on the **Add Authenticator** button:

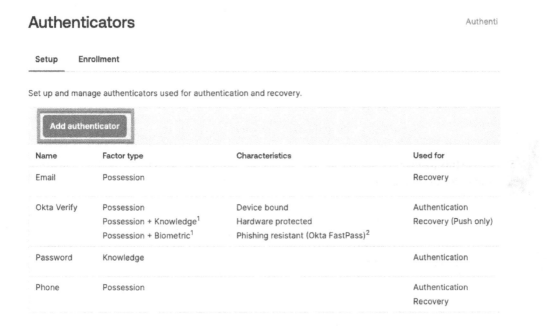

Authenticators

Authenti

Setup   Enrollment

Set up and manage authenticators used for authentication and recovery.

Add authenticator

Name	Factor type	Characteristics	Used for
Email	Possession		Recovery
Okta Verify	Possession Possession + Knowledge[1] Possession + Biometric[1]	Device bound Hardware protected Phishing resistant (Okta FastPass)[2]	Authentication Recovery (Push only)
Password	Knowledge		Authentication
Phone	Possession		Authentication Recovery

Figure 4.53 – Setup authenticators page

Before we select Duo, we need some information from the Duo dashboard.

3. Log into the Duo dashboard as an administrator, select **Applications**, and click on the **Protect an Application** button:

Figure 4.54 – Duo dashboard – the Applications page

4. Search for Okta and click on the **Protect** button:

Protect an Application
okta
Application
Protection Type
okta
Okta
2FA
Documentation
Protect

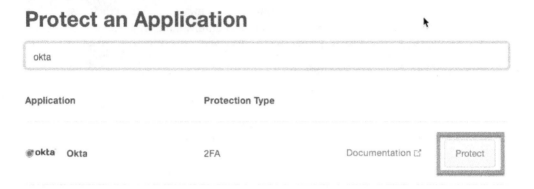

Figure 4.55 – Duo dashboard – protecting an application page

5. Leave this page open – you will need the **Integration Key**, **Secret Key**, and **API hostname** values for the authenticator configuration in Okta:

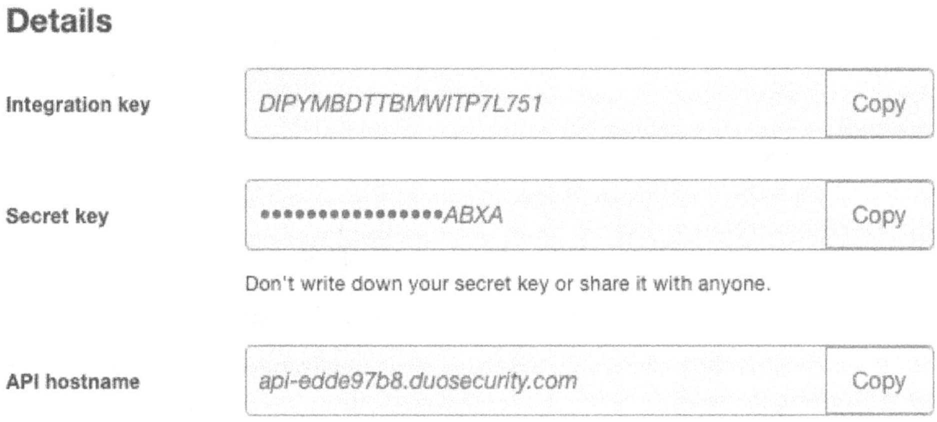

Details
Integration key
DIPYMBDTTBMWITP7L751
Copy
Secret key
••••••••••••••••ABXA
Copy
Don't write down your secret key or share it with anyone.
API hostname
api-edde97b8.duosecurity.com
Copy

Figure 4.56 – Okta configuration details

6. Back in the Okta dashboard, click on the Duo Security **Add** button:

Add Authenticator

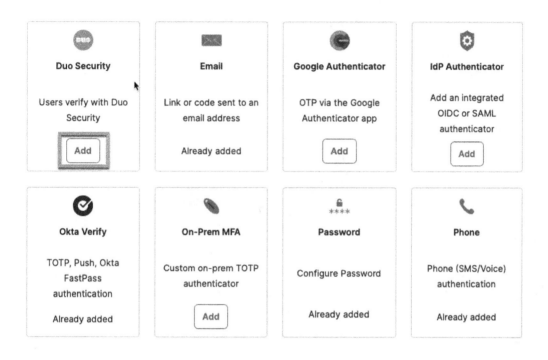

Figure 4.57 – Adding a new authenticator

7. Paste the **Integration Key**, **Secret Key**, and **API hostname** values into the authenticator configuration in Okta. Select **Email** as the Duo username format. Then, click on **Add**:

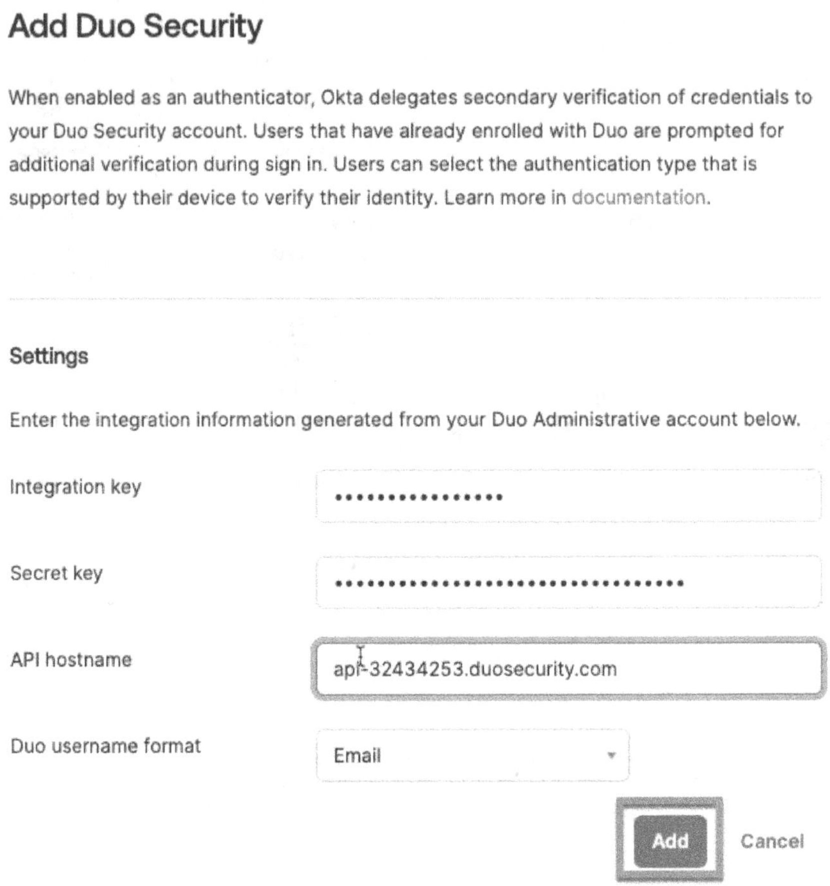

Add Duo Security

When enabled as an authenticator, Okta delegates secondary verification of credentials to your Duo Security account. Users that have already enrolled with Duo are prompted for additional verification during sign in. Users can select the authentication type that is supported by their device to verify their identity. Learn more in documentation.

Settings

Enter the integration information generated from your Duo Administrative account below.

Integration key	••••••••••••••••
Secret key	••••••••••••••••••••••••••••••••••
API hostname	api-32434253.duosecurity.com
Duo username format	Email ▾

Add   Cancel

Figure 4.58 – Adding Duo Security as an authenticator

Upon adding Duo as an authenticator, we changed the authentication configuration, which was *step 4* of the getting started list from Okta. To complete the final of the 5 steps that Okta recommends, we require MFA for all users when accessing Okta.

Requiring MFA to access Okta Workforce Identity apps

Okta Workforce Identity allows MFA to be configured at the organizational level or application level. We are going to require MFA for the whole organization. Okta also allows administrators to create more specific authentication policies per application.

Let's get started:

1. From the **Administrator Dashboard** area, on the **Getting started** page, click **Configure Policy**:

Figure 4.59 – Require MFA to access Okta

Okta provides a **Global Session Policy** that applies to the entire organization. We can use this policy to change the user's session length, create blocking rules, and enforce MFA. We are going to add a new rule, with higher priority than the **Default Rule** option, requiring MFA for all.

2. Click the **Add rule** button:

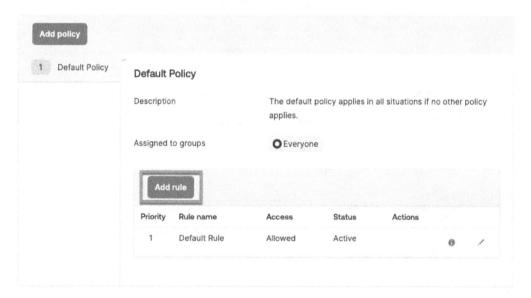

Figure 4.60 – Global Session Policy

3. Provide a descriptive name in the **Rule Name** field. Also, make sure that the **Multifactor Authentication (MFA) is** selection is set to **Required**. Click on the **Create Rule** button:

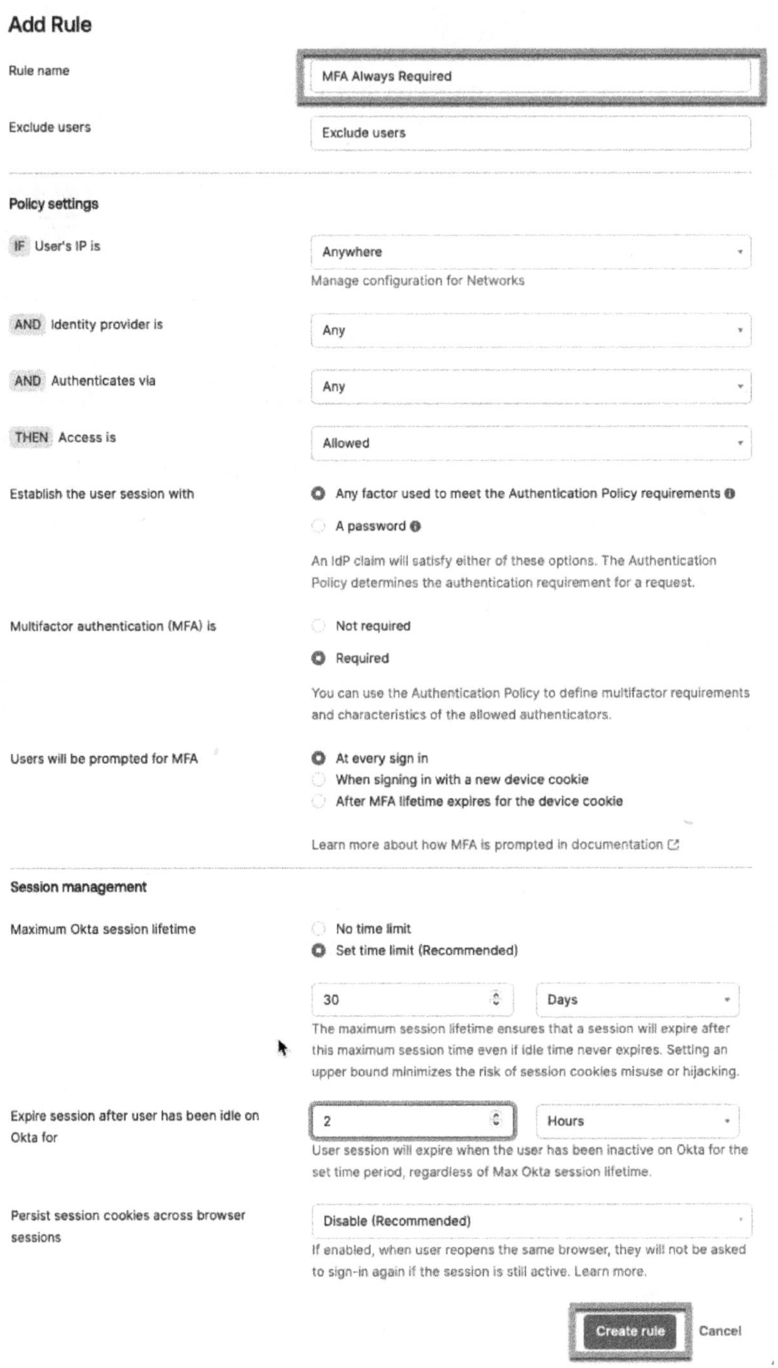

Add Rule

Rule name

MFA Always Required

Exclude users

Exclude users

Policy settings

IF   User's IP is

Anywhere ▾

Manage configuration for Networks

AND   Identity provider is

Any ▾

AND   Authenticates via

Any ▾

THEN   Access is

Allowed ▾

Establish the user session with

◉ Any factor used to meet the Authentication Policy requirements ⓘ

◯ A password ⓘ

An IdP claim will satisfy either of these options. The Authentication Policy determines the authentication requirement for a request.

Multifactor authentication (MFA) is

◯ Not required

◉ Required

You can use the Authentication Policy to define multifactor requirements and characteristics of the allowed authenticators.

Users will be prompted for MFA

◉ At every sign in
◯ When signing in with a new device cookie
◯ After MFA lifetime expires for the device cookie

Learn more about how MFA is prompted in documentation ☑

Session management

Maximum Okta session lifetime

◯ No time limit
◉ Set time limit (Recommended)

| 30 ⌄ | Days ▾ |

The maximum session lifetime ensures that a session will expire after this maximum session time even if idle time never expires. Setting an upper bound minimizes the risk of session cookies misuse or hijacking.

Expire session after user has been idle on Okta for

| 2 ⌄ | Hours ▾ |

User session will expire when the user has been inactive on Okta for the set time period, regardless of Max Okta session lifetime.

Persist session cookies across browser sessions

Disable (Recommended) ▾

If enabled, when user reopens the same browser, they will not be asked to sign-in again if the session is still active. Learn more.

Create rule   Cancel

Figure 4.61 – The Add Rule page

The result is a new priority 1 rule that will enforce MFA for all applications in the organization:

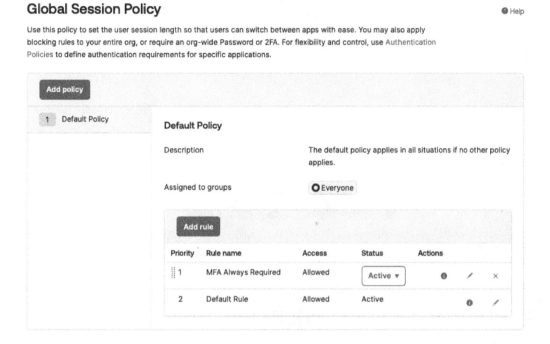

Global Session Policy

Use this policy to set the user session length so that users can switch between apps with ease. You may also apply blocking rules to your entire org, or require an org-wide Password or 2FA. For flexibility and control, use Authentication Policies to define authentication requirements for specific applications.

Add policy

| 1 | Default Policy |

Default Policy

| Description | The default policy applies in all situations if no other policy applies. |
| Assigned to groups | ⊙ Everyone |

Add rule

Priority	Rule name	Access	Status	Actions		
1	MFA Always Required	Allowed	Active ▾	❶	✎	✕
2	Default Rule	Allowed	Active		❶	✎

Figure 4.62 – Modified global session policy

We are now ready to test all the changes we've made to the default Okta configuration. We added a new authenticator, Duo, and we made MFA a requirement for all applications in the organization.

Let's log into the Okta dashboard as one of our users that's been imported from Microsoft Azure AD. If you did not import users from Microsoft Azure AD, log in as any non-admin user you created in your Okta organization:

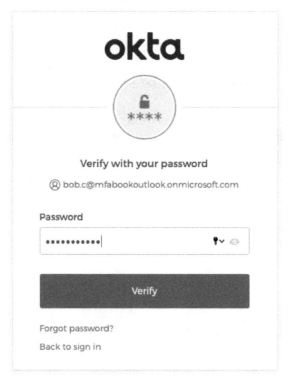

Figure 4.63 – Username and password prompts

This is the first time this user is performing a login that requires MFA. Because of that, Bob is prompted to set up any of the available authenticators for the organization, including Duo:

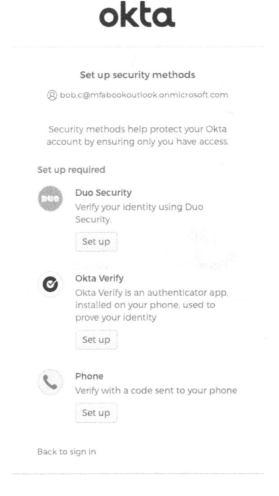

Figure 4.64 – Setting up the required MFA security methods

Complete the setup steps for Duo Security and Okta Verify and try logging in again. This time, instead of asking for a password after the username is entered, Okta allows Bob to choose among the available authenticators he has configured (not all the ones available for the organization):

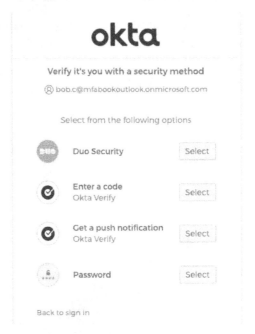

Figure 4.65 – Okta MFA selection

After selecting two factors to verify it's him, Bob is signed into the dashboard:

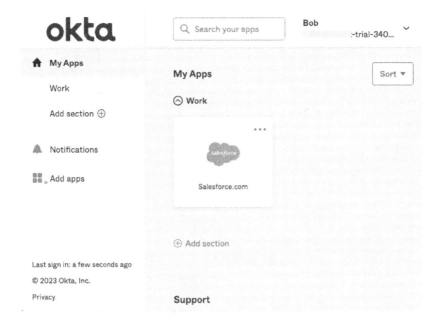

Figure 4.66 – Okta dashboard

Again, we can use SSO to log into Salesforce.com by selecting it from the menu:

Figure 4.67 – Selecting Salesforce.com from the list of apps

The dashboard can be seen in the following screenshot:

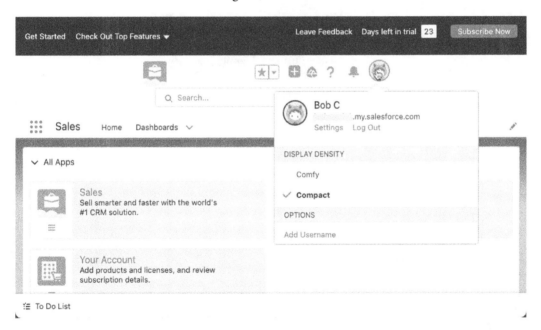

Figure 4.68 – SSO to Salesforce.com Okta dashboard

With that, we have configured MFA with the Okta Workforce Identity solution. In the second part of this chapter, we will look at a different solution from Okta – one created not for workforce users but for users external to the company.

Customer Identity with Okta

Acme also decided to try the Okta Customer Identity product as a solution for managing external customers. The main goal for Acme is to create frictionless registration and login for the existing and future cloud apps the company plans to develop. At the same time, the company wanted to keep strong security using Okta's adaptive MFA.

This step is optional and only required if you don't have an Okta Customer Identity account:

1. To create a Customer Identity trial account, go to `https://www.okta.com/free-trial/customer-identity/` and select **Customer Identity Trial** if it's not selected already:

okta
Workforce Identity Trial
Customer Identity Trial
Get Started in 15

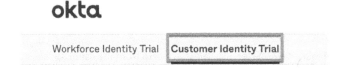

Figure 4.69 – Customer Identity Trial

2. Enter the required information and click **Get Started**:

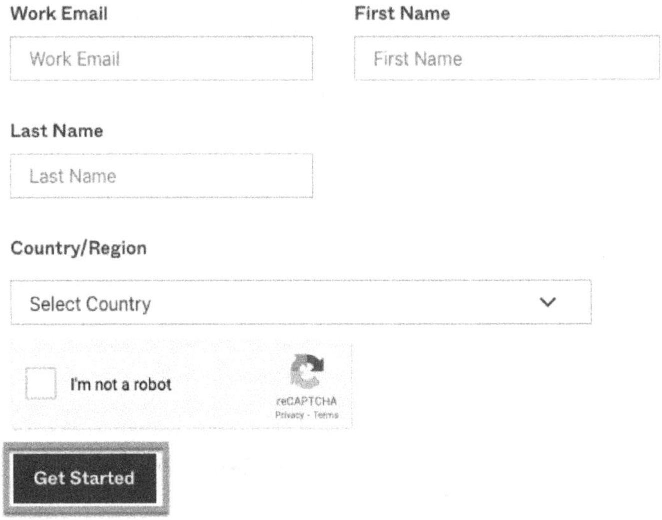
Work Email
Work Email
First Name
First Name
Last Name
Last Name
Country/Region
Select Country
I'm not a robot
reCAPTCHA
Privacy - Terms
Get Started

Figure 4.70 – Okta Customer Identity registration

Similar to Workforce Identity, this will initiate the process of creating a new account with Okta:

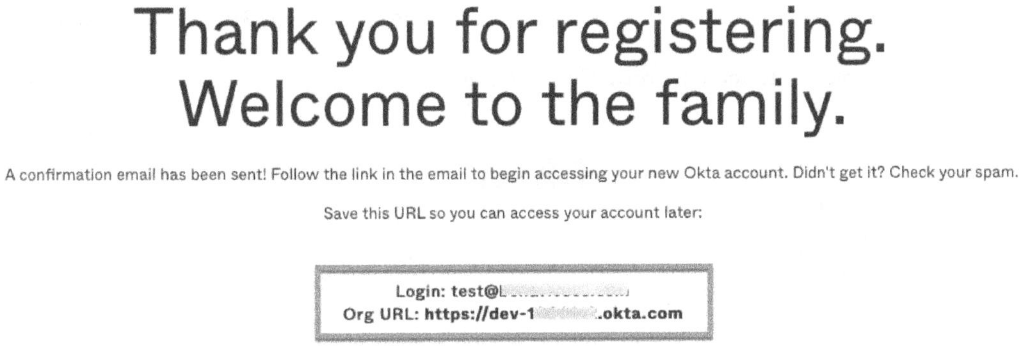

Thank you for registering.
Welcome to the family.
A confirmation email has been sent! Follow the link in the email to begin accessing your new Okta account. Didn't get it? Check your spam.
Save this URL so you can access your account later:
Login: test@
Org URL: https://dev-1 .okta.com

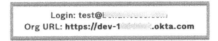

Figure 4.71 – Okta Customer Identity welcome page

3. Copy and save the **Login** and **Org URL** details. After confirming the account via the email sent by Okta, you can use that URL to open the Okta dashboard:

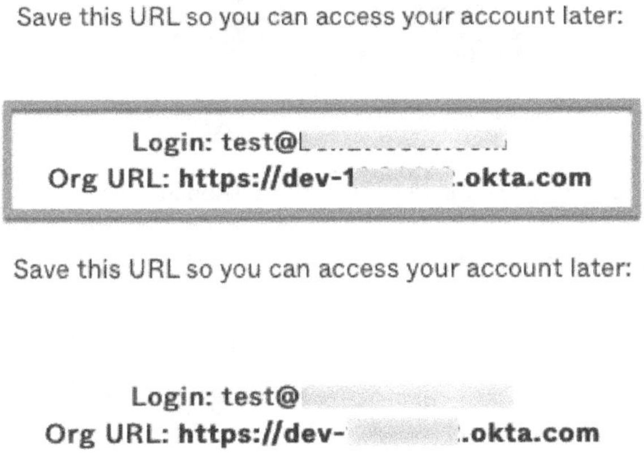

Save this URL so you can access your account later:
Login: test@
Org URL: https://dev-1 .okta.com
Save this URL so you can access your account later:
Login: test@
Org URL: https://dev- .okta.com

Figure 4.72 – Okta Customer Identity organization URL

With that, you have created the Okta Customer Identity organization. You can now use the specified username and organization URL to access the administrator dashboard.

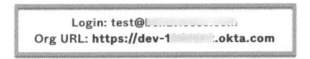

Customer Identity administration

Log into the Okta dashboard (this will look something like `https://dev-9999999.okta.com/`):

1. Enter your username and password. Click **Sign in**:

Figure 4.73 – Customer Identity dashboard sign in

You will find yourself on the Customer Identity dashboard page:

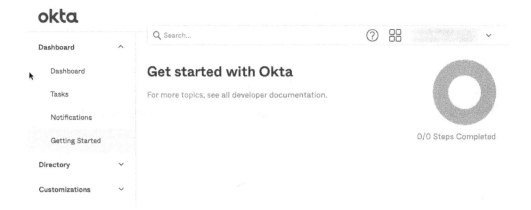

Figure 4.74 – Customer Identity dashboard

The first time we log into the Customer Identity dashboard, there will be no recommended tasks; we will see **0/0 Steps Completed**.

Testing Okta's Customer Identity solution

To start using the Customer Identity solution from Okta, we need two things: users and applications. In the Workforce scenario, we used Acme's users, imported from Microsoft Azure AD, as the test users for the workforce applications. In the Customer Identity scenario, end users should be able to register to the external applications themselves. Our first step is enabling user self-service.

User self-service

Okta uses a directory to store the identities for the solution. Directories contain users and groups that control access to the application. Because this is customer-facing, we need to allow customers (or test users) to register for the application:

1. Expand **Directory** and select **Self-Service Registration**:

Figure 4.75 – The Self-Service Registration menu

2. The **Self-Service Registration** window will open. Click **Edit** and tick the checkbox for **Add to Sign-In widget**:

Figure 4.76 – Self-service registration configuration

3.   At the bottom of the page, uncheck **User must verify email address to be activated** and click **Save**:

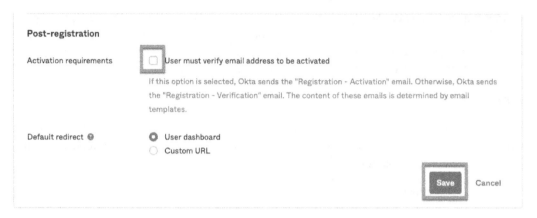

Figure 4.77 – Self-service registration configuration

> **Important note**
>
> Unchecking **User must verify email address to be activated** should only be used in a test environment. Performing the registration verification step is very important when exposing self-registration functionality to external customers to reduce fraud and will only occur when this checkbox is selected.

Now that we have allowed users to self-register, let's add one or more apps to the environment.

Adding applications

Suppose your company already has customer-facing **Software-as-a-Service** (**SaaS**) applications. In that case, you can use the Okta Customer Identity dashboard to manage those applications' integrations and user access. Using the dashboard, administrators can quickly add existing applications to the Okta platform and enable SSO, allowing users to access these applications without having to enter separate login credentials.

To add an application to SSO using the Okta Customer Identity dashboard, an administrator would typically need to perform the following steps:

1. Navigate to the Okta Customer Identity dashboard and select the **Applications** tab.

2. Click the **Add Application** button and choose the desired application from the available options.

3. Follow the on-screen instructions to configure the application for SSO and specify any additional settings, such as group or attribute mappings.

4. Save the changes and test the SSO integration to ensure it works as expected.

Acme decided not to use its production applications to test SSO but to create a sample app using Okta's CLI. The Okta CLI is a tool designed for developers. With this CLI tool, you can quickly set up an application with authentication in the language and framework you choose in just minutes without navigating any web-based admin consoles.

To use the Okta CLI, you must set it up on your machine. This native command-line application is compatible with macOS, Linux, and Windows operating systems.

To install the Okta client on different platforms, follow the instructions at `https://cli.okta.com/manual/#installation`.

After installing the Okta client, you need to log into your org. The Okta client uses API tokens, not passwords, to authenticate requests to Okta APIs. API tokens are created with the privilege level of the account used to create them. Creating service-level accounts to assign tokens to the specific privilege level needed is beyond the scope of this book, but it is a recommended security practice. To understand what an API token is and what the best practices are, one good place to go is `https://developer.okta.com/docs/guides/create-an-api-token/main/`.

Important note

Tokens are valid for 30 days. If a token is not used for 30 days, it will expire and a new one will need to be generated. The 30 days from creation or last use can't be changed. If a user account is deactivated in Okta, all API tokens created by that user account become invalid immediately.

Generating new tokens

We will need to log in using an API token. To create new tokens, from the Okta dashboard, do the following:

1. Expand **Security** and click on **API**:

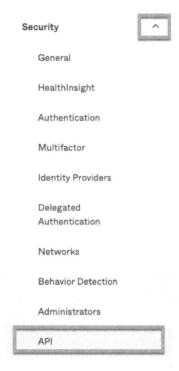

Figure 4.78 – Security – API

2. Select **Tokens**. Click on **Create Token**:

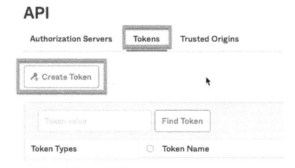

Figure 4.79 – The Tokens tab

3. Name your token. Then, click on **Create Token**:

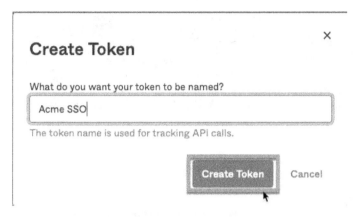

Figure 4.80 – The Create Token menu

4. Click on the **Copy to clipboard** button and copy this token to a safe location. This will be the only time that you will be able to see it in cleartext. Click on **OK, got it**:

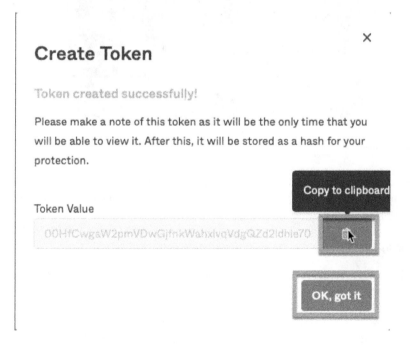

Figure 4.81 – The Create Token menu

The Okta client command-line tool will appear:

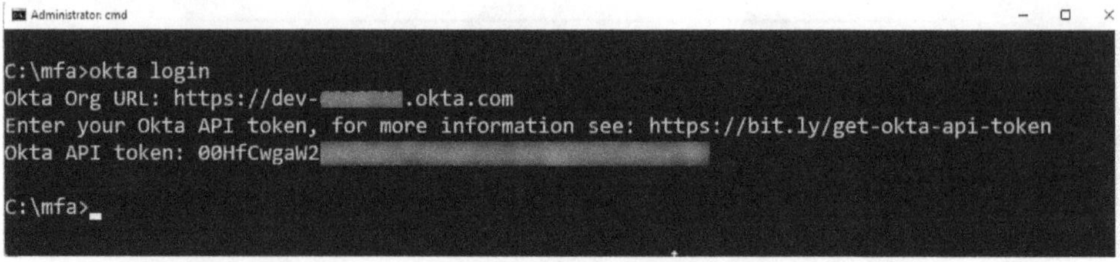

Figure 4.82 – Okta client command-line tool

5. Type okta start. Select the sample type or press *Enter* for the default (Spring Boot + Okta):

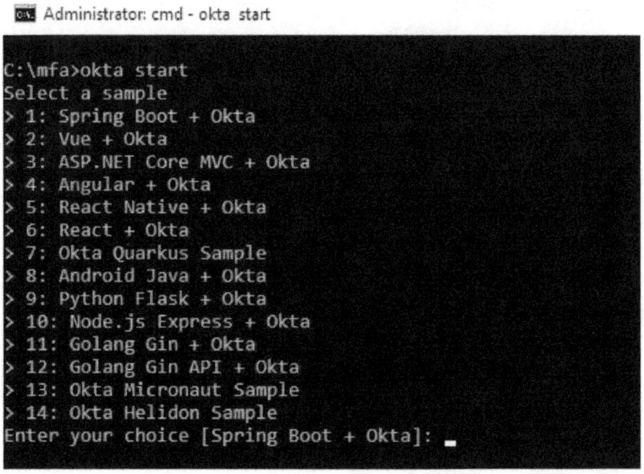

Figure 4.83 – Building an application

After the start command completes, follow the instructions provided by the okta start command to build and run the application. The okta start command will also register the application in your organization so that it's ready to be tested.

Testing the application

Now that we have applications and users registered to use them, we can validate that the okta start command created the application correctly:

1. Open a new browser window in InPrivate or incognito mode and browse to the URL for your application. It should look like http://localhost:9999/ (where 9999 is the port number your application is running on). In the case of the Spring Boot app, the URL will be http://localhost:8080.

2. You should be prompted to log in (if not, clear your cache and try again in a new window). In this case, do not enter a username and password but instead choose to **Sign up** as a new user:

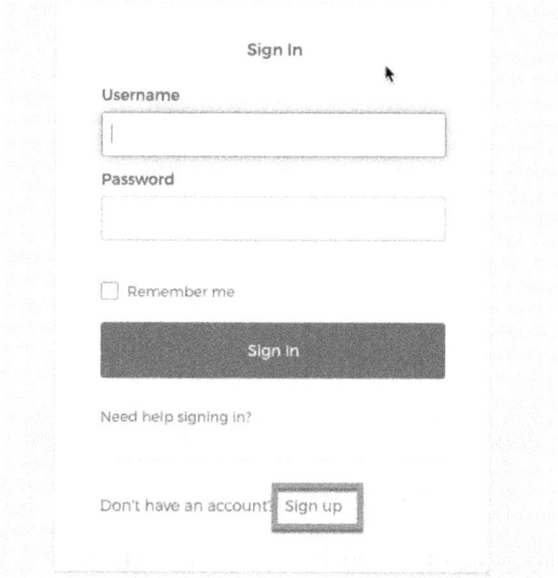

Figure 4.84 – Signing up as a new user

3. Enter the email, password, and first and last names of the new user, and click **Register**:

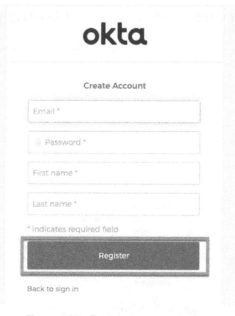

Figure 4.85 – Registering a new user

Because we selected not to require confirmation via email after registering the new user, as soon as the data for the user is validated, the user will be logged in to the application and the test will be successful:

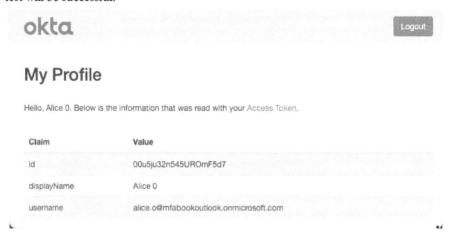

Figure 4.86 – Login successful

We just saw that protecting applications with Okta's Customer Identity solution is very similar to what we did in the first part of this chapter, with Okta's Workforce Identity. One major difference, however, is that for both dashboard and application authentication, the default with Okta Workforce Identity was to require MFA.

Requiring MFA to access Okta Customer Identity apps

Okta Customer Identity only has a password pre-defined as an authentication factor. To be able to configure MFA, we must first enable additional factors:

1.   Expand **Security** and click on **Multifactor**:

Figure 4.87 – Security – Multifactor

2. The **Multifactor** page will be displayed, as well as the default tab, **Factor Types**. We are going to enable **Okta Verify**:

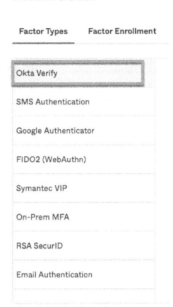

Figure 4.88 – Multifactor configuration

3. Change the status of **Okta Verify** from **Inactive** to **Active**. Optionally, you can also enable push notifications and require Touch ID or Face ID by clicking on **Edit** and making the appropriate selections:

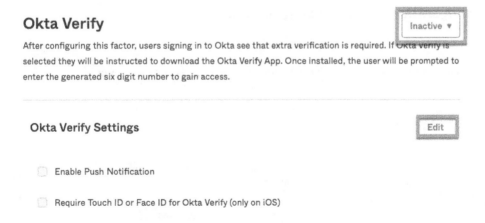

Figure 4.89 – Okta Verify configuration

4.   Once you've finished configuring Okta Verify, at the bottom of the page, click **Save**:

Figure 4.90 – Activating Okta Verify

At this point, if you try to log in to the dashboard or any configured SSO application, you will still not be prompted for a second factor of authentication. There are no rules that require it, so you don't get prompted to set up or use Okta Verify or any other factor you configured.

One option would be to follow similar instructions to those we used for Okta Workforce Identity and add a new rule for the whole organization, forcing MFA globally:

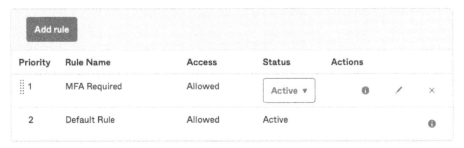

Figure 4.91 – Security – Authentication – Sign On – Add Rule

For external customers, it may not always make sense to require MFA for all applications. For this reason, we are not going to configure MFA for the whole organization, as shown earlier. Instead, we are going to configure MFA for selected applications.

MFA configuration for individual applications

Now, let's configure MFA for individual applications:

1.   Expand **Applications** and click on **Applications** via the dashboard menu:

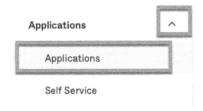

Figure 4.92 – Applications – Applications

2. From the list of applications shown, select and click on the application you want to configure MFA for:

Figure 4.93 – Security – Authentication – Sign On – Add Rule

3. On the selected application, change the tab to **Sign On** from **General**:

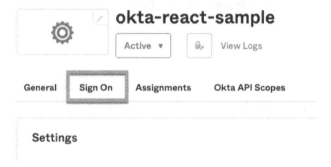

Figure 4.94 – The Sign On tab for an application

4. Click on + **Add Rule** to create a new rule:

Figure 4.95 – Adding a new rule

5. At the top of the new rule page, add a descriptive name for the rule in the **Rule Name** textbox:

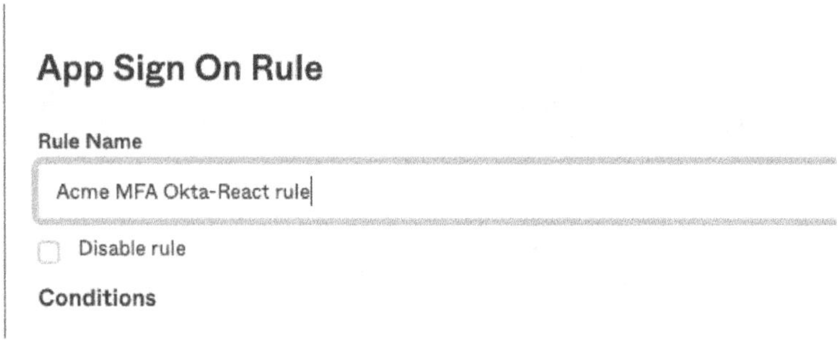

App Sign On Rule

Rule Name

Acme MFA Okta-React rule

☐ Disable rule

Conditions

Figure 4.96 – Adding a new app sign-on rule

6. At the bottom of the **New Rule** page, select the frequency for the MFA prompt. Then, click **Save**:

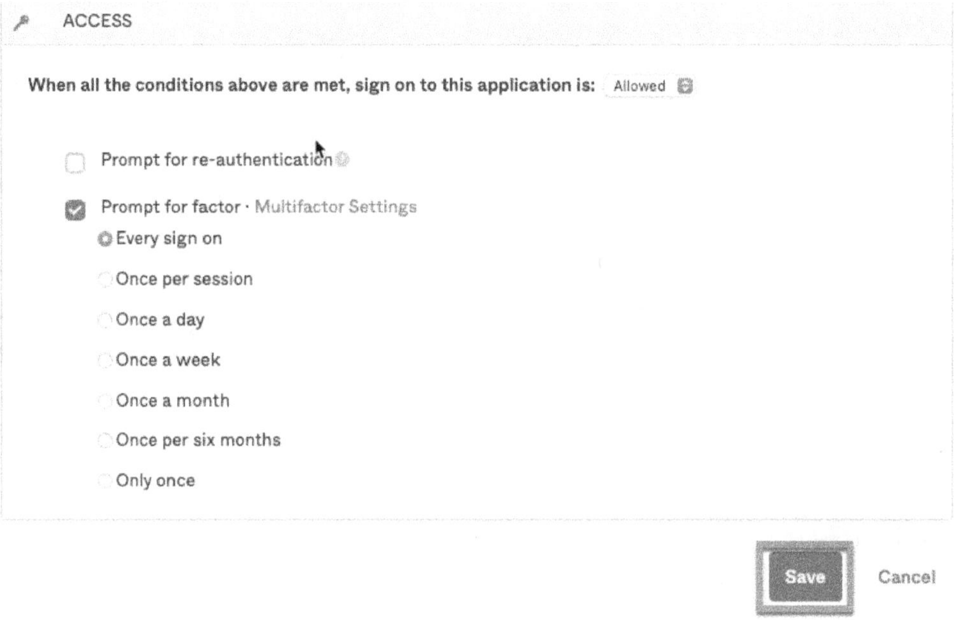

🔑 ACCESS

When all the conditions above are met, sign on to this application is: Allowed 🔲

☐ Prompt for re-authentication 🔘

☑ Prompt for factor · Multifactor Settings
  🔘 Every sign on
  ⦾ Once per session
  ⦾ Once a day
  ⦾ Once a week
  ⦾ Once a month
  ⦾ Once per six months
  ⦾ Only once

Save   Cancel

Figure 4.97 – Adding a new app sign-on rule

The resulting sign-on policy for the application now contains two rules. The first one guarantees that users will need to use MFA every time they sign into the app. The same rule does not apply to other apps, so this procedure must be repeated for each app where we want it to apply:

Sign On Policy

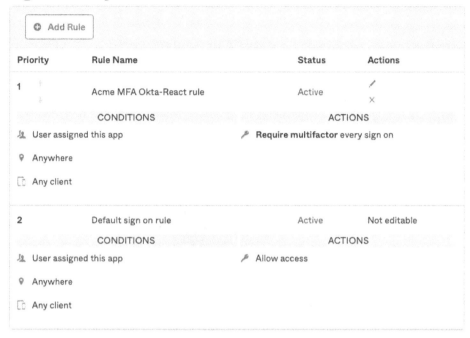

Figure 4.98 – Modified application sign-on policy

With that, we have created a rule that will force users to perform MFA every time they sign into specific applications in your organization. If that behavior should apply to all apps, the better option is to apply the MFA rule to the organization instead of to individual apps.

Testing MFA with the application

Now, we are ready to test the application again. For the applications we changed the configuration of, users will be prompted to enter MFA after they enter their username and password correctly:

1. Open a new browser window in InPrivate or incognito mode and browse to the URL for your application. It should look like `http://localhost:9999/` (where `9999` is the port number your application is running on). In the case of the React app, the URL will be `http://localhost:3000/profile`.

2.   Enter the username and password for a user and click **Sign In**:

Figure 4.99 – The Sign In page

3.   If this is the first time the user is using MFA, the setup page will be displayed. Click **Setup**:

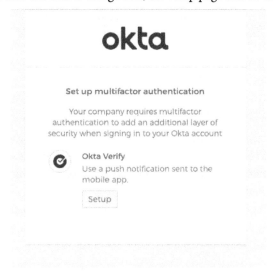

Figure 4.100 – Okta Verify setup prompt

If the user is accessing any MFA-configured application for the first time, the user will be prompted to set up an authentication factor. On subsequent logins, the user will just be prompted to confirm themselves with the authenticator. Okta Verify will be shown in this test.

After setup or confirmation of the authentication factor, the user's profile page will be displayed:

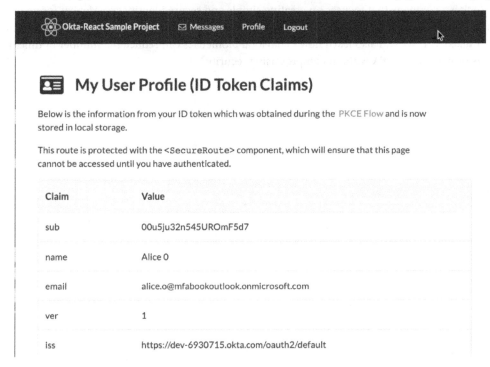

Figure 4.101 – React sample application profile page

This completes our test of using an application policy to enforce the use of a second factor of authentication when accessing selected apps in Okta's Customer Identity.

Okta also allows for more granular control of the rules required to access the organization or selected applications protected by Customer Identity.

Summary

In this chapter, we went over the process of importing users, enabling SSO for custom-built and Enterprise applications, and using different authentication factors for MFA with the use of organization-based policies, all to protect workforce users (employees, contractors, and temporary users) using the Okta Workforce Identity solution.

In the second part of this chapter, we learned how to enable self-registration and how to configure SSO and MFA using application-based policies with Okta Customer Identity. By deploying Okta Customer Identity, we enabled frictionless, secure access to custom-built apps for external users of the company.

In the next chapter, we will learn about another workforce and customer-oriented company, ForgeRock. ForgeRock is a company that focuses on creating simple and secure identity experiences for people at work or on the go, with no awkward registration processes and without worrying about fraudulent account takeovers. We will also learn how behavioral biometrics can reduce the number of times a user gets prompted for MFA without compromising security.

5

Access Management with ForgeRock and Behavioral Biometrics

In this chapter, we are going to explore ForgeRock **identity and access management** (**IAM**) products, and how Acme can use ForgeRock to provide a great customer experience for external users and empower and secure an agile workforce. We will also introduce an important new technology that will help us increase security and reduce friction during authentication: the behavior of the user login process.

We are going to cover the following main topics:

- Experiencing ForgeRock
- Introducing Authentication Trees
- What are behavioral biometrics?

Technical requirements

This chapter requires a ForgeRock software platform account. If required, a trial is available at `https://www.forgerock.com/trial-registration`.

For deploying ForgeRock's **Access Manager** (**AM**), Java **software development kit** (**SDK**) version 11 or above is required. We are going to use Apache's Tomcat web server to deploy the AM, but other web servers are also supported by ForgeRock.

An optional step is to create and deploy a Java-based application to demonstrate ForgeRock's access management capabilities.

An account with BehavioSec is required for the *What are behavioral biometrics?* section.

Experiencing ForgeRock

We will start by creating a ForgeRock account.

Creating a ForgeRock software platform account

This step is optional and only required if you don't have a ForgeRock software platform account, also known as a **backstage account**:

1. To create a ForgeRock software platform trial account, go to `https://www.forgerock.com/trial-registration`.

2. Enter the required information and click **REGISTER**:

Figure 5.1 – ForgeRock software platform account registration

This will initiate the process of creating a new trial account with ForgeRock:

Figure 5.2 – Backstage account welcome page

With that, your ForgeRock account will be created.

Signing into your backstage account for the first time

Log in with your backstage account at `https://sso.forgerock.com/UI/Login?goto=https%3A%2F%2Fbackstage.forgerock.com%2Fdownloads`. Then, follow these steps:

1. Enter your username and password. Click **Next**:

Figure 5.3 – Backstage login – username and password

2.   Choose one of the devices that you configured during registration – for example, **Push Auth**:

Please choose one of your previously registered devices to continue. This could be one of the following: WebAuthn, a hardware token (e.g. YubiKey) or biometric (e.g. Touch ID) authentication, Push Auth, a message based authentication with a mobile device, or One-time Password, a single-use, 6-digit, limited lifetime password generated by a mobile device

Figure 5.4 – Backstage login – MFA

3.   Click **Accept**:

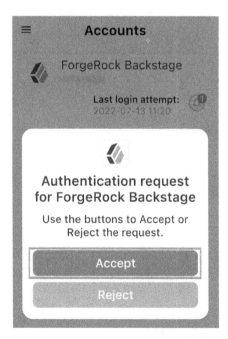

Figure 5.5 – Setting up security methods

Backstage is the place to go for support, education, and documentation, and for us, the place we can download the products we are going to use from ForgeRock. The products we will use are part of the ForgeRock Identity Platform.

ForgeRock also has a version of the Identity products that is cloud-based, known as **ForgeRock Identity Cloud**. We will not cover ForgeRock Identity Cloud in this book. You can find more information at https://www.forgerock.com/platform/identity-cloud:

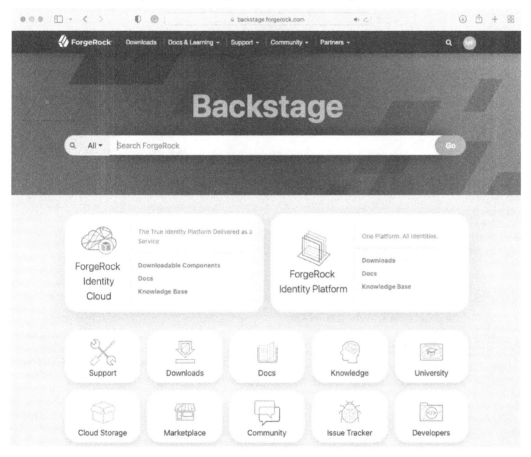

Figure 5.6 – ForgeRock backstage home page

The product we want to install from ForgeRock is Access Management. The latest version at the time of writing is 7.2.0.

Before we can install ForgeRock's Access Management product, we need to install the Java SDK and Apache's Tomcat. Instructions to install Java SDK version 11 can be found in *Appendix A*. Instructions to install Apache Tomcat version 9 can be found in *Appendix C*.

Installing ForgeRock Access Manager

Once Tomcat is up and running, we must install ForgeRock's AM.

Navigate to the **Access Management – Featured Downloads** page (`https://backstage.forgerock.com/downloads/browse/am/featured`) and select the **Access Management** ZIP file:

1. Click on **ZIP** (`https://backstage.forgerock.com/downloads/get/familyId:am/productId:am/minorVersion:7.2/version:7.2.0/releaseType:full/distribution:war`):

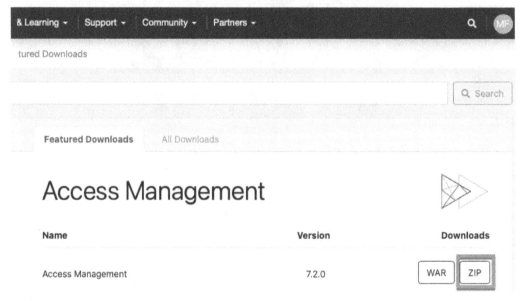

Figure 5.7 – Downloading the Access Management ZIP file

2. Click on **Download**:

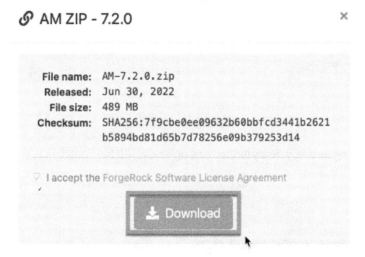

Figure 5.8 – Downloading AM

Save the downloaded ZIP file on the same server where Tomcat was installed:

Figure 5.9 – ZIP file

3. Unzip AM-7.2.0.zip using the unzip command (Mac or Linux) or by right-clicking the ZIP file and choosing **Extract All** (Windows):

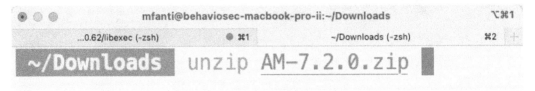

Figure 5.10 – Unzipping openam

After unzipping the file, a new subdirectory, openam, will be created. In it, you will find a file named AM-7.2.0.war:

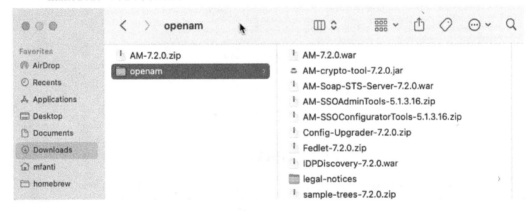

Figure 5.11 – Expanded openam ZIP file

4. Create an environment variable, $CATALINA_HOME, with the path of the Tomcat installation you want to use for openam:

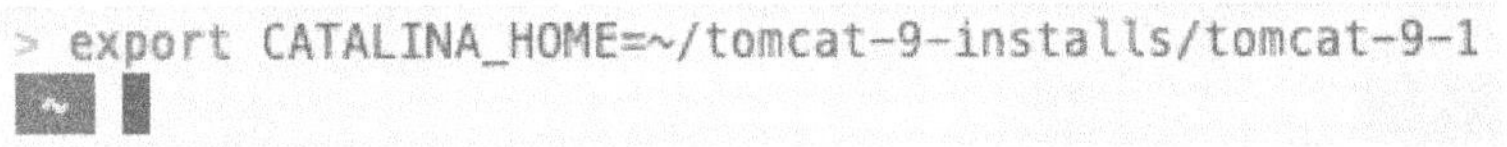

```
> export CATALINA_HOME=~/tomcat-9-installs/tomcat-9-1
```

Figure 5.12 – The $CATALINA_HOME environment variable

5. Stop Tomcat if it's running. Copy the WAR file in the openam subdirectory to the $CATALINA_HOME/webapps directory in Tomcat. Rename the file openam.war:

```
> cp ~/Downloads/openam/AM-7.2.0.war $CATALINA_HOME/webapps/openam.war
```

Figure 5.13 – The openam.war file copied to the webapps directory

6. Start Tomcat by running $CATALINA_HOME/bin/catalina.sh run:

> `$CATALINA_HOME/bin/catalina.sh run`

Figure 5.14 – Starting Tomcat

By adding openam.war to the webapps directory, we instructed to deploy the WAR file every time Tomcat is restarted. Tomcat uses the name of the WAR file without the extension as the local URL for openam.

Configuring ForgeRock's Access Manager (openam)

After deploying openam.war for the first time, the application needs to be configured before it can be used. Follow these steps:

1. Open a browser with the URL and port for your Tomcat instance and append /openam to the URL. In this example, the URL is http://openam.acme.local:18080/openam. Click on **Create Default Configuration**:

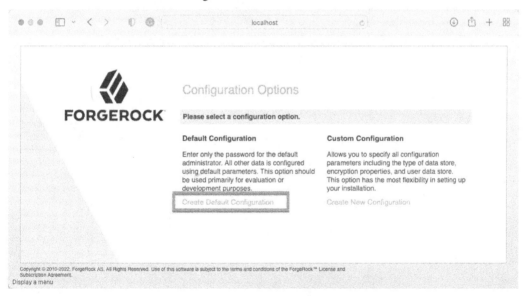

Figure 5.15 – The Configuration Options page

2.   Scroll down and accept the license agreement. Click **Continue**:

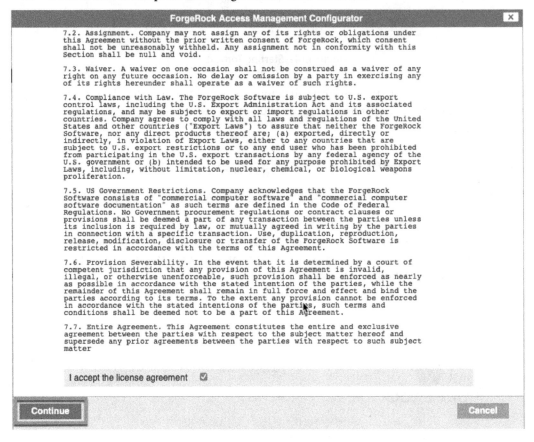

Figure 5.16 – License agreement page

3.   Enter the **Default User** password twice. Click **Create Configuration**:

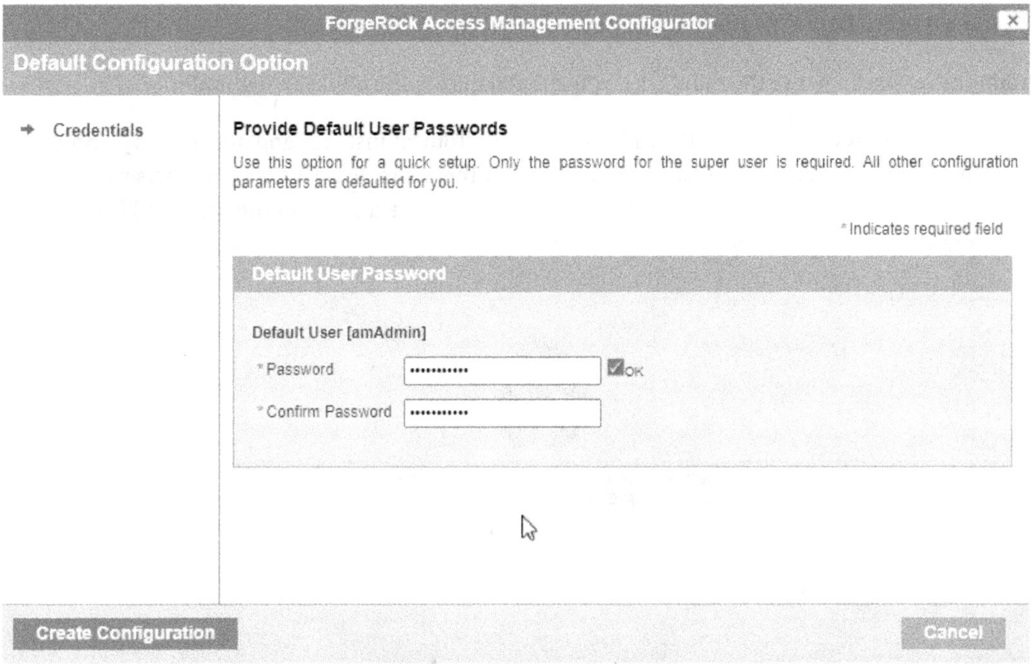

ForgeRock Access Management Configurator
Default Configuration Option
Credentials
Provide Default User Passwords
Use this option for a quick setup. Only the password for the super user is required. All other configuration parameters are defaulted for you.
* Indicates required field
Default User Password
Default User [amAdmin]
* Password
OK
* Confirm Password
Create Configuration
Cancel

Figure 5.17 – Default user password page

4. After the configuration is complete, click on **Proceed to Login**:

Configuration Complete!
Proceed to Login

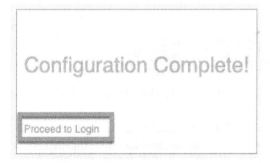

Figure 5.18 – The Configuration Complete! page

With that, openam has been installed and configured for use.

Using openam

To administer openam, use the same URL you entered during the initial configuration:

1. Open a browser with the URL and port for your Tomcat instance and append `/openam` to the URL. As in the previous section, the URL for this example is `http://openam.acme.local:18080/openam`. The default username is `amadmin` and the password is the one you chose during the initial configuration.

    Click **LOG IN**:

Figure 5. 19 – Initial login page

The main administration page will appear, along with the top-level realm. Realms in ForgeRock are administrative units that group configuration and identities together. We are going to be using the top-level realm in this chapter:

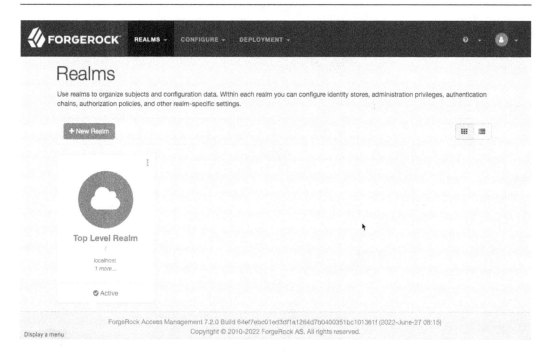

Figure 5. 20 – Main administration page

Before we continue using openam, let's deploy a test application on a separate Tomcat server. Instructions for deploying a custom web application on Tomcat can be found in *Appendix D*. You can also use the examples provided with Tomcat for this example.

Protecting a Java application using openam

ForgeRock provides components called Agents to protect web applications. For Tomcat and other Java-based web servers, ForgeRock created Java Agents. Java Agents enforce AM policies for web applications deployed on a Java container.

Java Agents intercept requests to web applications and ensure that clients have the required authentication and authorization.

To install a Java Agent in one of the web servers, we need to download the appropriate type of agent for the web server we are using. The web server cannot have openam deployed.

To download the AM Java Agent, follow these steps:

1. Login to `https://backstage.forgerock.com` and select **Downloads | Access Management | Featured Downloads | AM Java Agents** (`https://backstage.forgerock.com/downloads/browse/am/featured/java-agents`). Download the ZIP file for the **Tomcat** (not **Tomcat – Jakarta**) agent by clicking **ZIP**:

Figure 5.21 – Main administration page

2. Create a text file for the agent password. For production servers, you are strongly encouraged to generate secure passwords. You can change the password value and path shown in the following example:

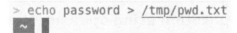

Figure 5.22 – Creating a password file

3. Make sure the file is protected:

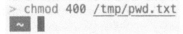

Figure 5.23 – Protecting the file from other users

Before installing the agent on the server, we need to go back and perform a few tasks in the openam administrator dashboard.

4. On the home screen, click on **Top Level Realm**:

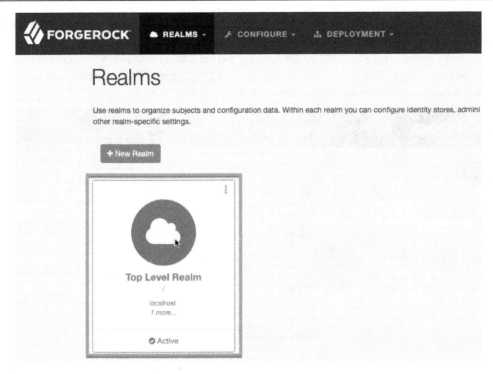

Figure 5.24 – Openam dashboard main screen

5.   Select **Applications | Agents | Java**. Click on + **Add Java Agent**:

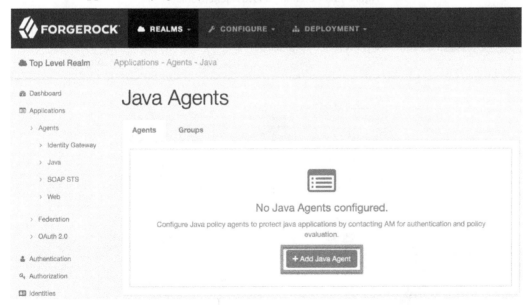

Figure 5.25 – Adding a new Java Agent profile

6.   Enter a unique **Agent ID**. **Agent URL** should be the URL of the Tomcat server where the agent is going to be installed, followed by `/agentapp`. In this example, the URL is `http://agent.acme.local:28080/agentapp`. **Server URL** is the full URL for openam – for example, `http://openam.acme.local:18080/openam`. Also, make sure that the password matches the value in the password file you created in *step 2*:

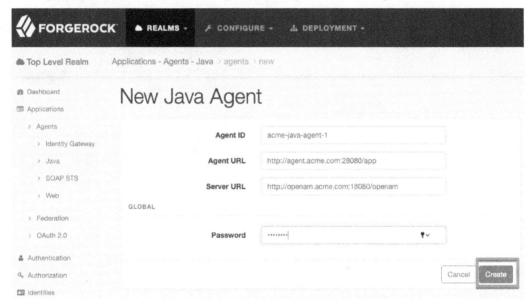

Figure 5.26 – Creating a Java Agent profile

After creating a new agent profile, the next step is to configure a new policy. Policies define the resources that should be protected by agents.

7. Select **Authorization | Policy Sets**. Click on **Default Policy Set**:

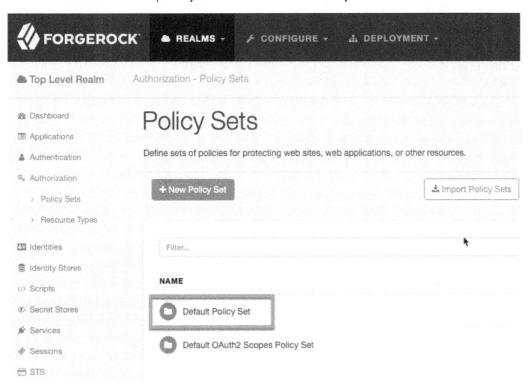

Figure 5.27 – Creating a Java Agent profile

8. Click on + **Add a Policy**:

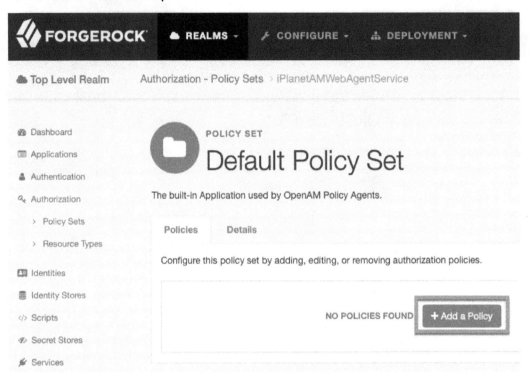

Figure 5.28 – Adding a new policy

9. Enter `acme-all-resources` for **Name**, URL for **Resource Type**, and `*://*:*/*` for **Resource pattern**. Leave the default for **Resource value**. Click **Add**:

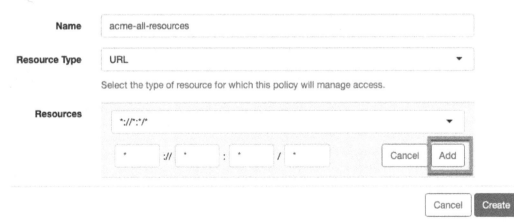

Figure 5.29 – Adding a new policy

10. Click on **Create**:

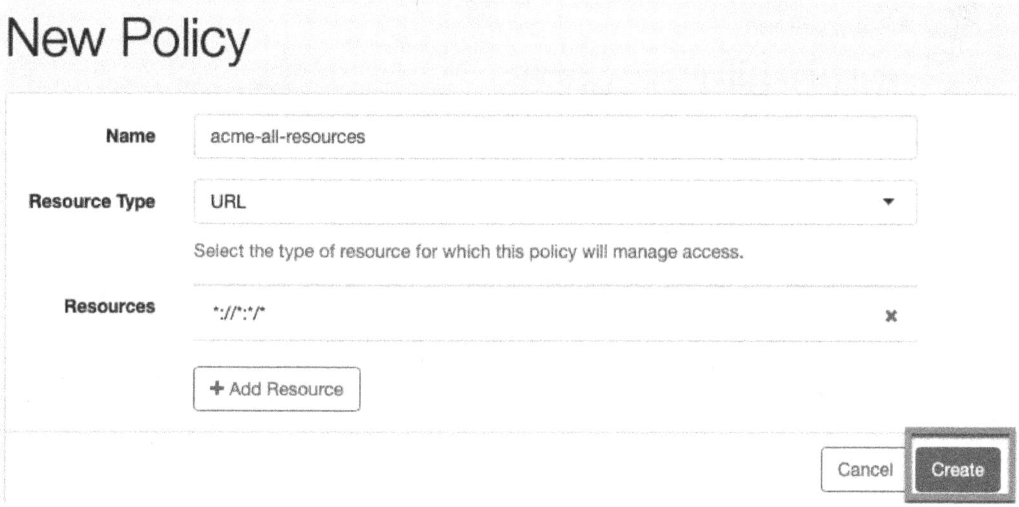

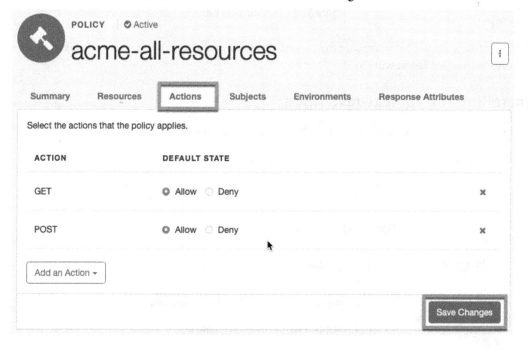

Figure 5.30 – Creating a new policy

11. In the **Actions** tab, allow **GET** and **POST**. Click **Save Changes**:

Figure 5.31 – The Actions tab

12. In the **Subjects** tab, select **All of…** and **Authenticated Users**. Click **Save Changes**:

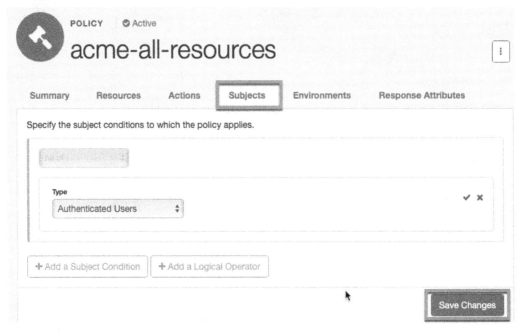

Figure 5.32 – The Subjects tab

Keep openam running but make sure the Tomcat server where the agent is to be installed is stopped; otherwise, the installation will fail.

Installing the Tomcat Java Agent

To install the Access Manager Tomcat Java Agent on a Tomcat server, do the following:

1. Unzip the Tomcat Java Agent file you downloaded in *step 1* in the previous section:

Figure 5.33 – Unzipping the Tomcat Java Agent file

2. Change directory to `./java_agents/tomcat_agent/bin`:

Figure 5.34 – Change to the bin directory

3. Create a response file called `acme1.txt` in the current directory with the following values (replacing $CATALINA_HOME with the real path to Java Agent Tomcat, not the openam Tomcat):

```
# Response File
CONFIG_DIR= $CATALINA_HOME/conf
AM_SERVER_URL= http://openam.acme.local:18080/openam
CATALINA_HOME= $CATALINA_HOME
AGENT_URL= http://agent.acme.local:28080/agentapp
AGENT_PROFILE_NAME= acme-java-agent-1
AGENT_PROFILE_REALM= /
AGENT_PASSWORD_FILE= /tmp/pwd.txt
```

This is how it would look:

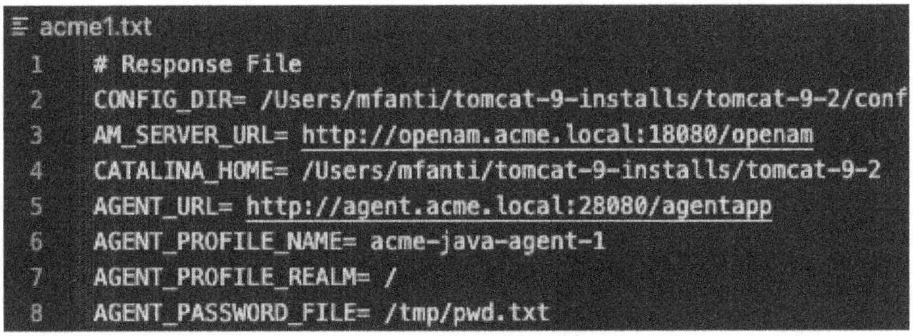

```
≡ acme1.txt
 1    # Response File
 2    CONFIG_DIR= /Users/mfanti/tomcat-9-installs/tomcat-9-2/conf
 3    AM_SERVER_URL= http://openam.acme.local:18080/openam
 4    CATALINA_HOME= /Users/mfanti/tomcat-9-installs/tomcat-9-2
 5    AGENT_URL= http://agent.acme.local:28080/agentapp
 6    AGENT_PROFILE_NAME= acme-java-agent-1
 7    AGENT_PROFILE_REALM= /
 8    AGENT_PASSWORD_FILE= /tmp/pwd.txt
```

Figure 5.35 – Example response file

4. Run `./agentadmin --install --acceptLicense --useResponse ./acme1.txt` to install the agent:

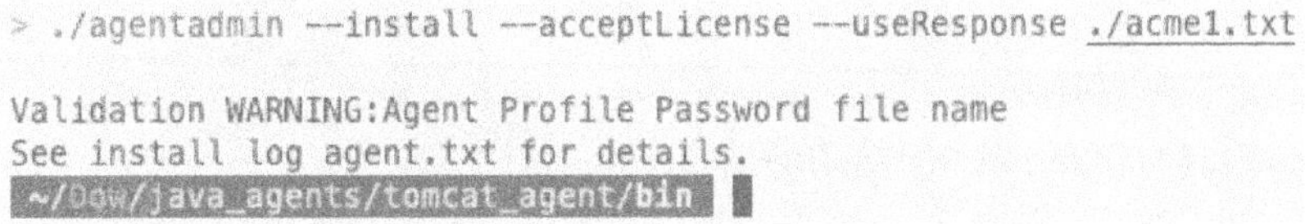

```
> ./agentadmin --install --acceptLicense --useResponse ./acme1.txt

Validation WARNING:Agent Profile Password file name
See install log agent.txt for details.
~/Dow/java_agents/tomcat_agent/bin  ▮
```

Figure 5.36 – The agentadmin install command

Now that the Java Agent has been installed on the container, we need to enable filters for each of the web applications in the container that we want the agent to protect.

Protecting a web application

In Apache Tomcat, all web applications are deployed in the $CATALINA_HOME/webapps directory. To protect the examples web application, we are going to modify the file located at $CATALINA_HOME/webapps/examples/WEB-INF/web.xml. As before, remember that in this case, $CATALINA_HOME is the real path to Java Agent Tomcat, not to the openam Tomcat:

1. Edit $CATALINA_HOME/webapps/examples/WEB-INF/web.xml.

2. After the <!-- Define example filters --> line, add the following:

    ```
    <filter>
      <filter-name>Agent</filter-name>
      <display-name>AM Agent</display-name>
      <description>AM Agent Filter</description>
      <filter-class>
    com.sun.identity.agents.filter.AmAgentFilter
      </filter-class>
    </filter>
    ```

    The preceding code should be added after the following line:

    ```
    <!-- Define example filters -->
    ```

    Figure 5.37 – The beginning of the <filter> section

3. After the <!-- Define filter mappings for the timing filters --> line, add the following:

    ```
    <filter-mapping>
      <filter-name>Agent</filter-name>
      <url-pattern>/*</url-pattern>
      <dispatcher>REQUEST</dispatcher>
      <dispatcher>INCLUDE</dispatcher>
      <dispatcher>FORWARD</dispatcher>
      <dispatcher>ERROR</dispatcher>
    </filter-mapping>
    ```

    The preceding code should be added after the following line:

    ```
    <!-- Define filter mappings for the timing filters -->
    ```

    Figure 5.38 – The beginning of the <filter-mapping> section

4. Save the file and restart Java Agent Tomcat.

After restarting the agent, we are ready to test if the Java Agent was installed correctly and that the examples application is protected by the agent.

Testing the Java Agent

To test the Java Agent's configuration, we can use any of the pages in the examples application. Other web applications on the same Tomcat server of the agent are not protected because the agent is present in the filter configuration for those web applications:

1. Open a private (or incognito) browser window and go to the examples application – for example, `http://agent.acme.local:28080/examples/servlets/servlet/CookieExample`.

2. Enter `amadmin` for the username and the corresponding password to log in:

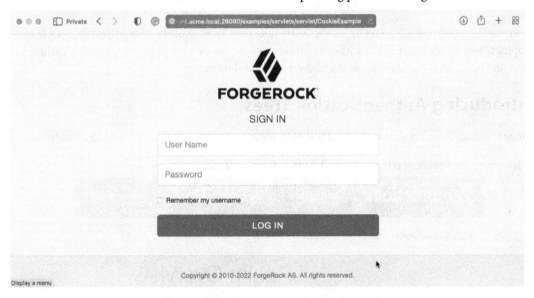

Figure 5.39 – Login prompt for a protected application

After logging in, the page is displayed. Other pages on the examples web application can be browsed without logging in again:

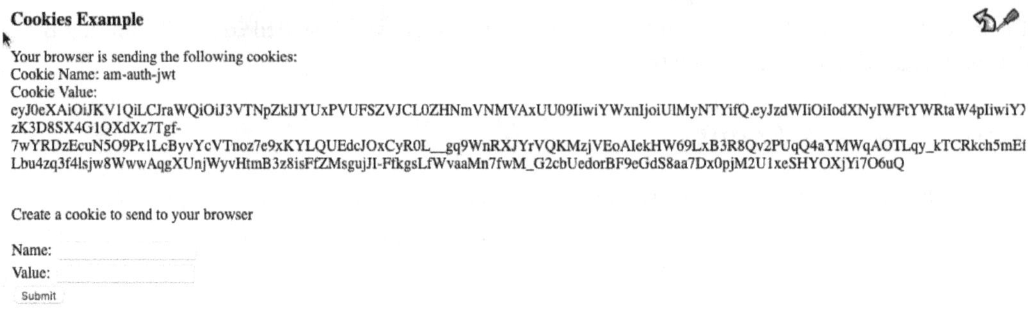

Cookies Example

Your browser is sending the following cookies:
Cookie Name: am-auth-jwt
Cookie Value:
eyJ0eXAiOiJKV1QiLCJraWQiOiJ3VTNpZklJYUxPVUFSZVJCL0ZHNmVNMVAxUU09IiwiYWxnIjoiUlMyNTYifQ.eyJzdWIiOiIodXNyIWFtYWRtaW4pIiwiY)
zK3D8SX4G1QXdXz7Tgf-
7wYRDzEcuN5O9Px1LcByvYcVTnoz7e9xKYLQUEdcJOxCyR0L__gq9WnRXJYrVQKMzjVEoAIekHW69LxB3R8Qv2PUqQ4aYMWqAOTLqy_kTCRkch5mEf
Lbu4zq3f4lsjw8WwwAqgXUnjWyvHtmB3z8isFfZMsgujJI-FfkgsLfWvaaMn7fwM_G2cbUedorBF9eGdS8aa7Dx0pjM2U1xeSHYOXjYi7O6uQ

Create a cookie to send to your browser

Name:
Value:
Submit

Figure 5.40 – Cookies example page

We just finished protecting a web application with a Java Agent in Tomcat using a username and a password, which is the default for AM.

In the next section, we are going to look at Authentication Trees. Authentication Trees are an easy, graphical way of configuring the authentication workflow in openam. That configuration includes MFA and behavioral biometrics, among many other possibilities.

Introducing Authentication Trees

Authentication Trees are part of the configuration tools for openam. Let's learn how to access them:

1. From the **Quick Start** menu, select **Authentication Trees**:

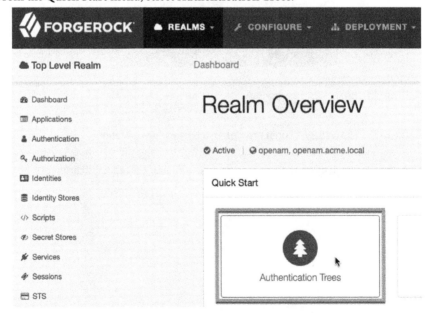

Figure 5.41 – The Quick Start menu – Authentication Trees

2. **Authentication Trees** can also be accessed from the left menu. Select **Authentication**, then **Trees**:

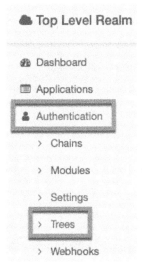

Figure 5.42 – Authentication sub-menu

Several trees are pre-configured in openam. **PlatformLogin** is used for the default login in openam. We are going to use that as a model for our custom Authentication Tree:

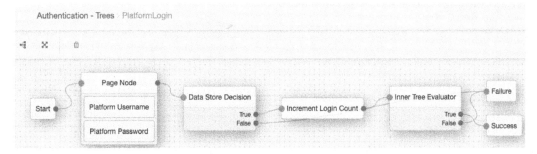

Figure 5.43 – PlatformLogin Authentication Tree

3.   Click on + **Create Tree** to create a new Authentication Tree:

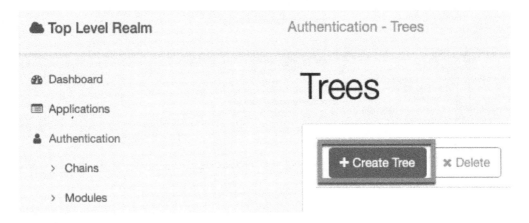

Figure 5.44 – Creating a new Authentication Tree

4.   Enter a name and click on **Create** to create a new tree:

Figure 5.45 – Naming the new Authentication Tree

5. In the new Authentication Tree, drag a **Data Store Decision**, a **Password Collector**, and a **Username Collector**:

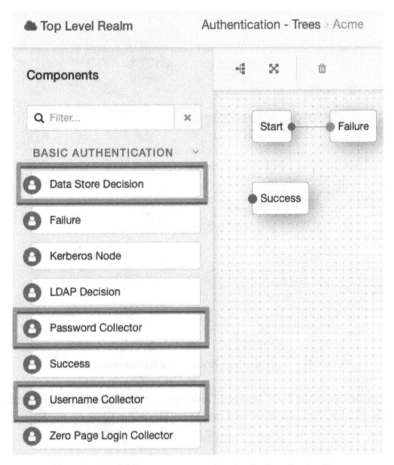

Figure 5.46 – Adding components to the Authentication Tree

6. Search for and add a **Page Node** to the tree:

Figure 5.47 – Adding a Page Node to the Authentication Tree

7. Add **Username Collector** and **Password Collector** to **Page Node**. Arrange the nodes so that **Page Node** follows **Start**, followed by **Data Store Decision**. Click on the **Save** button when you're done:

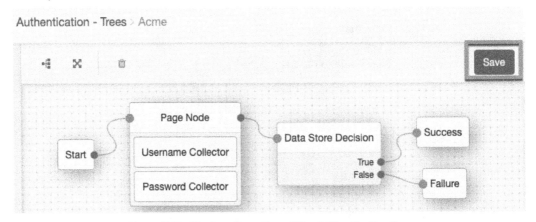

Figure 5.48 – Final configuration of the Authentication Tree

8. In the **Page Node** configuration, click on the **Add** button for **Page Header – Optional**:

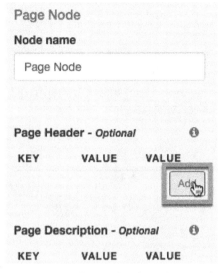

Figure 5.49 – Adding a header to Page Node

9. Enter en under **KEY** and Sign In under **VALUE**. Keys for languages other than English can also be used. Click on +:

Figure 5.50 – Adding an English header to Page Node

Optionally, do the same thing for **Page Description – Optional**.

After saving the Authentication Tree, log out of openam before testing it. openam allows you to test individual trees by following the openam URL with /XUI/#login&service=<Tree name>, where Tree Name should be the name of the tree being tested:

1. In our example, go to http://openam.acme.local:18080/openam/XUI/#login&service=Acme in your browser.

2. Confirm that you are testing the correct Authentication Tree by checking the header and description you added in the previous steps:

Figure 5.51 – Login page for Acme's Authentication Tree

After logging in, you should see the profile of the user, confirming that the user was authenticated correctly using Acme's Authentication Tree:

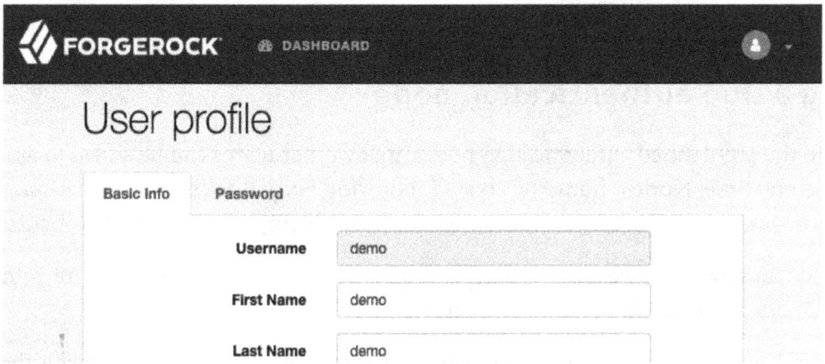

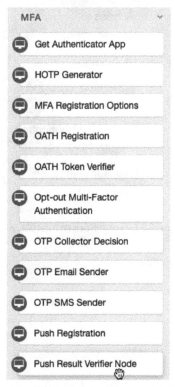

Figure 5.52 – Authenticated user profile page

Next, we are going to add MFA to the Authentication Tree. Even though you can find some pre-defined MFA nodes in the menu already, we are going to add a new one, from ForgeRock's marketplace. ForgeRock's marketplace is the place where ForgeRock, its partners, and others can add custom integrations with ForgeRock's products. To avoid installing and configuring a new MFA product in this chapter, we are going to use Duo Security, which we used in the previous two chapters:

Figure 5.53 – Default MFA nodes

Next, let's see how to install a Duo authentication node.

Installing a Duo authentication node

In addition to the pre-defined authentication nodes, many other nodes can be added to an openam installation at any time. Nodes should be created following ForgeRock's documentation, which is available at `https://backstage.forgerock.com/docs/am/7.1/auth-nodes`.

For this section and the next, we are not going to create custom nodes. Instead, we are going to use pre-existing nodes published in ForgeRock's marketplace.

To search for and download pre-existing nodes published in ForgeRock's marketplace, follow these steps:

1. Open a browser window. The URL for ForgeRock's marketplace is `https://backstage.forgerock.com/marketplace`:

Figure 5.54 – ForgeRock marketplace

2. Using the search bar, search for Duo:

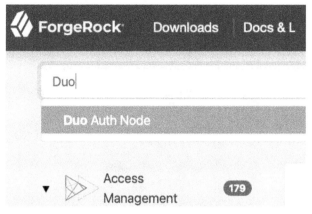

Figure 5.55 – ForgeRock marketplace search bar

Follow the instructions to download and install the JAR file needed for **Duo Auth Node**.

3. Download the JAR file by clicking **here**:

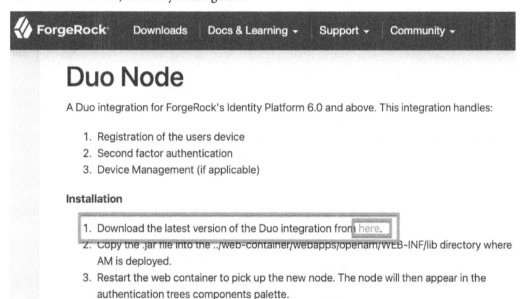

Figure 5.56 – The Duo Node marketplace page

4. Copy the downloaded file to $CATALINA_HOME/webapps/openam/WEB-INF/lib.

5. Restart Tomcat:

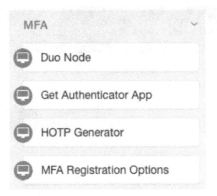

Figure 5.57 – MFA menu with Duo Node

As in *Figure 5.57*, you should now see a new entry on the Authentication Tree's MFA menu for Duo Node.

Configuring authentication with a Duo authentication node

For our second Authentication Tree example, we are going to add Duo to the Authentication Tree we created previously. You may want to test other functionality without Duo, so it is recommended that you create an Authentication Tree with a different name – for example, Duo:

1. Follow the steps we used to create the **Acme** Authentication Tree to add a page node, username and password collectors, and a data store decision node to the **Duo** tree. When adding the headers to the page node, make them different than the Acme tree so that you can differentiate them when testing them. Add a new **Duo Node** to the tree. The final configuration should look exactly like a copy of the **Acme** Authentication Tree, with an additional Duo Node:

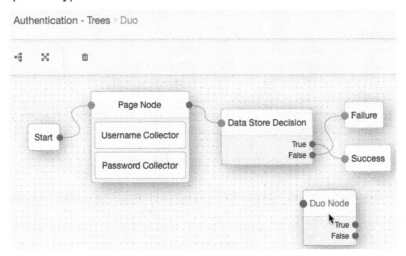

Figure 5.58 – Authentication tree with a Duo Node

2. Create a new Web SDK application in the **Duo Security** administration dashboard, following the instructions on the **Duo Node** page.

3. Complete the information in the Duo Node. Add an **Integration Key** (client ID), **Secret Key** (client secret), and **API Host Name** for the Duo Web SDK application you created in the previous step:

Figure 5.59 – Duo Node parameters

4. Connect the **True** outcome of **Data Store Decision** to **Duo Node** and the **True** outcome from **Duo Node** to **Success**, as in the following figure:

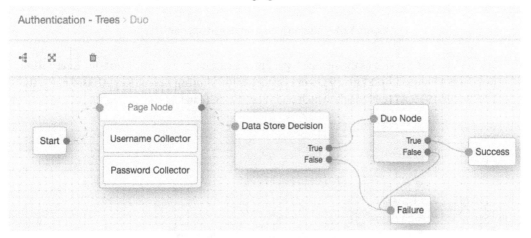

Figure 5.60 – Final configuration for the Duo Authentication Tree

Before we can test the Duo Authentication Tree, we need users that have Duo accounts. Administrators can create new users in ForgeRock openam. If configured, end users can also use self-registration to create new accounts in ForgeRock.

Configuring self-registration in openam

To configure self-registration in openam, do the following:

1. To configure self-registration in openam, select **Services**, then + **Add a Service**:

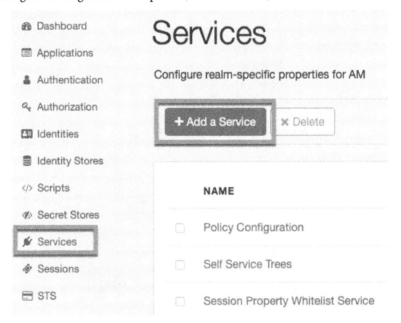

Figure 5.61 – Adding a new service in openam

2. Select **User Self-Service** as the service type. Enter `selfserviceenctest` for **Encryption Key Pair Alias** and `selfservicesigntest` for **Signing Secret Key Alias**. Enable the **USER REGISTRATION** toggle switch and then click **Create**:

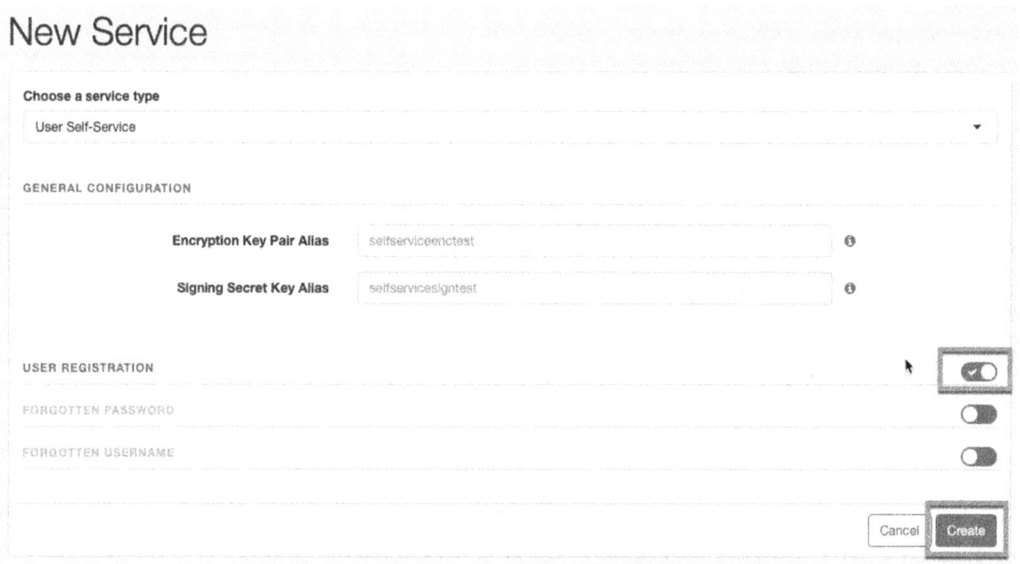

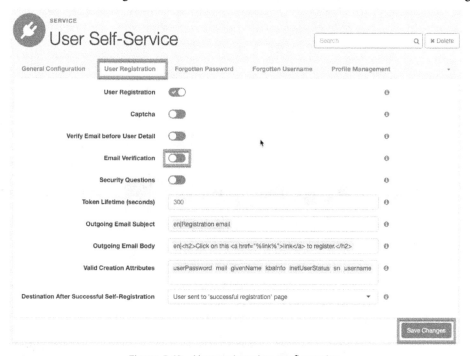

Figure 5.62 – Self-service configuration

3.   Click on the **User Registration** tab, disable **Email Verification**, and click on **Save Changes**:

Figure 5.63 – User registration configuration

After creating the new service and enabling self-registration, log out of openam before testing the Duo Authentication Tree.

Testing self-registration and MFA

As seen before, the format of the URL that's used for testing the new tree is /XUI/#login& service=<Tree name>, where Tree name should be the name of the tree being tested:

1. In our example, open http://openam.acme.local:18080/openam/ XUI/#login&service=Duo in your browser. Again, confirm you are testing the correct Authentication Tree by checking the header and description that were created for the **Duo** tree. Click on **Create an account** to create an account in ForgeRock that matches an existing account in Duo Security:

Figure 5.64 – Duo Authentication Tree

2. Enter the new account information and click **SAVE**:

Figure 5.65 – ForgeRock self-registration page

3. After the account is successfully created, click on the **Return to Login Page** button:

Figure 5.66 – Successful self-registration

4.   Enter the new username and password you created in ForgeRock and click **LOG IN**:

Figure 5.67 – Logging in with a new account

5.   Follow the steps to set up the account with Duo. When you're finished, try logging in again:

Figure 5.68 – ForgeRock Duo setup

6. When logging in with an account that's been set up properly with Duo, you can choose an authenticated method:

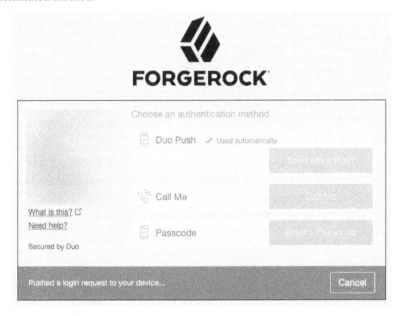

Figure 5.69 – Duo prompt

If the method is approved in Duo, you will be redirected to the profile page for that user:

Figure 5.70 – Profile page after logging in with Duo MFA

We just finished the basic configuration of Duo MFA with ForgeRock and tested two different Authentication Trees, with and without Duo. However, when logging into ForgeRock or protected applications, those trees are not used.

> **Important note**
>
> Unless you tested the custom tree with `amadmin` or with another user that has administrator privileges, do not use that tree in the organization authentication configuration. You may be locked out of openam and may need to reinstall or restore from backup. If you are locked out of openam, try logging in using `amadmin` and a specific custom tree that doesn't use MFA.

To use one of the custom trees to log in to the protected application, click on **Authentication |
Settings | Core** and select one of the custom Authentication Trees for **Organization Authentication
Configuration**. We don't recommend using a tree that includes MFA unless that tree has been tested
with `amadmin` or another user with administrator privileges (see the preceding *Important note*).
This example uses the Acme tree:

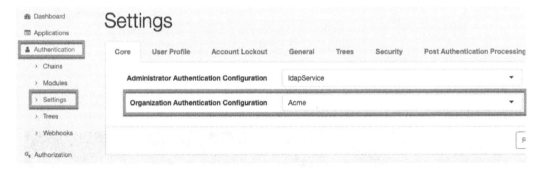

Figure 5.71 – Core Organization Authentication Configuration

Now, every time you log in to openam or a protected application, you will be prompted to log in
with that custom Authentication Tree. Remember that you can still bypass the default tree using
the `/XUI/#login&service=<Tree name>` custom URL. On the other hand, this can also
be used in production, so don't create custom trees that can be used in production to bypass the
minimum security configurations of your environment.

As we have seen so far, MFA can easily be added to improve the security of openam, including external
users. But how can you have your cake and eat it too? How can you have the benefits of MFA without
having to use an app on your phone every time you log in to an application?

The next section discusses behavioral biometrics, which claims we can do just that – that is, eliminate
the friction of multiple authentication factors while still keeping everything secure. In the next section,
we are going to see if that's possible.

What are behavioral biometrics?

Behavioral biometrics uses hardware such as a keyboard and mouse to analyze behavioral inputs and recognize users based on their interactions. The analysis does not affect the user's experience or force the user to do anything differently. In this section, we are going to use BehavioSec behavioral biometric products to improve security using continuous authentication.

BehavioSec, just like Duo, has integration with ForgeRock in the marketplace. We are going to use that integration to verify if the user appears to be the same one before deciding if Duo is required or not on each login.

Installing BehavioSec

BehavioSec's ForgeRock integration can be found at `https://cloud.behaviosec.com/dashboard/sdks`. If an account is needed, one can be obtained at `https://cloud.behaviosec.com/dashboard/signup`:

1. To install BehavioSec in ForgeRock, click on **DOWNLOAD**:

Figure 5.72 – Downloading BehavioSec

2.   Copy the downloaded file to `$CATALINA_HOME/webapps/openam/WEB-INF/lib`.

3.   Restart Tomcat:

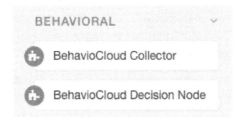

Figure 5.73 – The BEHAVIORAL menu with BehavioCloud

As in *Figure 5.74*, you should now see a new entry on the Authentication Tree's MFA menu for Duo Node.

Configuring authentication with BehavioSec

Our final Authentication Tree example will add BehavioSec to the Authentication Tree we created to test Duo. As before, we suggest that you recreate the Duo authentication and give it a different name – for example, `BehavioSecDuo`:

1.   Follow the steps we used to create the **Duo** Authentication Tree to add a page node, username and password collectors, a data store decision node, and a **Duo Node**. When adding the headers to **Page Node**, make sure you indicate that this includes BehavioSec. Add a **BehavioCloud Collector** and a **BehavioCloud Decision node**. The final configuration should look exactly like a copy of the **Duo** Authentication Tree, with an additional **BehavioCloud Collector** and a **BehavioCloud Decision Node**:

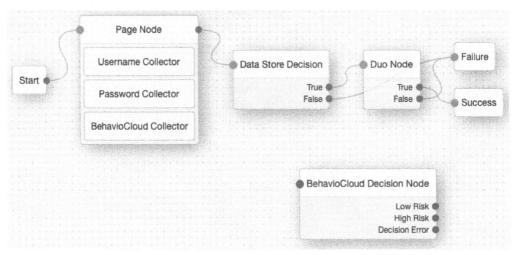

Figure 5.74 – BehavioSecDuo Authentication Tree

2. Configure **BehavioCloud Collector** by copying the **API Key** and **API Secret** properties from the BehavioSec dashboard:

Figure 5.75 – BehavioCloud Collector configuration

3. Connect the **True** outcome of **Data Store Decision** to **BehavioCloud Decision Node** and the **High Risk** and **Decision Error** outcomes from **BehavioCloud Decision Node** to **Duo Node**, as in the following figure:

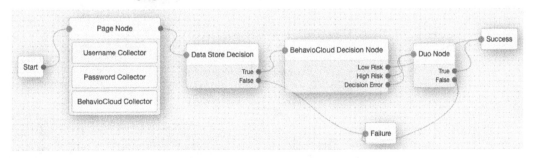

Figure 5.76 – Final configuration of the BehavioSecDuo Authentication Tree

Now, we are ready to test the Authentication Tree with BehavioSec and Duo.

Testing authentication with BehavioSec and Duo

The interesting thing you will see when testing the BehavioSecDuo Authentication Tree is that it looks exactly like the Duo Authentication Tree. After you log in a couple of times, you will still be prompted to authenticate with Duo:

1. In our example, go to `http://openam.acme.local:18080/openam/XUI/ - login&service=BehavioSecDuo` in your browser. Type (*do not copy and paste*) your username and password. Click on **LOG IN**:

Figure 5.77 – BehavioSecDuo login page

Nothing should change the first few times you log in as a new user. The user will still be prompted for Duo MFA. After a few logins, however, the user will be trained, and the risk will be low enough that BehavioSec will bypass Duo during authentication. You can verify the risk of each authentication by going to the BehavioSec dashboard (`https://cloud.behaviosec.com/dashboard/`):

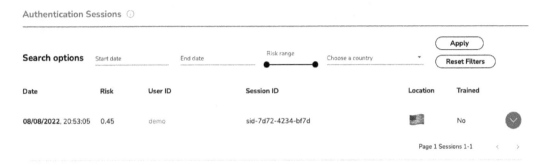

Figure 5.78 – BehavioSec dashboard

This completes our configuration of ForgeRock using MFA with Duo, as well as behavioral biometrics from BehavioSec.

This configuration shows how risk-based authentication from BehavioSec can be used to evaluate the username and password of the user to reduce friction while increasing security during login while maintaining a good customer experience.

Another huge benefit is that behavioral biometrics can detect bots and remote access. With BehavioSec, your users don't have to see a CAPTCHA challenge response ever again.

Summary

This chapter explored ForgeRock IAM products and how Acme used ForgeRock, combined with behavioral biometrics from BehavioSec, to provide an excellent, frictionless customer experience for the workforce and external users without reducing security.

In the next chapter, we are going to explore two new products: PingFederate and 1Kosmos. These products can provide workforce users with passwordless, strong identity-based authentication.

6

Federated SSO with PingFederate and 1Kosmos

In this chapter, we are going to explore user authentication and **single sign-on** (SSO) with PingFederate. We are also going to see how 1Kosmos can provide a more frictionless and secure experience for workforce users.

We are going to cover the following main topics:

- Introducing PingFederate and federated SSO
- What is passwordless authentication?
- Introducing 1Kosmos and passwordless **multifactor authentication** (MFA) with verified identity

Technical requirements

This chapter requires a Ping Identity account. If required, a trial is available at https://www.pingidentity.com/en/try-ping.html.

For deploying PingFederate, Java **software development kit** (SDK) version 11 or above is required. Alternatively, the Ping Identity software can be deployed using container technologies. Although not covered in this chapter, information about running Ping Identity products in containerized environments is available at https://devops.pingidentity.com.

An account with 1Kosmos is required for passwordless authentication using BlockID. A 1Kosmos free trial is available, if needed, at https://www.1kosmos.com/free-trial/.

Experiencing Ping Identity's PingFederate

PingFederate from Ping Identity provides SSO, **federated identity management (FIM)**, and **customer identity and access management (CIAM)**. PingFederate can extend employee, customer, and partner identities across domains without passwords. SAML, WS-Federation, and OAuth are some of the identity standard protocols used in PingFederate.

More information about the standards used in PingFederate can be found at the following links:

- **SAML**: `http://docs.oasis-open.org/security/saml/Post2.0/sstc-saml-tech-overview-2.0.html`

- **WS-Federation**: `http://docs.oasis-open.org/wsfed/federation/v1.2/os/ws-federation-1.2-spec-os.html`

- **OAuth**: `https://oauth.net/2/`

- **OpenID Connect**: `https://openid.net/connect/`

Before we can install PingFederate, we need to install the Java SDK. Instructions to install Java SDK version 11 can be found in *Appendix A*.

Installing PingFederate

We will start off by downloading and installing PingFederate, as follows:

1. Log in with your Ping Identity account at `https://www.pingidentity.com/en/account/sign-on.html`. Select **Support and Community** from the **Select Account** drop-down list. Enter your email address and password. Click **Sign In**:

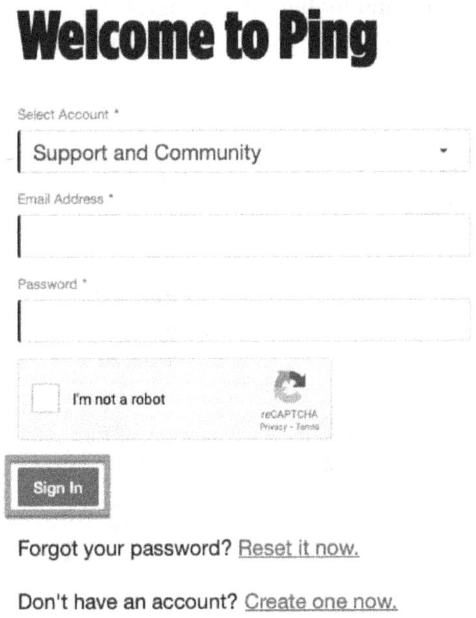

Welcome to Ping

Select Account *

| Support and Community | ▼ |

Email Address *

Password *

I'm not a robot — reCAPTCHA
Privacy - Terms

Sign In

Forgot your password? Reset it now.

Don't have an account? Create one now.

Figure 6.1 – Ping Identity Support and Community login

Support and Community is the account type to use for Ping Identity support, education, and documentation and provides access to the **Download** page for the products we are going to use from Ping Identity.

2. Click the **Resources** drop-down button and select **Downloads**:

| Developer | Support   Partner Portal

nmunity   Training   Resources

           Product Status

           Downloads

           Product Releases

any certifications.   **Integration Directory**   st
                                    rts, customers an
                        community.

Figure 6.2 – Downloads sub-menu

3. Ping Identity has a few cloud and mobile apps, **PingID** and **PingOne MFA**. We will not use those in this book. Instead, scroll down, and in the **PingOne Advanced Services** section, select **View Now** under **PingFederate**:

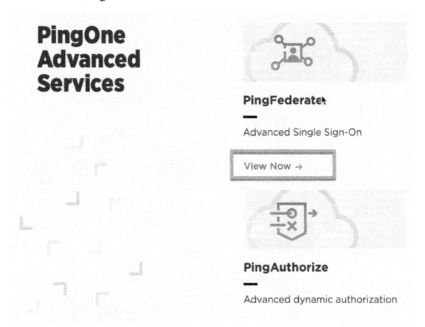

Figure 6.3 – PingOne Advanced Services download page

The product we want to download from Ping Identity is PingFederate. The latest version at the time of writing is 11.1.1. The download page for PingFederate also contains links to the installation guide and DevOps resources (for cloud deployment) and for requesting a license key.

4. Click on **Request a license key.**:

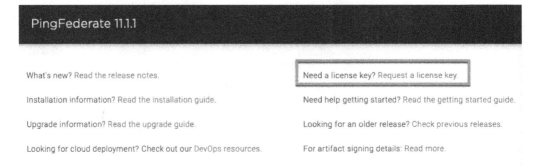

Figure 6.4 – PingFederate license key request

5.   Select **PingFederate**. Accept the license agreement and click **Submit**:

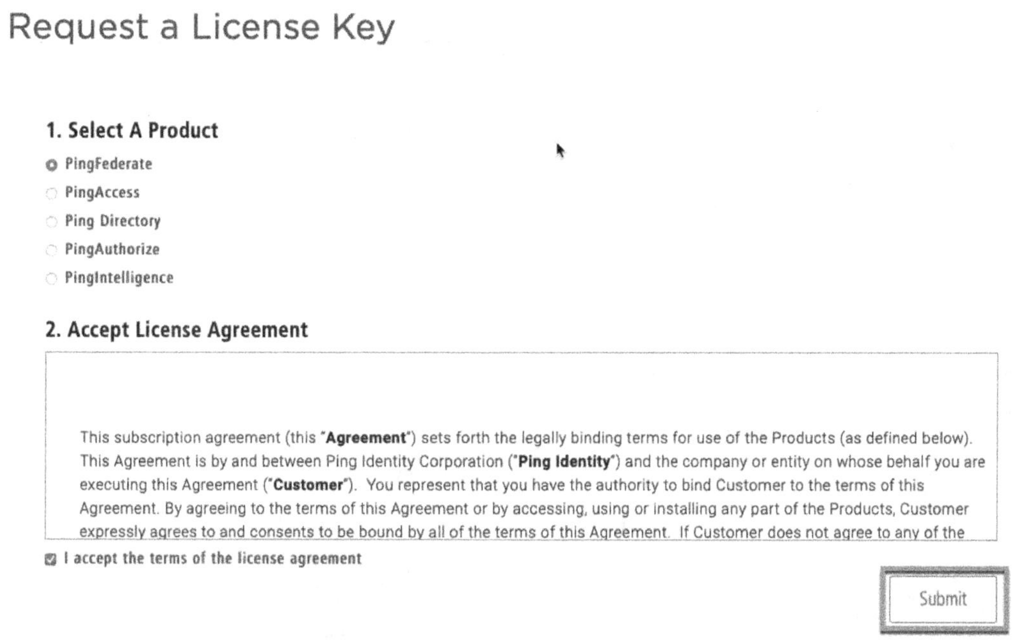

Request a License Key

1. Select A Product

- ○ PingFederate
- ○ PingAccess
- ○ Ping Directory
- ○ PingAuthorize
- ○ PingIntelligence

2. Accept License Agreement

This subscription agreement (this "**Agreement**") sets forth the legally binding terms for use of the Products (as defined below). This Agreement is by and between Ping Identity Corporation ("**Ping Identity**") and the company or entity on whose behalf you are executing this Agreement ("**Customer**"). You represent that you have the authority to bind Customer to the terms of this Agreement. By agreeing to the terms of this Agreement or by accessing, using or installing any part of the Products, Customer expressly agrees to and consents to be bound by all of the terms of this Agreement. If Customer does not agree to any of the

☑ I accept the terms of the license agreement

    Submit

Figure 6.5 – License key request

After requesting a license key, go back to the PingFederate **downloads** page and scroll to the **Downloads** section.

6.   Click the **PRODUCT DISTRIBUTION (ZIP)** button:

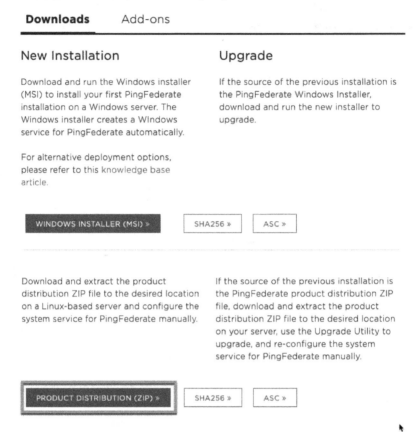

Figure 6.6 – Selecting the download type

We are going to use the ZIP version, which can be used on Linux servers, Windows, and Mac servers as well. On Windows, a Windows installer is available that creates a Windows server for PingFederate as part of the installation product.

7.   Save the downloaded ZIP file:

Figure 6.7 – ZIP file

8. Unzip `pingfederate-11.1.1.zip` using the `unzip` command (Mac or Linux) or by right-clicking the ZIP file and choosing **Extract All** (Windows). The complete PingFederate software is included in the ZIP file:

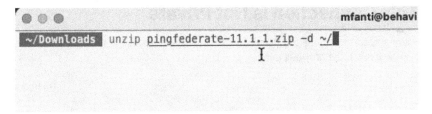

Figure 6.8 – Unzipping PingFederate

9. After unzipping the file, a new subdirectory, `pingfederate-11.1.1`, is created. Change to the `pingfederate-11.1.1/pingfederate` directory and run the `run.sh` file (or `run.bat` for Windows) in the `bin` directory:

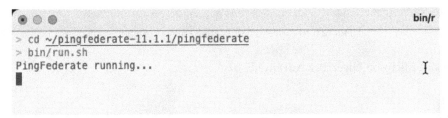

Figure 6.9 – Running PingFederate

When a `PingFederate running...` message is displayed, the installation is complete, and we can configure PingFederate for standalone use.

Configuring Ping Identity's PingFederate

After deploying PingFederate for the first time, it needs to be configured before it can be used. Here's how to do that:

1. Open a browser with the URL for your PingFederate instance with port `9999`. Remember to use `https`, not `http`. In this example, the URL is `https://localhost:9999`.

    Depending on the browser used, a different message may appear when using an HTTPS connection where the certificate used doesn't match the URL used (`localhost`). Choose the appropriate response to ignore the warning and continue browsing.

2. If using Safari, click **visit this website** to continue:

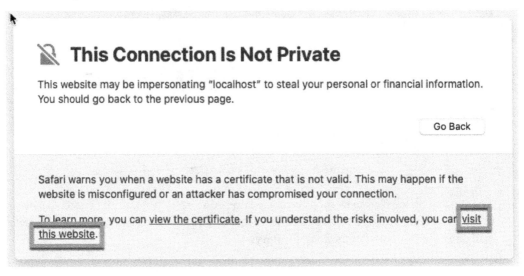

Figure 6.10 – Invalid certificate warning in Safari

3. If using Microsoft Edge, click **Advanced**:

Figure 6.11 – Private connection warning in Microsoft Edge

4. Then, click **Continue to localhost (unsafe)**:

Figure 6.12 – Ignoring warning in Microsoft Edge

The PingFederate administrative console (or the administrative API) only accepts **Secure Sockets Layer (SSL)** connections. When deploying in production, a valid certificate should be configured to avoid the warnings shown when using the default certificate. More information on managing PingFederate certificates can be found at `https://docs.pingidentity.com/bundle/pingfederate-111/page/yij1564002984865.html`.

5.   Click **Next** after agreeing to the terms and conditions:

PingFederate Setup   ×

Terms and Conditions

Subscription Agreement

This subscription agreement (this "Agreement") sets forth the legally binding terms for use of the Products and with respect to Order Forms (each as defined below). This Agreement is by and between Ping Identity Corporation ("Ping Identity") and the company or entity on whose behalf you are accepting this Agreement ("Customer"). You represent that you have the authority to bind Customer to the terms of this Agreement. By agreeing to the terms of this Agreement, by accessing, using or installing any part of the Products, or by executing an Order Form that references this Agreement, Customer expressly agrees to and consents to be bound by all of the terms of this Agreement. If Customer does not agree to any of the terms of this Agreement, Customer is prohibited from downloading, installing, activating or using the Products or executing an Order Form that references this Agreement. The effective date of this Agreement is the date set forth on an Order Form (if applicable) or otherwise the date on which Customer downloads, installs, activates or uses the Products (the "Effective Date"). Collectively, Ping Identity and Customer may be referred to as the "Parties" or in the singular as a "Party."

For good and valuable consideration, the receipt and sufficiency of which is hereby acknowledged, the Parties agree as follows: 1. Definitions.

"Administrator" is an individual who has been granted administrative permissions by Customer to the Service.

 I agree to the terms and conditions

Figure 6.13 – PingFederate terms and conditions

6. Click **Next** to accept the base URL:

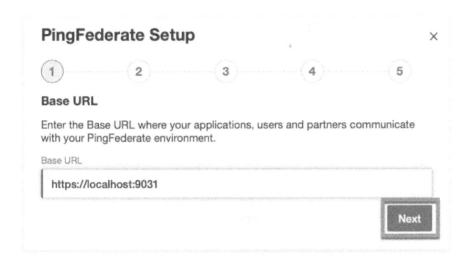

Figure 6.14 – PingFederate base URL

7. We are *not* going to connect to PingOne. Click **Next**:

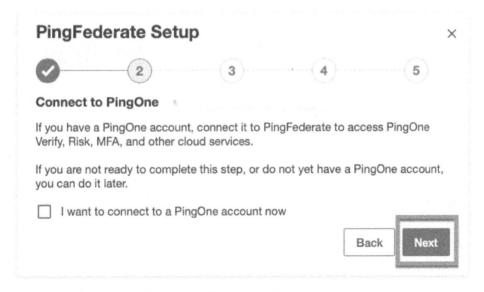

Figure 6.15 – PingFederate PingOne setup

8. Click **Choose file** to select a valid license for PingFederate:

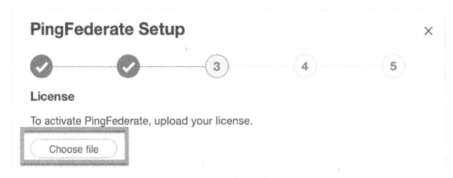

Figure 6.16 – PingFederate license setup

9. After uploading a valid license file, click **Next**:

Figure 6.17 – PingFederate license

10. Create a username and password for administering PingFederate. Click **Next**:

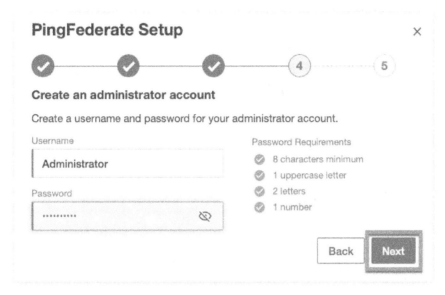

Figure 6.18 – PingFederate administrator account

11. After the configuration is complete, click **Finish**:

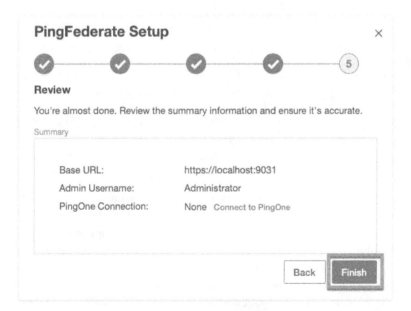

Figure 6.19 – PingFederate setup review

At this point, OpenAM is installed and configured for use.

12. Click on **Start** to take a tour:

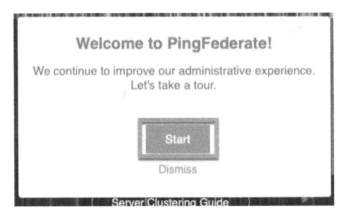

Figure 6.20 – Configuration complete page

To navigate to the main page on the administration dashboard, click on the **PingFederate** logo at the top left:

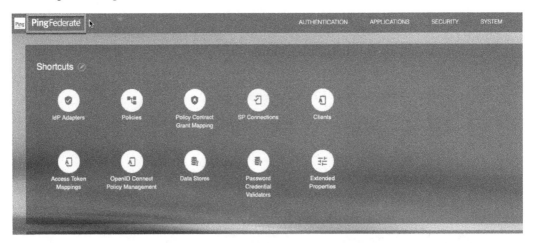

Figure 6.21 – Main PingFederate dashboard

The basic PingFederate installation does not include any sample applications. To understand the functionality of PingFederate without going into the details, we are going to use one of the samples provided by Ping Identity. The configuration of the samples can be done manually or automatically, as we will see in the sample. Details about the manual administration of PingFederate are beyond the scope of this book. More information about training for PingFederate can be found at https://www.pingidentity.com/en/support/training/course-details/388.html.

Deploying sample applications in PingFederate

We are going to deploy sample applications using the PingFederate Agentless Integration Kit. The Agentless Integration Kit allows PingFederate to perform SSO for **identity provider** (**IdP**) and **service provider** (**SP**) applications without the need for an agent. An IdP is a function of PingFederate that manages identity information for principals and also provides authentication services for SSO. PingFederate also allows for SP applications to be configured and to use the SSO functionality provided by the IdP. Ping Identity provides many samples for different products at `https://github.com/pingidentity/`. Proceed as follows:

1. To build the sample applications, download the ZIP file or clone the repository at `https://github.com/pingidentity/pf-agentless-ik-sample-java`:

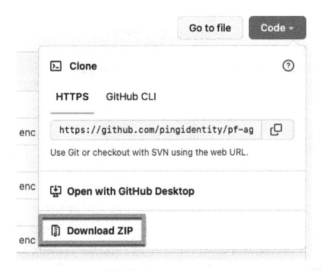

Figure 6.22 – Downloading sample applications' code

2. Extract the contents of the downloaded file:

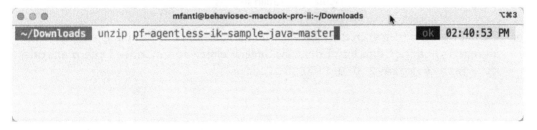

Figure 6.23 – Unzipping the sample project

3. Change to the root directory (`pf-agentless-ik-sample-java-master`) of the project:

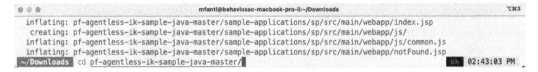

Figure 6.24 – Changing to the root directory

4. Use Maven to build the project (`mvn clean install`):

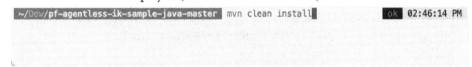

Figure 6.25 – Building the sample project

5. When the build finishes, change to the `assembly/target` directory (`cd assembly/target`):

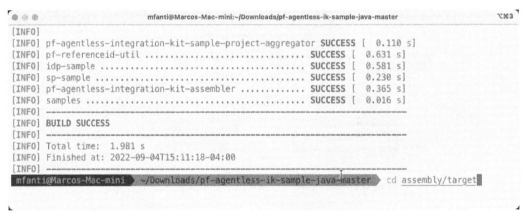

Figure 6.26 – Build success

6. The build process (*step 4*) creates a ZIP file with the contents of the sample application in the `assembly/target` directory. Extract the contents of `pf-agentless-integration-kit-java-sample-2.0.0-SNAPSHOT.zip`:

Figure 6.27 – Unzipping the file

7. Stop PingFederate:

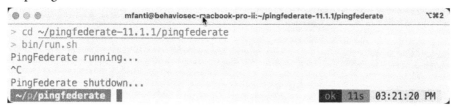

Figure 6.28 – Stopping PingFederate

8. Copy the two WAR files in the `dist` directory to `<pf_install>/pingfederate/server/default/deploy`:

Figure 6.29 – Duo node parameters

9. Restart PingFederate:

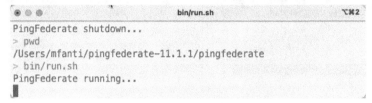

Figure 6.30 – Restarting PingFederate

After deploying the two sample applications, we proceed to import the configuration data that was created in the preceding steps:

1. In the dashboard, go to **SYSTEM | Server | Configuration Archive** and select **IMPORT**:

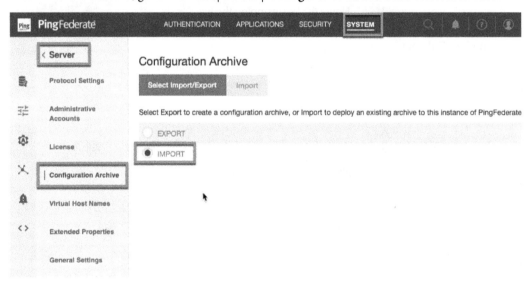

Figure 6.31 – Server/Configuration Archive page

2. Click **Next:**

Configuration Archive

Select Import/Export	Import

Select Export to create a configuration archive, or Import to deploy an existing archive to this instance of PingFederate.

○ EXPORT

● IMPORT

Next

Figure 6.32 – Configuration Archive/Import panel

3. Click **Choose File:**

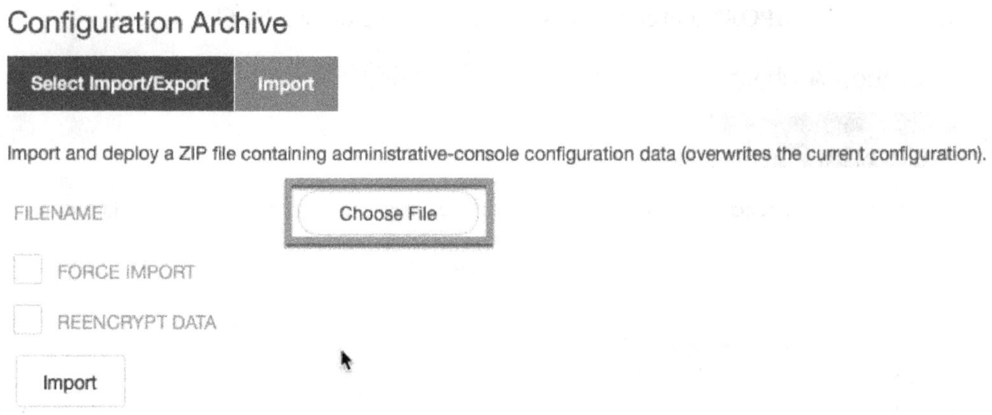

Configuration Archive

Select Import/Export	Import

Import and deploy a ZIP file containing administrative-console configuration data (overwrites the current configuration).

FILENAME   Choose File

☐ FORCE IMPORT

☐ REENCRYPT DATA

Import

Figure 6.33 – Configuration Archive/Choose File panel

4. Select data.zip from the unzipped project file:

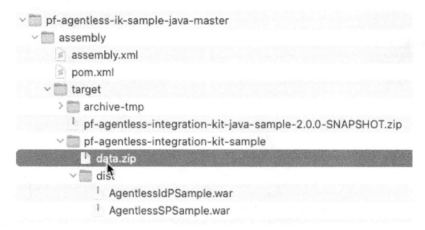

Figure 6.34 – data.zip file

5. Select **FORCE IMPORT** and click **Import** after uploading the archive file:

Figure 6.35 – Configuration Archive/Import panel

6. Click **Done** after the **The configuration archive was successfully imported.** message is displayed:

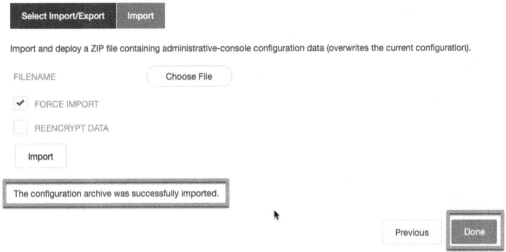

Figure 6.36 – Forcing import of new configuration archive

We just finished importing a new configuration for PingFederate and deployed two sample applications. To confirm the configurations were deployed correctly, we can go back to the dashboard.

7. Click **AUTHENTICATION** and then **IdP Connections**:

Figure 6.37 – AUTHENTICATION page

8. Confirm **Sample IdP** is configured:

Figure 6.38 – IdP Connections page

9. Click **APPLICATIONS** and then **SP Connections**:

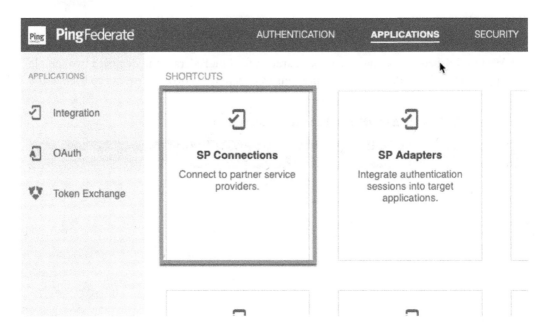

Figure 6.39 – APPLICATIONS page

10. Confirm that **Sample SP** is configured:

Figure 6.40 – SP Connections page

The final step is confirming the applications were deployed successfully.

11. Open a new browser window at `https://localhost:9031/AgentlessIdPSample/app`:

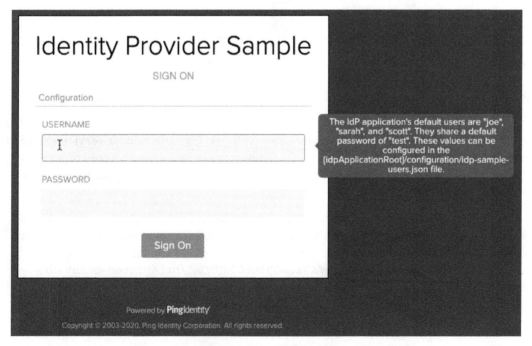

Figure 6.41 – IdP sample app

You should be able to log in using the pre-defined users `joe`, `sarah`, and `scott`. For all three, the password is `test`.

After logging in, you can validate the functionality of the application and log out:

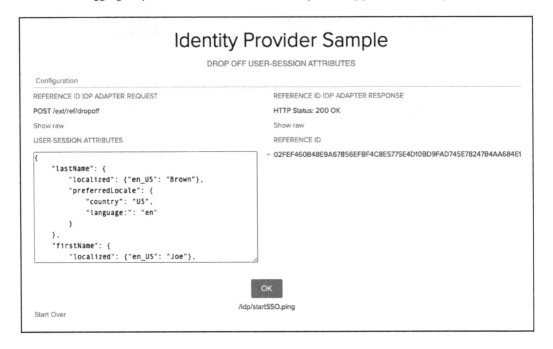

Figure 6.42 – IdP sample

You can also try logging in using the SP sample application (`https://localhost:9031/AgentlessSPSample/app`).

In this section, we deployed two sample applications. One provided a service called `AgentlessSPSample`. Another provided an authentication mechanism used to access the services of `AgentlessSPSample`. We also confirmed that we could log in to the application using the credentials provided.

As with ForgeRock in the previous chapter, PingFederate also integrates with Duo and BehavioSec's behavioral biometric product. The integration files, as well as detailed instructions on integration with BehavioSec and PingID, can be found in Ping's marketplace at `https://support.pingidentity.com/s/marketplace-integration/a7i1W00000006UhQAI/behaviosec-behavioral-biometrics`:

BehavioSec Behavioral Biometrics

Ping Identity and BehavioSec integrate to provide a risk based authentication solution based on behavioral biometrics. PingFederate will serve the javacode on a login page to capture the user's behavioral biometric signals. PingFederate then sends this information to BehavioSense to compare to the user's baseline. If the patterns deviate from the baseline, the PingFederate policy can either deny authentication or step up authentication to PingID.

This gives a frictionless second factor to determine if strong authentication is required.

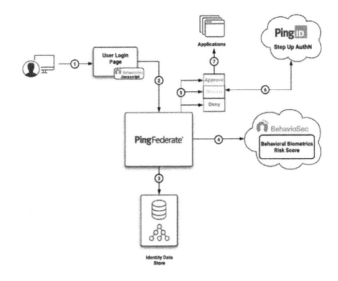

1. User is presented with a Login Form with the embedded BehavioSec Javascript which executes on the page load. The execution of the BehavioSec Javascript creates a browser fingerprint. Also, the user's behavioral analytics (typing cadence, mouse movement) is collected.
2. PingFederate performs the authentication by validating the user's identifier and password in the identity data store.
3. PingFederate calls the BehavioSec adapter and passes it the browser fingerprint and the behavioral analytics.
4. The Iovation adapter calls the BehavioSec REST API and uploads the browser fingerprint blackbox and behavioral analytics to the BehavioSec platform. The user's browser fingerprint and behavioral analytics will be evaluated by the Behavioral engine and will the return the user's risk score. The BehavioSec-PingFederate adapter translates the risk score into the required Approved, Review, and Deny result.
5. The PingFederate Policy Tree evaluates the result from the BehavioSec adapter to determine if Approved access or Step-Up authentication is required or Denied access is granted.
6. If the Policy Tree determines Step-Up authentication is required, the user is redirected to PingID. After, the successful Step-Up authentication, PingID redirects back to PingFederate.
7. PingFederate creates an authentication session and redirects the user to the requested target application.

Figure 6.43 – Ping Identity marketplace page

PingID is a cloud MFA product from Ping Identity, like Duo.

In this chapter, we are going to explore a different type of MFA product: BlockID from 1Kosmos.

What is passwordless MFA?

BlockID from 1Kosmos provides a frictionless authentication experience, with strong identity proofing.

Employees can self-enroll and use the mobile application for advanced biometric authentication. For more details about all the features of BlockID for the workforce, go to `https://www.1kosmos.com/blockid-workforce-user-journey/`.

Integrating BlockID and PingFederate

A BlockID tenant is required for the integration. A free BlockID tenant can be obtained at `https://www.1kosmos.com/free-trial/`. With a free trial, 1Kosmos will create a tenant URL, a tenant tag, and a license key, which are going to be used in the integration. The PingConnector package can also be obtained by contacting 1Kosmos. Proceed as follows:

1. To install the PingFederate BlockID connector, shut down the PingFederate server:

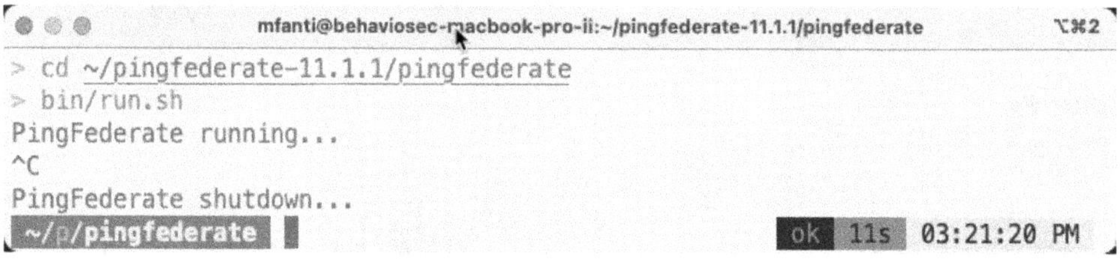

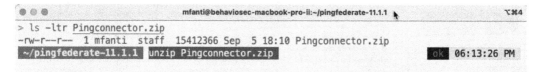

Figure 6.44 – Stopping PingFederate

2. Extract the `Pingconnector.zip` file contents:

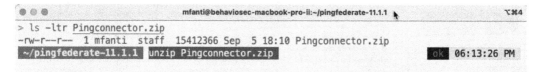

Figure 6.45 – Downloading BehavioSec

3. Copy the `BlockIDAdapter.jar` file to the `$PF_HOME/server/default/deploy/` directory:

```
● ● ●                    mfanti@behaviosec-macbook-pro-ii:~/pingfederate-11.1.1                    ⌥⌘4
> ls -ltr Pingconnector
total 31872
-rw-rw-r--  1 mfanti  staff     4646 Apr  8  2021 BlockIDDecrypt.jar
-rw-rw-r--  1 mfanti  staff     4368 Apr  8  2021 BlockIDAdapter.jar
-rw-rw-r--  1 mfanti  staff   880303 Apr  8  2021 bcpkix-jdk15on-166.jar
-rw-rw-r--  1 mfanti  staff  9533091 Apr  8  2021 BlockIDSDK.war
-rw-rw-r--  1 mfanti  staff  5884134 Apr  8  2021 bcprov-jdk15on-166.jar
~/pingfederate-11.1.1  cp Pingconnector/BlockIDAdapter.jar pingfederate/server/default/deploy/█
```

Figure 6.46 – Copying BlockIDAdapter.jar

4. Copy the `BlockIDSDK.war` file to the `$PF_HOME/server/default/deploy/` directory:

```
● ● ●                    mfanti@behaviosec-macbook-pro-ii:~/pingfederate-11.1.1                    ⌥⌘4
> ls -ltr Pingconnector
total 31872
-rw-rw-r--  1 mfanti  staff     4646 Apr  8  2021 BlockIDDecrypt.jar
-rw-rw-r--  1 mfanti  staff     4368 Apr  8  2021 BlockIDAdapter.jar
-rw-rw-r--  1 mfanti  staff   880303 Apr  8  2021 bcpkix-jdk15on-166.jar
-rw-rw-r--  1 mfanti  staff  9533091 Apr  8  2021 BlockIDSDK.war
-rw-rw-r--  1 mfanti  staff  5884134 Apr  8  2021 bcprov-jdk15on-166.jar
~/pingfederate-11.1.1  cp Pingconnector/BlockIDAdapter.jar pingfederate/server/default/deploy/█
```

Figure 6.47 – Copying BlockIDSDK.war

5. Copy the `BlockIDDecrypt.jar`, `bcpkix-jdk15on-166.jar`, and `bcprov-jdk15on-166.jar` files to the `$PF_HOME/server/default/lib/` directory:

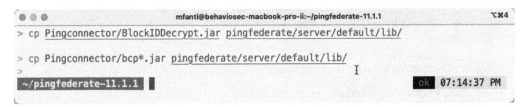

```
● ● ●                    mfanti@behaviosec-macbook-pro-ii:~/pingfederate-11.1.1                    ⌥⌘4
> cp Pingconnector/BlockIDDecrypt.jar pingfederate/server/default/lib/

> cp Pingconnector/bcp*.jar pingfederate/server/default/lib/
>                                                           I
~/pingfederate-11.1.1  █                              ok  07:14:37 PM
```

Figure 6.48 – Copying BlockIDDecrypt.jar, bcpkix-jdk15on-166.jar, and bcprov-jdk15on-166.jar

6. Restart PingFederate:

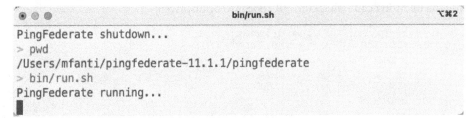

```
● ● ●                           bin/run.sh                           ⌥⌘2
PingFederate shutdown...
> pwd
/Users/mfanti/pingfederate-11.1.1/pingfederate
> bin/run.sh
PingFederate running...
█
```

Figure 6.49 – Restarting PingFederate

After copying all the required files, the second part of the integration requires changing the default login page for PingFederate (or modifying a copy of it) and adding the QR code that is going to be used by BlockID for authentication.

Remember—we are eliminating the need for usernames and passwords. Follow these steps:

1. Using a text editor, open the `html.form.login.template.html` file in the `$PF_HOME/server/deploy/conf/template/` directory.

2. Import the following JavaScript code:

```
<script type="text/javascript"
src="https:// <pingfedhostname>/BlockIDSDK/js/blockid.js"></
script>
```

3. Add the following JavaScript code to the head of the page:

```
<script type="text/javascript">
function createSession() {
            createNewSession("Fingerprint", "did,userid",
"qrcode",null
                , function(result, error) {client_
dataReceived(result)}
            )
        }
        function client_dataReceived(result) {
let str = JSON.stringify(result, null, 4);
        var obj = JSON.parse(str);
    }
createSession();
</script>
```

4. Replace the existing form on the page with this:

```
< <form id="loginbid" name="loginbid" method="POST"
action="$URL">
 <input type="hidden" name="blockidAuthn" id="blockidAuthn"
value="true" />
 <input type="hidden" name="payload" id="payload"  /> </form>
```

5. Add the following code where the QR code is to be displayed:

```
<div id="qrcode" style="z-index:
1;position:relative;margin:10px"></div>
```

That completes the changes to the login page. The remainder of the configuration is performed using the PingFederate dashboard. Proceed as follows:

1. Log in to the PingFederate admin console.

2. Select **AUTHENTICATION | IdP Adapters**:

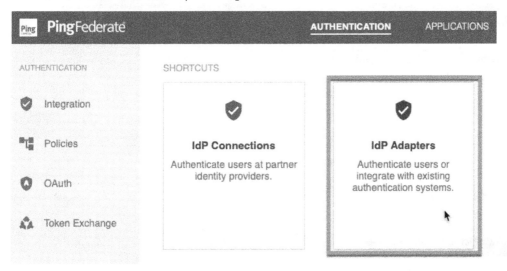

Figure 6.50 – AUTHENTICATION/IdP Adapters

3. Give the new adapter a name and an instance ID. The type should be **BlockID Adapter v 1.0.1**. Click **Next**:

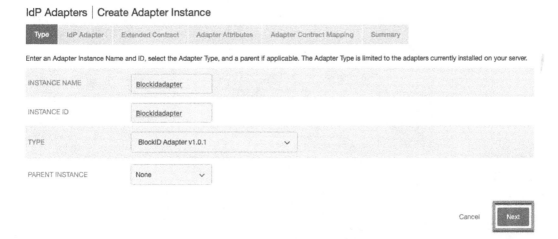

Figure 6.51 – Type of adapter

4. Click **Next**:

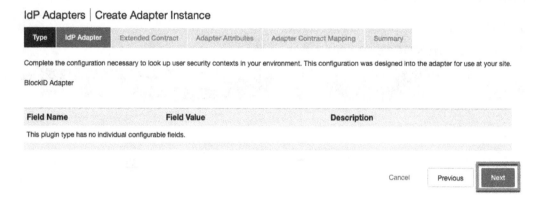

Figure 6.52 – Configuration fields

5. Click **Next**:

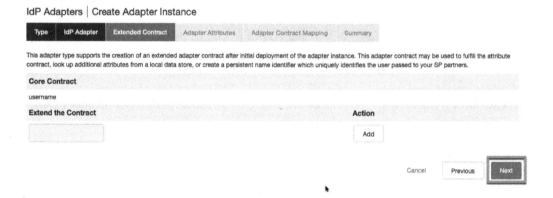

Figure 6.53 – Extending the contract

6. Check the **Pseudonym** field for username. Click **Next**:

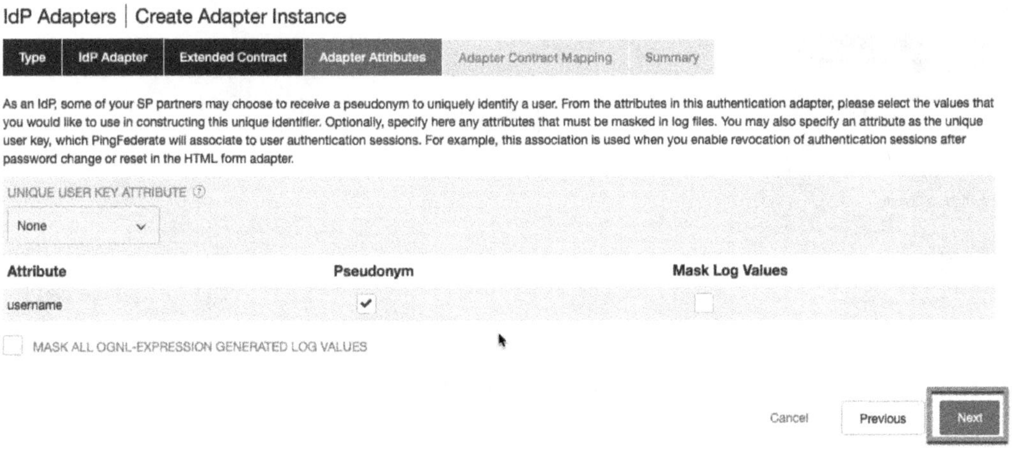

Figure 6.54 – Adapter attributes

7. Click **Configure Adapter Contract**:

IdP Adapters │ Create Adapter Instance

| Type | IdP Adapter | Extended Contract | Adapter Attributes | Adapter Contract Mapping | Summary |

An Adapter Contract may be used to fulfill the Attribute Contract passed to your SP partners. By default, the adapter contract is fulfilled by the adapter itself. Optionally, additional attributes from local data stores can be used to fulfill the contract.

Configure Adapter Contract

Cancel   Previous   Next

Figure 6.55 – Adapter contract mapping

8. Click **Next**:

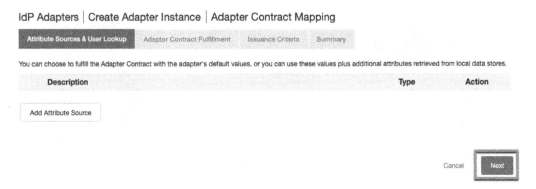

Figure 6.56 – Attribute source

9. Click **Next**:

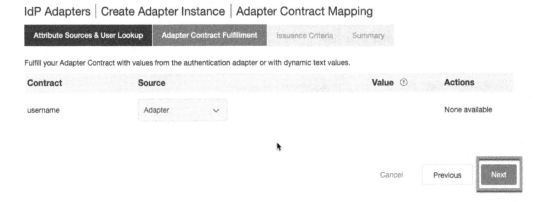

Figure 6.57 – Configuration fields

10. Click **Next**:

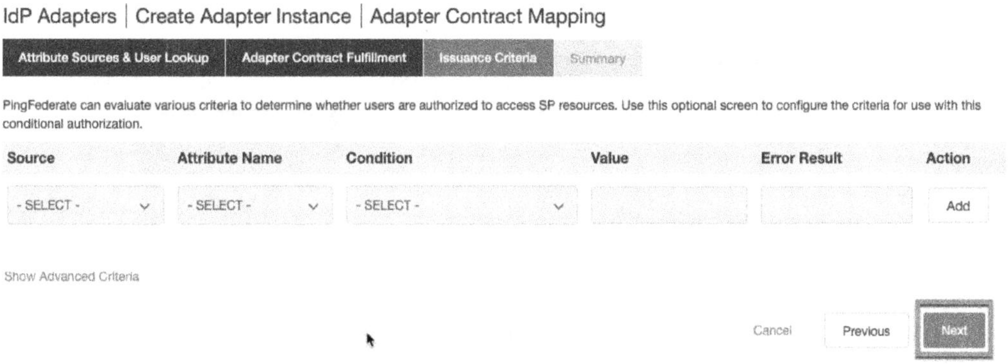

IdP Adapters | Create Adapter Instance | Adapter Contract Mapping

| Attribute Sources & User Lookup | Adapter Contract Fulfillment | Issuance Criteria | Summary |

PingFederate can evaluate various criteria to determine whether users are authorized to access SP resources. Use this optional screen to configure the criteria for use with this conditional authorization.

Source	Attribute Name	Condition	Value	Error Result	Action
- SELECT -	- SELECT -	- SELECT -			Add

Show Advanced Criteria

Cancel   Previous   Next

Figure 6.58 – Issuance criteria

11. This completes the configuration of the adapter contract mapping. Click **Done**:

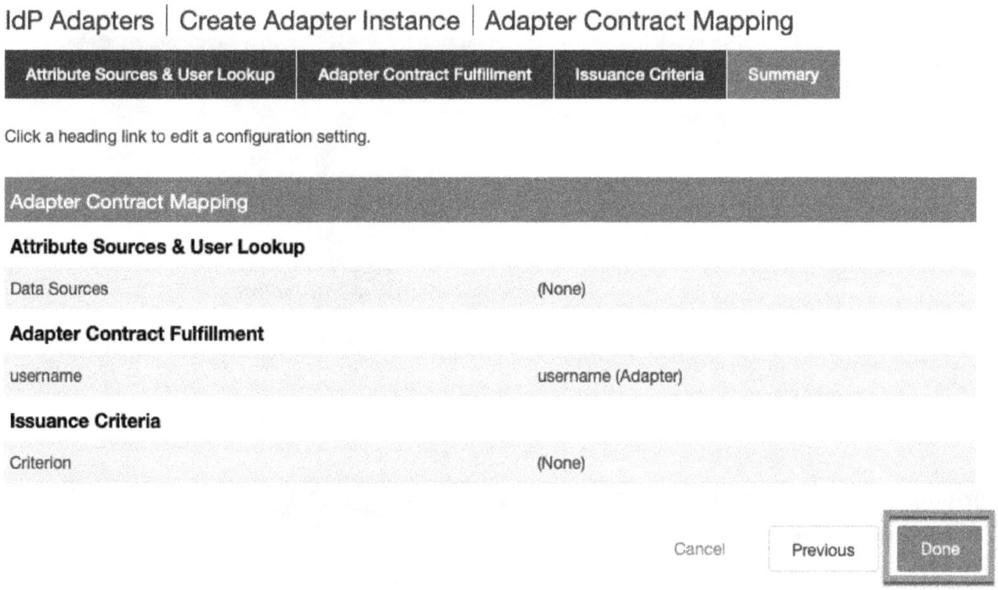

IdP Adapters | Create Adapter Instance | Adapter Contract Mapping

| Attribute Sources & User Lookup | Adapter Contract Fulfillment | Issuance Criteria | Summary |

Click a heading link to edit a configuration setting.

Adapter Contract Mapping

Attribute Sources & User Lookup

| Data Sources | (None) |

Adapter Contract Fulfillment

| username | username (Adapter) |

Issuance Criteria

| Criterion | (None) |

Cancel   Previous   Done

Figure 6.59 – Adapter contract mapping summary

12. Click **Next**:

Figure 6.60 – Adapter contract mapping

13. Check all the values on the IdP adapter summary page. Click **Save**:

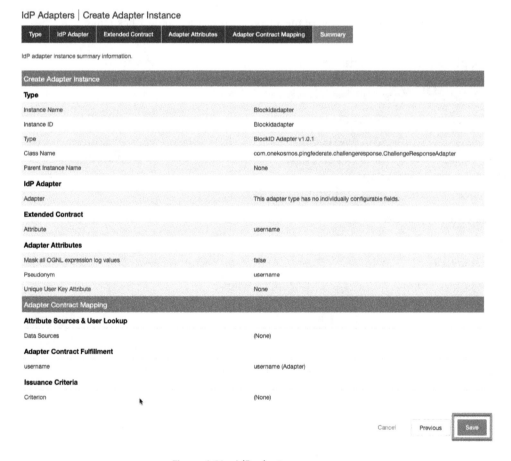

Figure 6.61 – IdP adapter summary

We have finished creating an IdP connector. The connector will be used to authenticate users using the BlockID adapter. The next step is creating an authentication selector that will branch the authentication policy during login. Here's how to do that:

1.   Select **AUTHENTICATION | Selectors**:

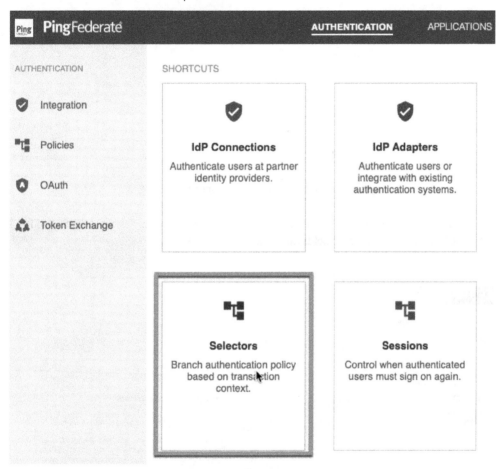

Figure 6.62 – AUTHENTICATION/Selectors

2.   Give the selector a name and instance ID. The type should be **HTTP Request Parameter Authentication Selector**. Click **Next**:

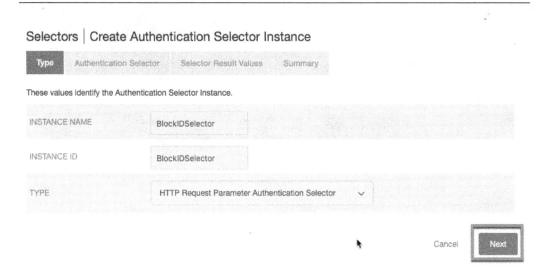

Figure 6.63 – Authentication selector instance type

3. For **HTTP REQUEST PARAMETER NAME**, use `blockidAuthn`. Select **CASE-SENSITIVE MATCHING**. Leave all other fields unchecked. Click **Next**:

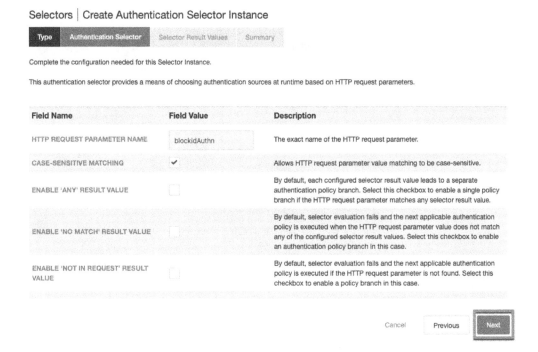

Figure 6.64 – Authentication selector instance configuration

4. Add a new attribute, `true`. Click **Save**:

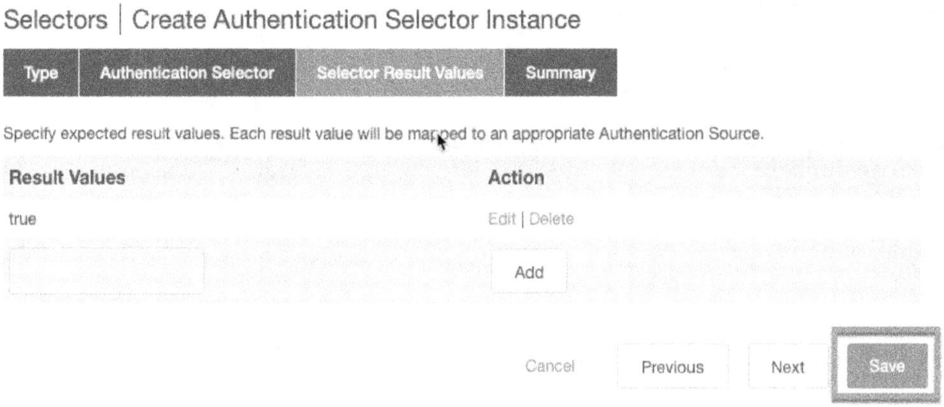

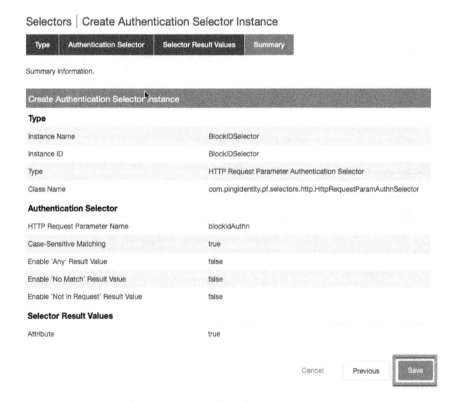

Figure 6.65 – Authentication selector instance result

5. Validate the authentication selector instance configuration and click **Save**:

Figure 6.66 – Authentication selector instance summary

We are going to create a new policy in PingFederate to define which path the authentication will take for our users. A policy contract is required before we can create a new policy, so this is what we'll do:

1. With the authentication selector instance created, let's create a new policy contract. Select **AUTHENTICATION | Policies**:

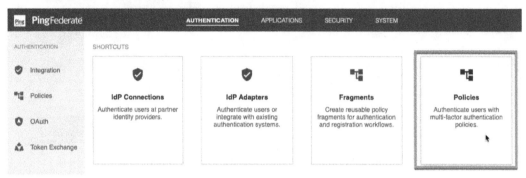

Figure 6.67 – AUTHENTICATION/Policies menu

2. Select **Policy Contracts** and **Create New Contract**:

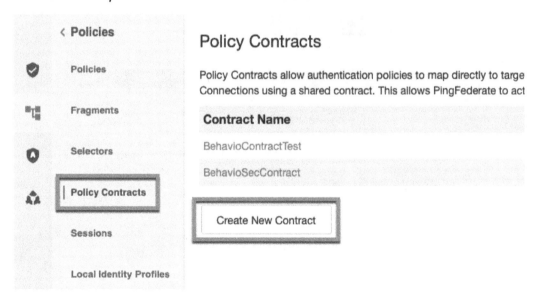

Figure 6.68 – Creating a new policy contract

3. Add a name for the new contract and click **Next**:

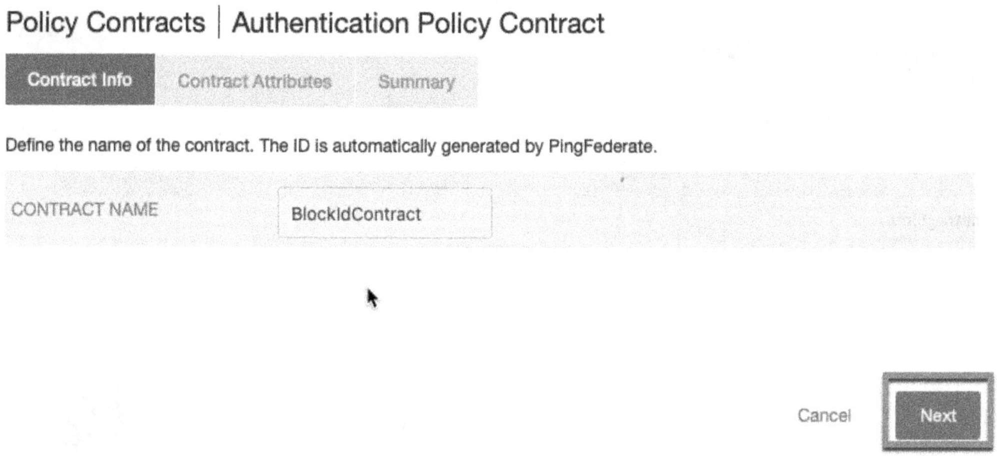

Policy Contracts | Authentication Policy Contract

Contract Info	Contract Attributes	Summary

Define the name of the contract. The ID is automatically generated by PingFederate.

CONTRACT NAME	BlockIdContract

Cancel   Next

Figure 6.69 – Policy contract info

4. There's no need to add new attributes. Click **Next**:

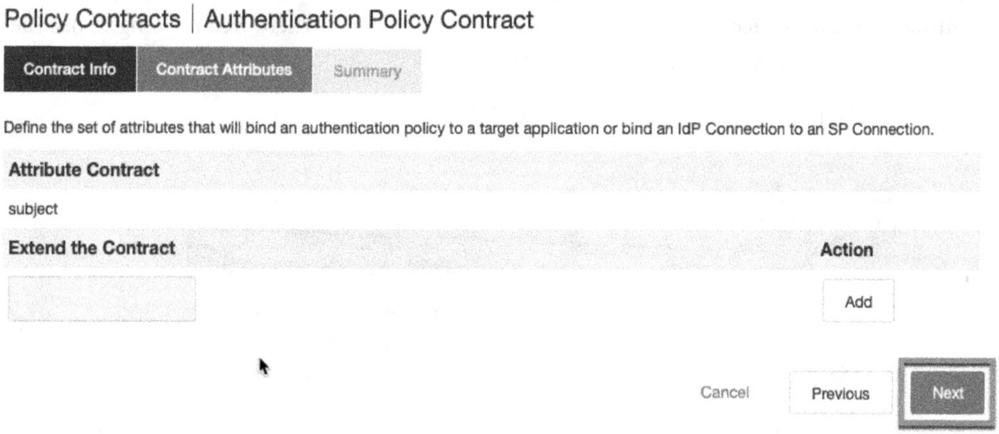

Policy Contracts | Authentication Policy Contract

Contract Info	**Contract Attributes**	Summary

Define the set of attributes that will bind an authentication policy to a target application or bind an IdP Connection to an SP Connection.

Attribute Contract

subject

Extend the Contract	**Action**
	Add

Cancel   Previous   Next

Figure 6.70 – Policy contract attributes

5.   Click **Save** to finish the contract creation:

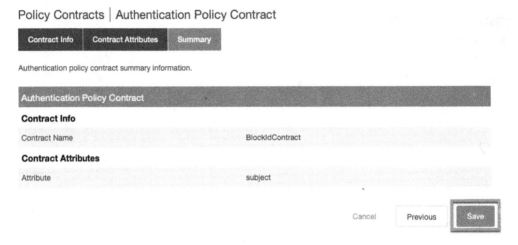

Figure 6.71 – Policy contract summary

The final elements before we test authenticating to PingFederate using BlockID are an authentication policy and an IdP connection. As the name implies, the authentication policy defines the path taken by the authentication. A successful path ends with a policy contract where we are going to use the policy contract we just created. Proceed as follows:

1.   Select **AUTHENTICATION | Policies** and click **Add Policy**:

Figure 6.72 – New authentication policy

2. Add a name and an ID to the new policy:

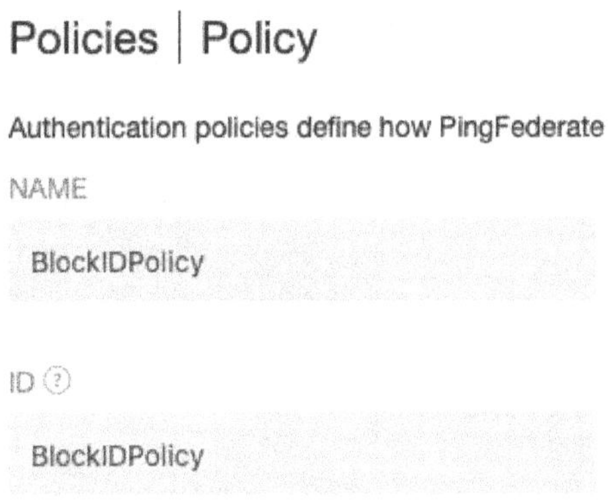

Policies | Policy

Authentication policies define how PingFederate authenticates users. Selectors

NAME

BlockIDPolicy

ID ⑦

BlockIDPolicy

Figure 6.73 – New authentication policy name and ID

3. Configure the new policy as in the following screenshot. Click **Done**:

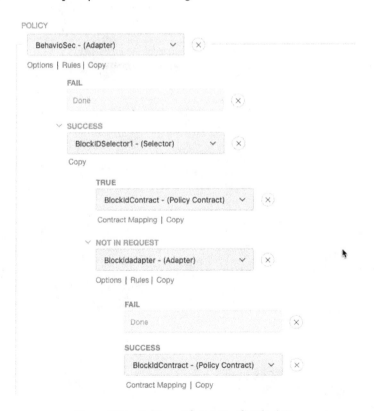

Figure 6.74 – Policy configuration for BlockID

Finally, we create an IdP connection to test the SSO login, as follows:

1. Select **AUTHENTICATION** | **Integration** | **IdP Connections** and click **Create Connection**:

Figure 6.75 – New IdP connection

2. If you have an existing application protected by PingFederate, enter the information for that application. The information for the application deployed in the *Deploying sample applications in PingFederate* subsection can also be used. Click **Done**:

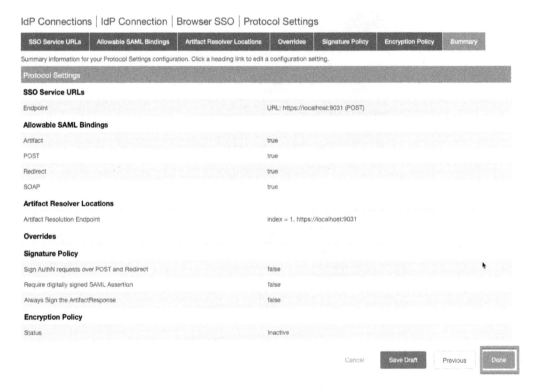

Figure 6.76 – Finished IdP connection

This completes our configuration of PingFederate with BlockID. After registering your account on the BlockID mobile app, we are ready to start testing.

Testing authentication with BlockID

The interesting thing you will see when testing BlockID is that there is no input performed on the page by the user. All the MFA process is done with the mobile app. Proceed as follows:

1. Open `https://localhost:9031/idp/startSSO.ping?PartnerSpId=1KosmosSPConnection`:

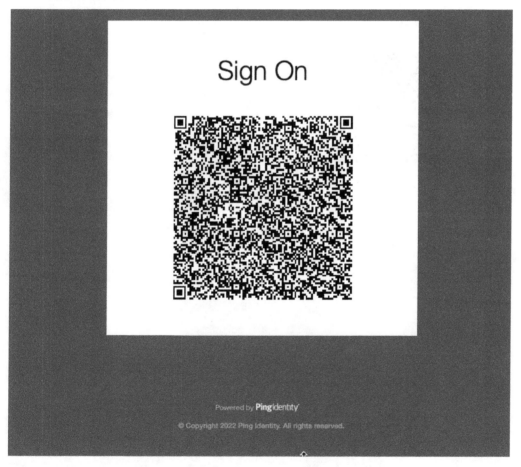

Figure 6.77 – BlockID login page and QR code

2.   Another option is to use the `https://www.1kosmos.com/demo/` 1Kosmos-hosted website:

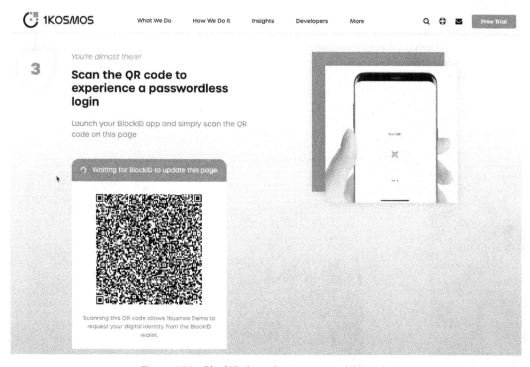

Figure 6.78 – BlockID demo login page and QR code

3. After scanning the QR code, approve the login on the app for a passwordless login:

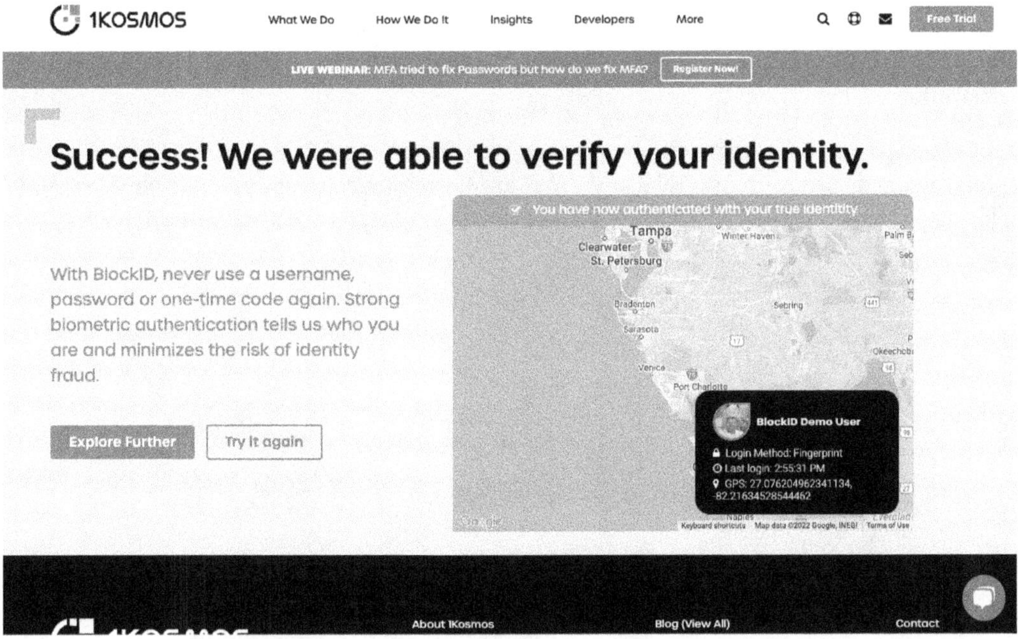

Figure 6.79 – Successful BlockID login

This completes our configuration and test of PingFederate using passwordless MFA with BlockID. As we can see from *Figure 6.79*, the verified information of the user is available for the application to use and can be used in the contract created in PingFederate.

The configuration also shows how passwordless authentication from 1Kosmos can be used to securely eliminate passwords from the authentication process while creating an excellent experience for the user.

Solutions such as 1Kosmos offer a comprehensive approach to digital security by protecting against synthetic identity fraud and account takeover and implementing passwordless authentication mechanisms. Synthetic identity fraud refers to creating fake identities using a combination of genuine and falsified information, which can be used to commit various types of fraud. Account takeover is a form of identity theft where a malicious actor gains unauthorized access to a user's online account and uses it for personal gain or to cause harm.

Summary

This chapter explored Ping's IAM products and how Acme used Ping Identity combined with passwordless authentication from 1Kosmos to provide an excellent, frictionless, verified experience for the workforce.

In the next chapter, we are going to explore **Amazon Web Services** (**AWS**) IAM services for protecting the workforce and customers. We will discuss AWS IAM, workforce user integration with AWS, and employing Amazon Cognito for customer and end user signup, sign-in, and authorization.

7

MFA and the Cloud – Using MFA with Amazon Web Services

Like most companies, Acme wants to use a cloud platform to develop and deploy products and services for its customers. In this chapter, we are going to see how Acme can use the **Identity and Access Management (IAM)** services from **Amazon Web Services (AWS)** to enable its workforce and customers.

We are going to cover the following topics:

- AWS IAM
- AWS and Workforce users
- Using Amazon Cognito for sign-up, sign-in, and authorization of Acme's customers and end users

Technical requirements

For this chapter, you are going to need an AWS root account. If needed, an account can be created at https://portal.aws.amazon.com/billing/signup#/start/email. With a new account, you should be able to follow most, if not all, of the examples in this chapter using the free tier (https://aws.amazon.com/free/). An optional step for creating and deploying an application to demonstrate Amazon Cognito capabilities requires Java **Software Development Kit (SDK)** version 11 or above.

AWS IAM

Before cloud services like AWS, Azure, or Google Cloud were available, cyber security was very different. Companies needed to build their own data centers or rent space with hosting companies to deploy their own servers somewhere else. Virtual security for those enterprise data centers was based on one fundamental idea: keep the bad actors out.

As seen in *Figure 7.1*, the main security perimeter for an enterprise was created using a firewall between the cloud (everything external to the enterprise) and the enterprise servers. Within the enterprise data center, an internal firewall would also separate the trusted part of the network from the **Demilitarized Zone (DMZ)**, where the public-facing servers, such as web and mail servers, would reside:

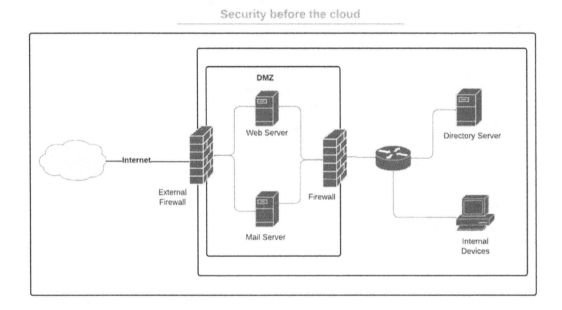

Figure 7.1 – Security before the cloud

In a typical scenario, a user would try to access a web application from Acme and connect to its web server. The user would then log in to the web server and get an encrypted token, which would be used on every request for services from the user – to see her bank balance on request on a banking application, for example, Alice would ask for the balance on the application. The request would go to the web server and include the token obtained during login. The web server would decrypt the cookie, verify that the session was still valid, and pass Alice's name in the clear to the application. The application would call a microservice using a generic system ID for authorization, and not Alice's name.

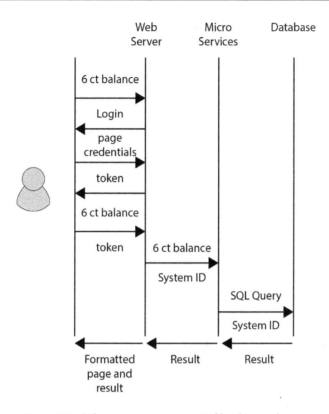

Figure 7.2 – Balance request on a typical banking web app

This model also assumes trust between zones, quite the opposite of the zero-trust approach advocated today. Any user inside the security perimeter of the enterprise could create their own request.

Another issue with this model was accountability. After the encrypted token was accepted by the web server, all calls from the applications the user would access would be made using internal system IDs, not the original user.

The database server or the microservices would not know that Alice was the original caller to the web application; they would only see a generic system account name.

What happened when a company moved from a private data center to the cloud? All those services that were inside the trusted perimeter were now shared by all users. But if the security perimeter vanishes, what that means is you need a pervasive application security model.

In addition to security, another important change that happened when data centers moved to the cloud was how companies paid for using the servers and services. Companies that were responsible for all the costs associated with their data centers now wanted to pay only for the services they used. After all, this was one of the main benefits of utilizing AWS and other cloud services.

Figure 7.2 shows an AWS user calling AWS services such as Cloud 9, Lambda functions, or an Aurora database. Of course, AWS users can also use other services:

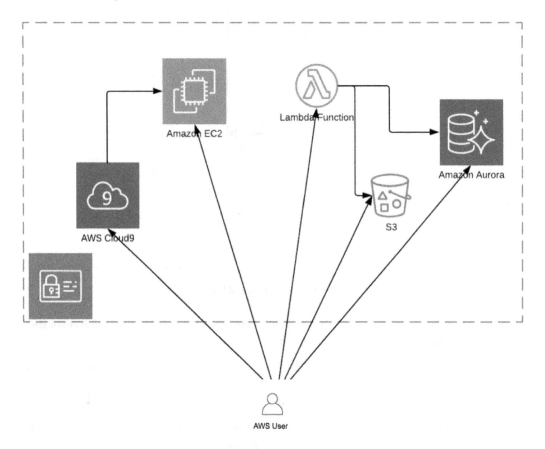

Figure 7.3 – Security on AWS

If we used the old security model in the cloud, users could call any service at any time and there would be no way to charge users only for the services they used. To solve this problem, Amazon created a model where security is everywhere. The model not only requires authentication and authorization with every service accessed but it also has a cost associated with each service.

In AWS, this model is implemented by the AWS IAM web service. AWS IAM is the service that controls access to every AWS resource. IAM controls who can sign in and has the appropriate permissions to access each resource.

In the *Technical requirements* section for this chapter, I only listed an AWS account – more specifically, an AWS root user account.

Every time you create an AWS account, you start with a user that has complete access to every service and resource that is available to that particular account. This is also known as an AWS root user account and it is accessed by logging in with the email address and password used during the creation of the account.

> **Important note**
>
> You should not use the root user for everyday tasks. Instead, the root user credentials should be stored in a safe place, and the account should be used to perform the tasks only the root user can perform.
>
> To enhance security, it is advised to set up **multifactor authentication** (**MFA**) to safeguard your AWS resources. MFA can be enabled for both the AWS account root user and IAM users. Activating MFA for the root user applies only to the root user's credentials. IAM users within the account are separate identities, each possessing unique credentials and individual MFA configurations. You can associate up to eight MFA devices, in any mix of the presently supported MFA types, with your AWS account root user and IAM users.

An AWS account is not a user account

If an AWS root user account is not to be used for everyday tasks, what should we use instead? For AWS, those are IAM users.

To avoid using the AWS root account, let's create our first IAM user. First, we need to enable billing account access:

1. Go to `https://aws.amazon.com/console/`. Select **Root user**, enter the root user email address, and click **Next**:

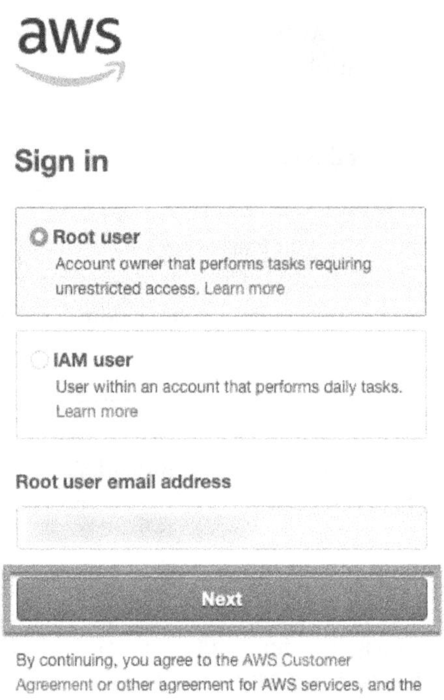

Sign in

○ **Root user**
Account owner that performs tasks requiring
unrestricted access. Learn more

○ **IAM user**
User within an account that performs daily tasks.
Learn more

Root user email address

[   ]

[**Next**]

By continuing, you agree to the AWS Customer
Agreement or other agreement for AWS services, and the

Figure 7.4 – AWS root user login

2.   Enter the password and click **Sign in**:

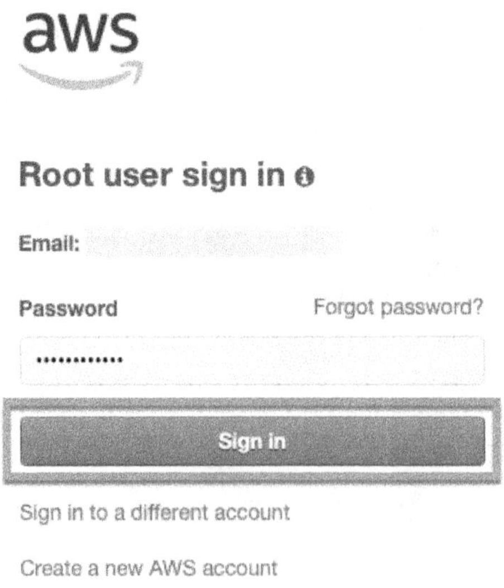

Root user sign in ❶

Email:

Password   Forgot password?

[············]

[**Sign in**]

Sign in to a different account

Create a new AWS account

Figure 7.5 – AWS root user login

3. If MFA is already enabled for the account, enter your **MFA code**. Click **Submit**:

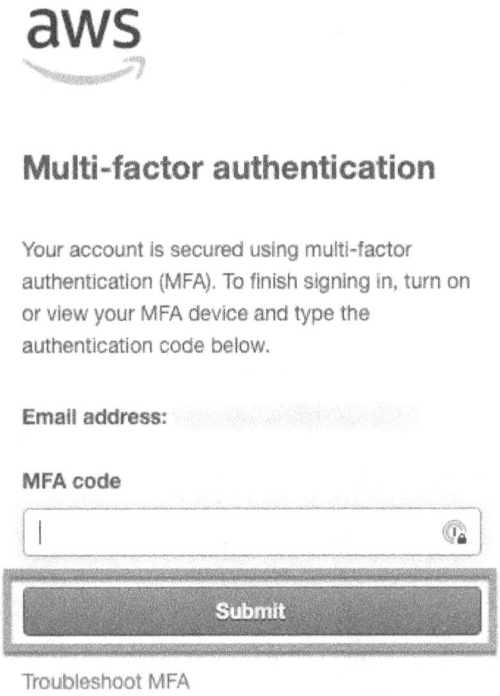

aws
Multi-factor authentication
Your account is secured using multi-factor authentication (MFA). To finish signing in, turn on or view your MFA device and type the authentication code below.
Email address:
MFA code
Submit
Troubleshoot MFA

Figure 7.6 – AWS root user MFA

4. On the navigation bar, choose your account name at the top left, and then choose **Account** from the drop-down list:

gs, docs, and more
[Option+S]
N. Virginia
mfabook
Reset to default layout
+ Add widg
Account ID: 7294-4593-7419
Account
Organization
Service Quotas
Billing Dashboard
Security credentials

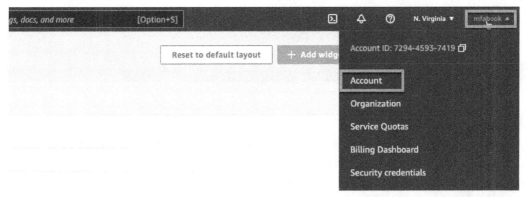

Figure 7.7 – The Account menu

5.   In the **IAM User and Role Access to Billing Information** section, click **Edit**:

▼ IAM User and Role Access to Billing Information

Use the **Activate IAM Access** setting to allow IAM users and roles access to pages of the Billing and Cost Management console. This setting alone doesn't grant IAM users and roles the necessary permissions for these console pages. In addition to activating IAM access, you must also attach the required IAM policies to those users or roles. For more information, see Granting access to your billing information and tools.

If this setting is deactivated, then IAM users and roles in this account can't access the Billing and Cost Management console pages, even if they have administrator access or the required IAM policies.

The **Activate IAM Access** setting does not control access to:

- The console pages for AWS Cost Anomaly Detection, Savings Plans overview, Savings Plans inventory, Purchase Savings Plans, and Savings Plan cart
- The Cost Management view in the AWS Console Mobile Application
- The Billing and Cost Management SDK APIs (AWS Cost Explorer, AWS Budgets, and AWS Cost and Usage Report APIs)
- The Customer Carbon Footprint Tool on the Cost & Usage Reports console page

IAM user/role access to billing information is deactivated.

Figure 7.8 – IAM User and Role Access to Billing Information

6.   Check the **Activate IAM Access** checkbox and click **Update**:

▼ IAM User and Role Access to Billing Information

Use the **Activate IAM Access** setting to allow IAM users and roles access to pages of the Billing and Cost Management console. This setting alone doesn't grant IAM users and roles the necessary permissions for these console pages. In addition to activating IAM access, you must also attach the required IAM policies to those users or roles. For more information, see Granting access to your billing information and tools.

If this setting is deactivated, then IAM users and roles in this account can't access the Billing and Cost Management console pages, even if they have administrator access or the required IAM policies.

The **Activate IAM Access** setting does not control access to:

- The console pages for AWS Cost Anomaly Detection, Savings Plans overview, Savings Plans inventory, Purchase Savings Plans, and Savings Plan cart
- The Cost Management view in the AWS Console Mobile Application
- The Billing and Cost Management SDK APIs (AWS Cost Explorer, AWS Budgets, and AWS Cost and Usage Report APIs)
- The Customer Carbon Footprint Tool on the Cost & Usage Reports console page

 Activate IAM Access

Cancel

Figure 7.9 – Activating IAM access

Now that we have updated the billing access, let us create our first user:

1.   On the navigation bar, click **Services**. In the left menu, select **Security, Identity, & Compliance** and then **IAM** to return to the IAM console:

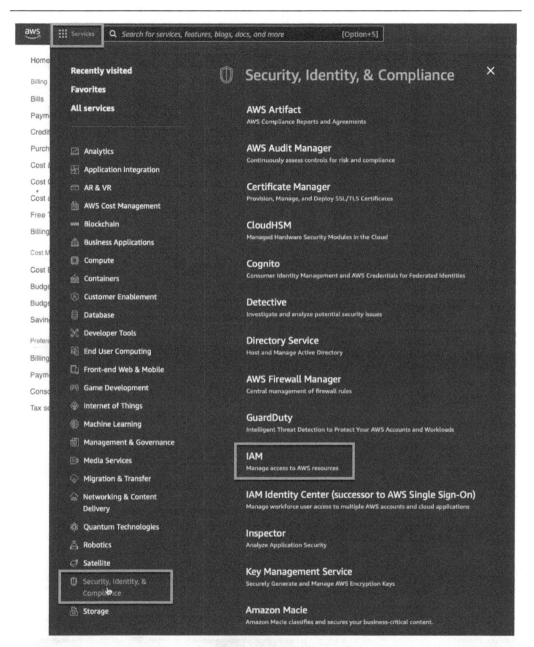

Figure 7.10 – The IAM menu

2. In the left navigation pane, select **Users**, and then click **Add users**:

Figure 7.11 – Users menu

3. On the **Add user** page, add a given **User name**, select **Password - AWS Management Console Access** as the AWS credential type, and click **Next: Permissions**:

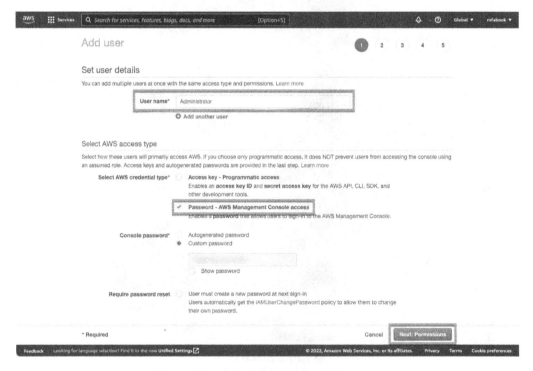

Figure 7.12 – Add user – Set user details page

4. On the **Set permissions** page, click on **Create group**:

Figure 7.13 – Add user – Set permissions page

5. In the **Create group** window that opens, check the checkbox for the **AdministratorAccess** job function. Click **Create group**:

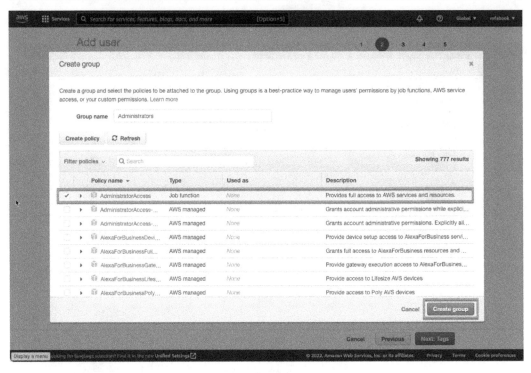

Figure 7.14 – Create group page

6. Click **Next: Tags**:

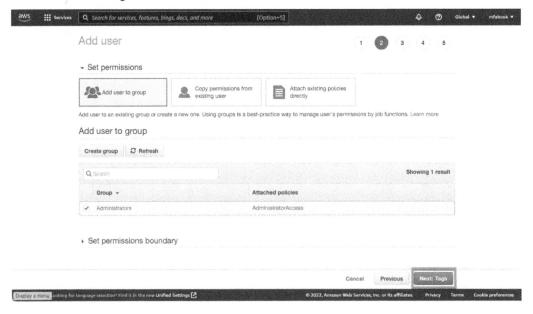

Figure 7.15 – Add user – Set permissions page

7. Click **Next: Review**:

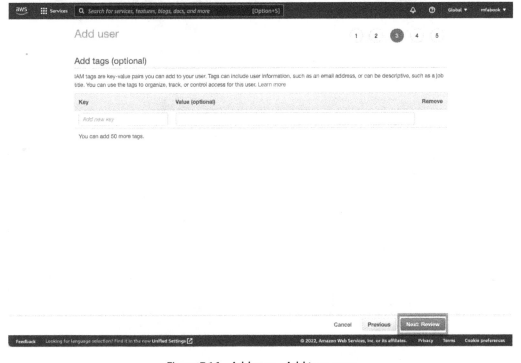

Figure 7.16 – Add user – Add tags page

8. Click **Create user**:

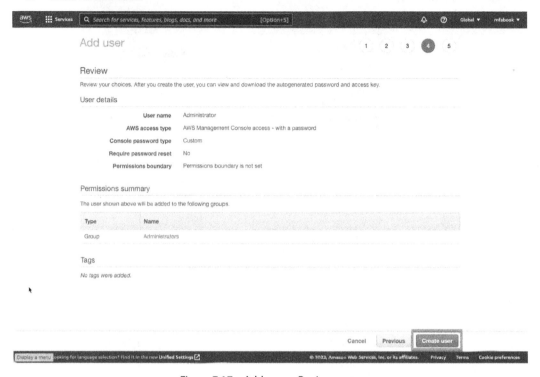

Figure 7.17 – Add user – Review page

9. Save the URL at the bottom of the **Success** green box. Click **Close** and then sign out of the AWS console:

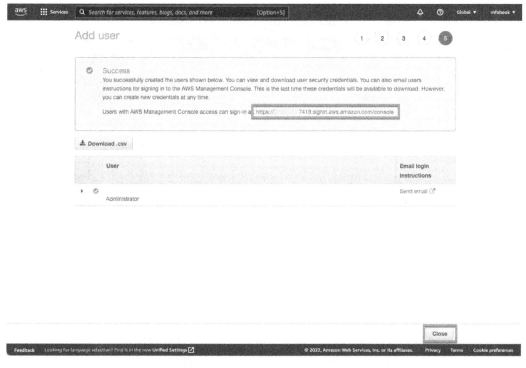

Figure 7.18 – Add user – Success page

Now that we have a user account (non-root user), let us use it:

1. Open the URL saved in the previous step and sign in with the Administrator ID and password. You should now see **Administrator@accountname** logged in on the dashboard:

Figure 7.19 – Console home

With such broad permissions, let's make sure MFA is enabled for the Administrator user account.

2.   On the navigation bar, click on the username. Then, select **Security credentials**:

Figure 7.20 – Security credentials menu

3. Click **Assign MFA device**:

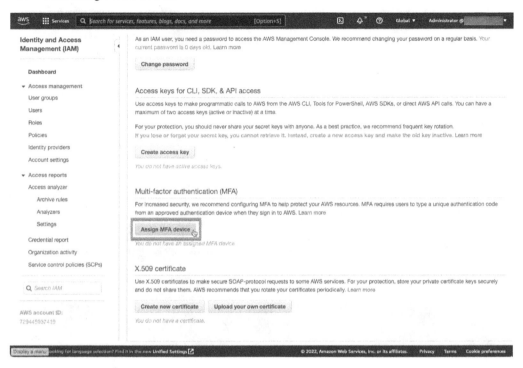

Figure 7.21 – Assign MFA device

4.   Select the type of MFA device to use and click **Continue**:

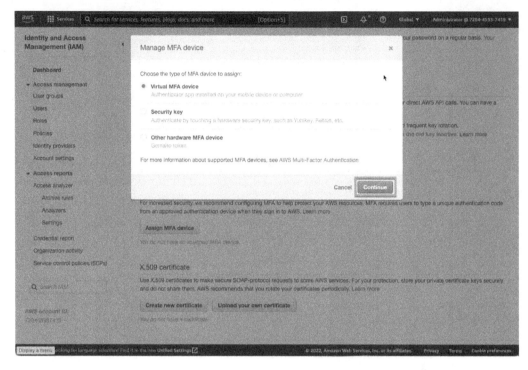

Figure 7.22 – MFA device selection

5. Enter the required information to set up your virtual device and click **Assign MFA**:

Figure 7.23 – Assign MFA

This concludes the configuration of our first IAM user.

With this user, we can avoid using the AWS root account except when required to change the root account settings. Changes to the account name, email address, root user password, and root user access keys require the root user credentials. Changes to contact information, AWS Regions, and payment currency preferences don't need root user credentials. Closing an account also requires root user credentials.

Workforce identities on AWS

In addition to having an all-powerful user, Acme needs other IAM user accounts for all other employees that need to use AWS. As the company grows, we must continue creating user accounts for new users with the minimum amount of effort. And, of course, we need to remove access when the user leaves or changes jobs with the company.

As we create user accounts and associated credentials, they are immediately available for authentication. Those users, however, will not have any default authorization associated with their accounts. All they can do is log in and set up their second-factor authentication.

IAM policies, a JSON formatted document of them that explicitly lists the permissions, must be created and assigned to a user. Only then will the user be allowed to perform the actions described in the policies assigned to them.

IAM policies do not need to be assigned directly to users. Instead, they may also be assigned to an IAM role. An IAM role is an AWS identity that determines what the identity is authorized to do in AWS.

The main difference between user accounts and roles is that a role does not have any long-term defined credentials, passwords, or associated access. Instead, roles are a way to dynamically create access keys, which are provided to the user temporarily and delegate access to users, applications, or services that don't usually have access to those resources.

AWS IAM Identity Center (successor to AWS Single Sign-on)

As a company grows, the manual management of accounts, policies, roles, and overall access to AWS can become complicated. Instead, companies are better served by using AWS IAM Identity Center to create or connect Workforce identities and manage access centrally to multiple applications. Identity Center is a free AWS product.

IAM Identity Center can connect to any of the major cloud identity providers that support SAML 2.0 and SCIM protocols, including Okta, Ping Identity, Azure AD, JumpCloud, CyberArk, and OneLogin, as well as Microsoft Active Directory Domain Services.

Identity Center can also be used to manage user assignments to SAML-based cloud applications such as Salesforce, Box, and Microsoft 365. With the user permissions configured, users have one-click access to AWS accounts and their assigned applications, simplifying the user experience.

Let's start using AWS IAM Identity Center:

1. Open `https://signin.aws.amazon.com/console`:

Figure 7.24 – AWS console login

2. Open `https://console.aws.amazon.com/singlesignon`. Click **Enable** to enable IAM Identity Center:

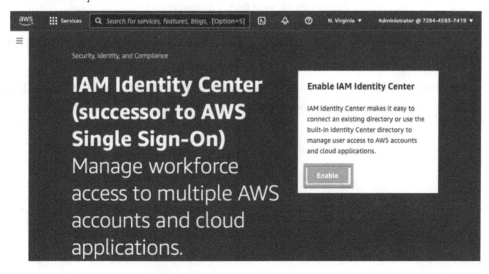

Figure 7.25 – Enable Identity Center

3. If prompted, create an AWS organization for use with the AWS account. Click **Create AWS organization**:

Enable IAM Identity Center

 IAM Identity Center requires AWS Organizations
We detected that your AWS account does not currently use this service. After you create an organization, you cannot join this account to another organization until you delete its current organization.

AWS Organizations provides the following benefits:

1. Enables single payer and centralized cost tracking

2. Lets you create and invite other AWS accounts

3. Allows you to apply policy-based controls

4. Helps you simplify organization-wide management of AWS services

Would you like us to create an AWS organization for you now?
We will also enable IAM Identity Center as part of this process.

Cancel   Create AWS organization

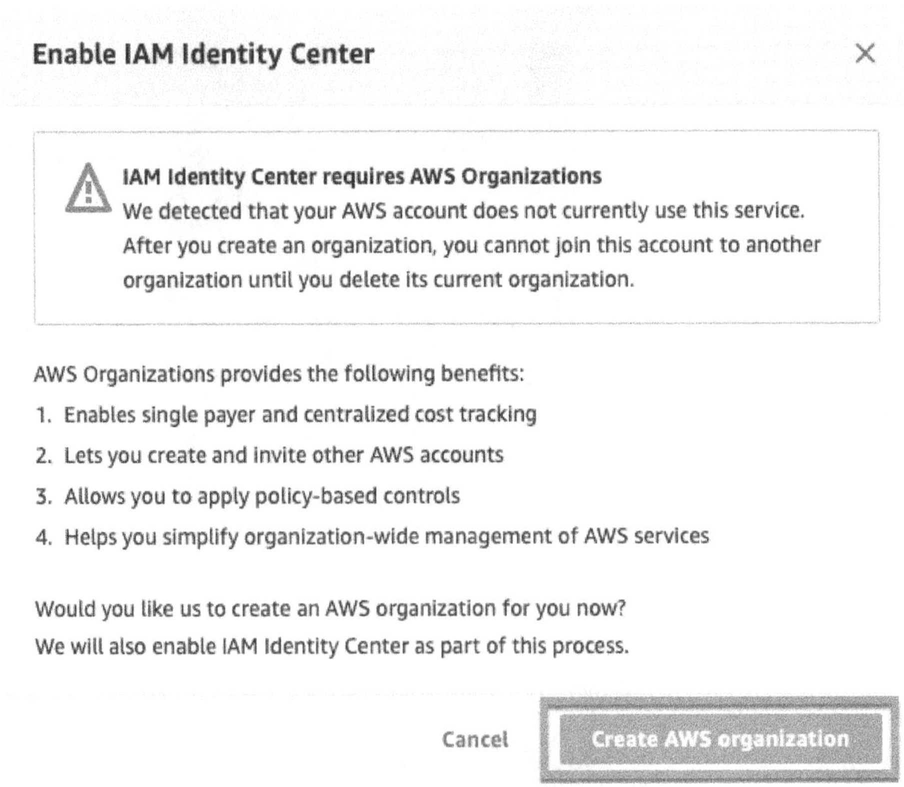

Figure 7.26 – Create AWS organization

When the IAM Identity Center console is used for the first time, the recommended first step is to choose your identity source:

1. If you are going to use an existing identity source (Okta, Ping Identity, Azure AD, JumpCloud, or other supported ones), click **Choose your identity source**:

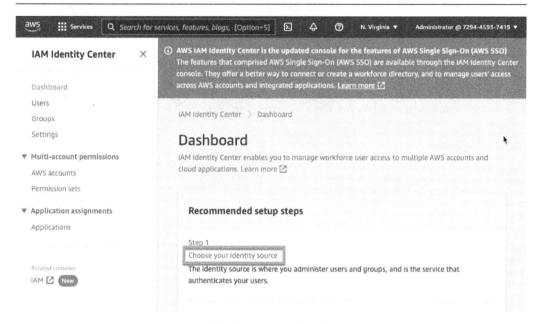

Figure 7.27 – Choose your identity source

2. In the **Choose identity source** section, select **Active Directory** or **External identity provider**, click **Next**, and follow the remaining steps for that directory. To keep **Identity Center directory**, click **Cancel**:

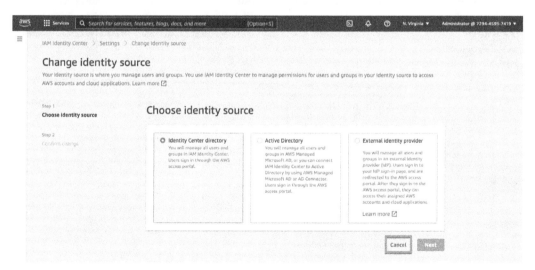

Figure 7.28 – Choose identity source

Next, we are going to configure MFA for all Workforce users:

1. Select **Settings** from the left menu, scroll down to the **Multi-factor authentication** section, and click **Configure**:

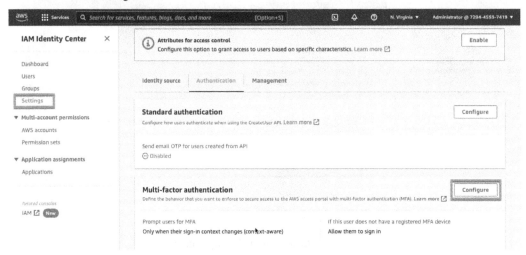

Figure 7.29 – Multi-factor authentication configuration

2. Select one of the first two options for **Prompt users for MFA**. Selecting the third option disables MFA. Select **Require them to register an MFA device at sign in**. Click **Save changes**:

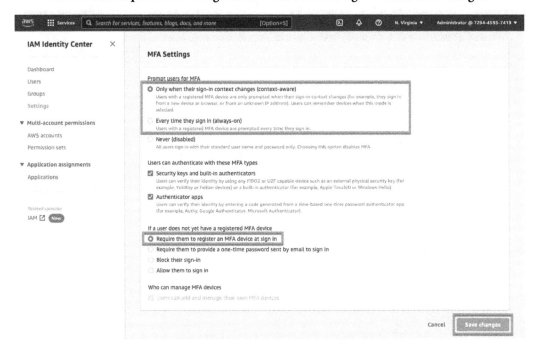

Figure 7.30 – Multi-factor authentication configuration

Let's try to use Identity Center to create a new user with mandatory MFA:

1. On the left menu, select **Users** and click **Add user**:

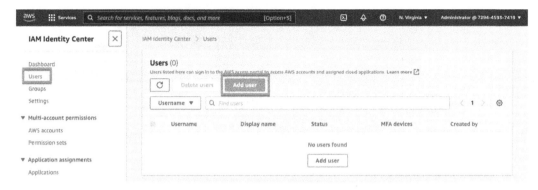

Figure 7.31 – Adding a new user

2. Enter a unique username and the other required details:

Figure 7.32 – New user details

3. Scroll down and click **Next**:

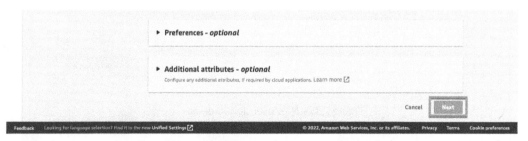

Figure 7.33 – New user details – Click Next

4. Verify the information and click **Add user**:

Figure 7.34 – Add user

5. Copy the **AWS access portal URL** and **One-time password** details and click **Close**:

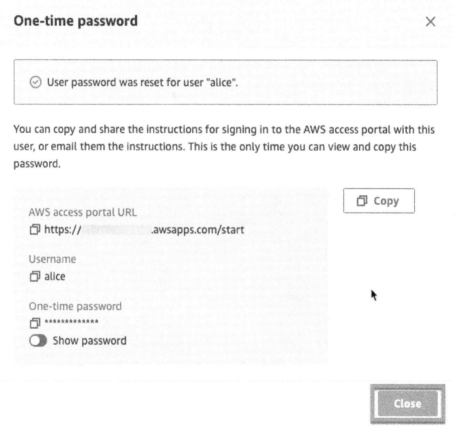

Figure 7.35 – New user sign-in details

Save the **AWS access portal URL** and **One-time password** details, as they will be used to test the user.

The list of users, including the newly created one, is displayed on the dashboard.

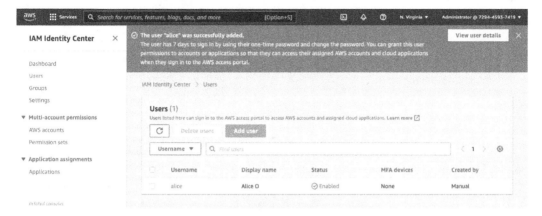

Figure 7.36 – New user created

Let's test the new user we just created:

1. Open the AWS access portal URL. Enter the username for the new user. Click **Next**:

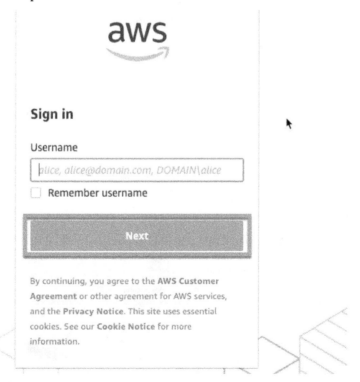

Figure 7.37 – New user sign-in

2.   Enter the one-time password generated for the user. Click **Sign in**:

Figure 7.38 – New user sign-in

3.   Select the MFA device to register. Click **Next**:

Figure 7.39 – Register MFA device

4.   Click **Done**:

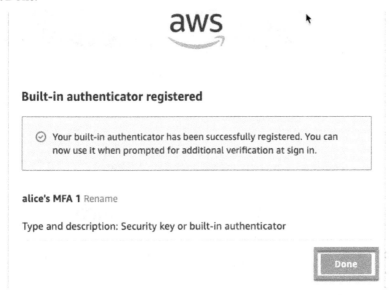

Figure 7.40 – MFA device registered

5.   Set a new password to replace the one-time password created initially. Click **Set new password**:

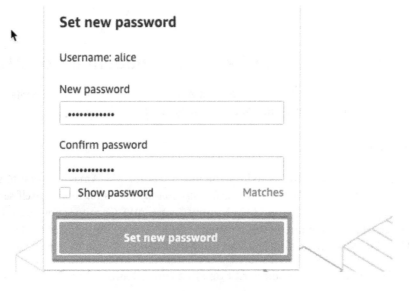

Figure 7.41 – Set new password

After login, the AWS user portal is displayed. In this case, the user doesn't have any applications:

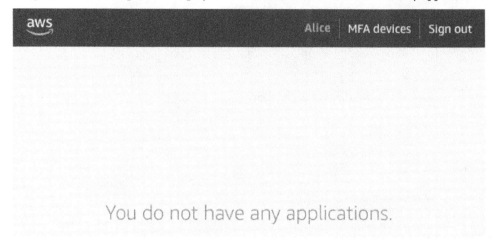

Figure 7.42 – AWS user portal

Adding applications to AWS SSO is beyond the scope of this book. More information about additional configurations in AWS IAM Identity Center can be found at https://docs.aws.amazon.com/singlesignon/latest/userguide/get-started-enable-identity-center.html.

This concludes the Workforce IAM section. We just learned how to centrally manage users using AWS IAM Identity Center and how to enforce MFA for all Workforce users.

Customer Identity and Access Management on AWS

Now that we have covered Workforce IAM on AWS, let's discuss the requirements and how can we implement **Customer Identity and Access Management (CIAM)** on AWS.

AWS Cognito

AWS Cognito is the tool to use to implement CIAM on AWS. Amazon Cognito satisfies all the requirements presented in *Chapter 1, On the Internet, Nobody Knows You're a Dog* (self-service, scalability, ease of use, and SSO). Unlike AWS IAM Identity Center, AWS Cognito is not free. It is a pay-as-you-use service. For the examples in this chapter, we will try to use as many of the free tier services as possible. With this introduction, let's start using the Cognito user pools to manage our customers:

1. Sign in to the AWS console at https://signin.aws.amazon.com/console:

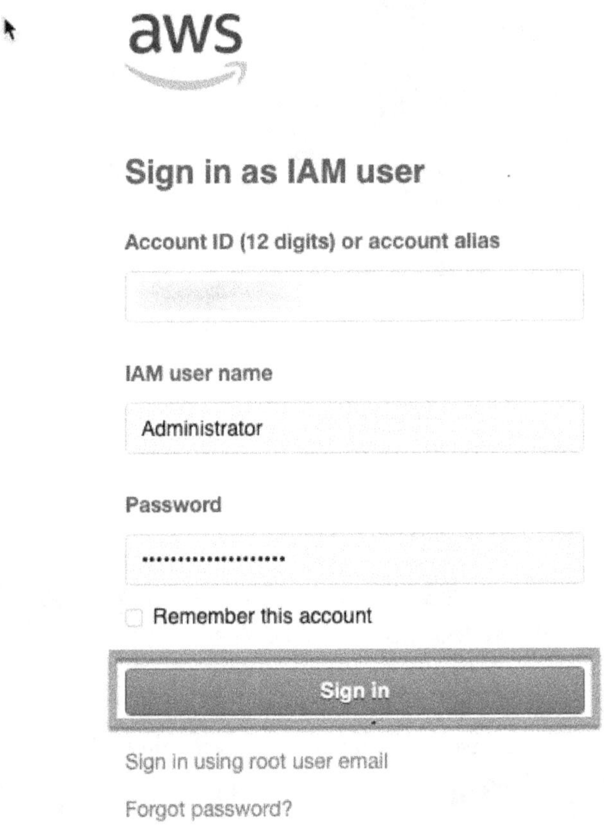

Sign in as IAM user

Account ID (12 digits) or account alias

IAM user name

Administrator

Password

....

☐ Remember this account

Sign in

Sign in using root user email

Forgot password?

Figure 7.43 – AWS console sign-in

2. Search for **Cognito**:

Figure 7.44 – Searching for Cognito

3. Click **Create user pool**:

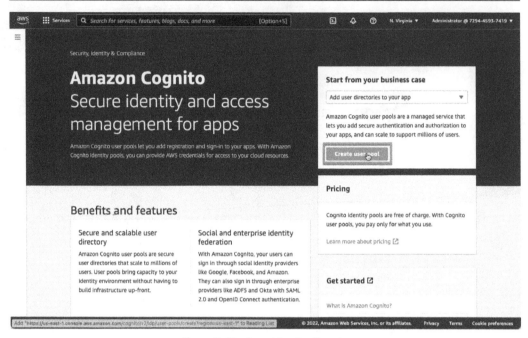

Figure 7.45 – Searching for Cognito

We are going to create a new Cognito user pool.

4.   In the **Provider types** section, leave the default, **Cognito user pool**. Click **Next**:

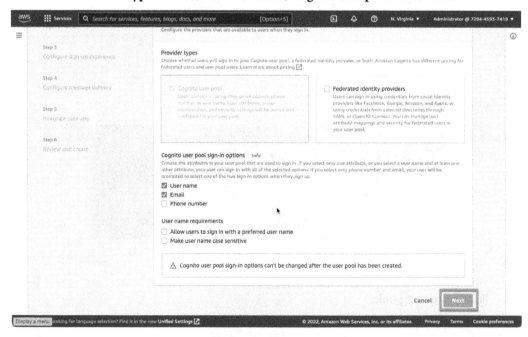

Figure 7.46 – Creating a user pool

5. Leave the default for **Password policy mode**:

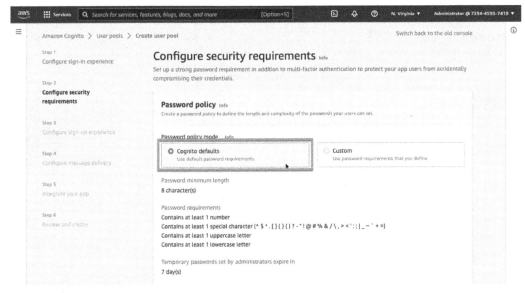

Figure 7.47 – Configure security requirements

6. For **MFA enforcement**, select **Require MFA - Recommended**. Uncheck the **SMS message** checkbox to avoid charges:

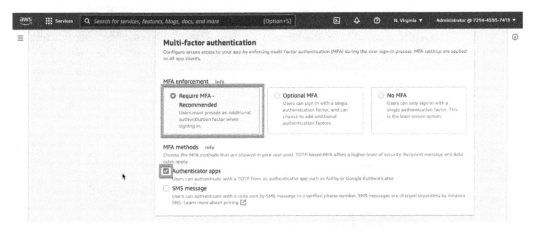

Figure 7.48 – Creating the user pool

7.   In **User account recovery**, select **Enable self-service account recovery - Recommended**. Click **Next**:

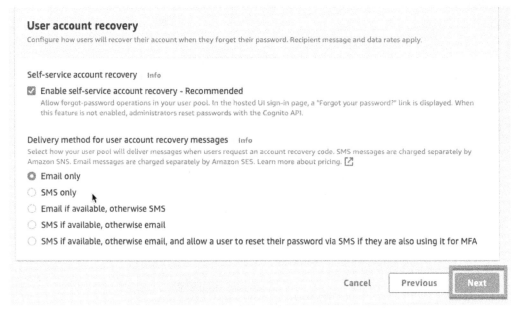

Figure 7.49 – Creating the user pool

8.   The next step configures the sign-up experience. Select **Enable self-registration**:

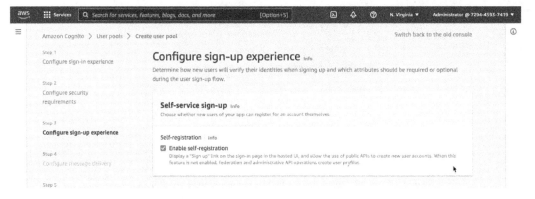

Figure 7.50 – Configure sign-up experience

9. Uncheck **Allow Cognito to automatically send messages to verify and confirm - Recommended**:

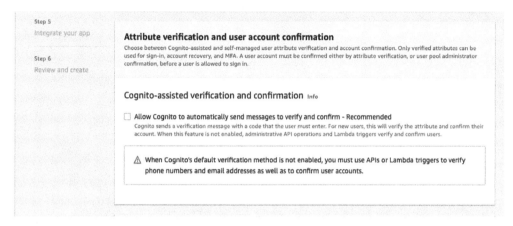

Figure 7.51 – Configure the sign-up experience

10. Click **Next** to finish configuring the sign-up experience:

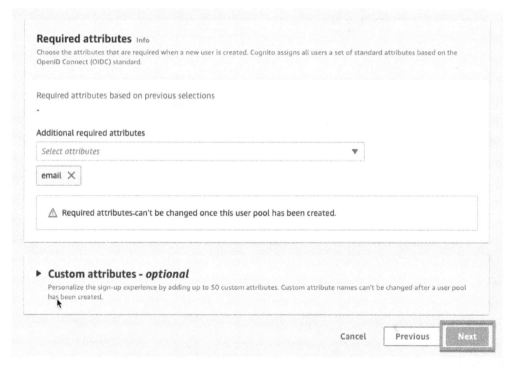

Figure 7.52 – Finishing configuring the sign-up experience

11. On the **Configure message delivery** screen, select **Send email with Cognito**. Click **Next** to finish **Step 4**:

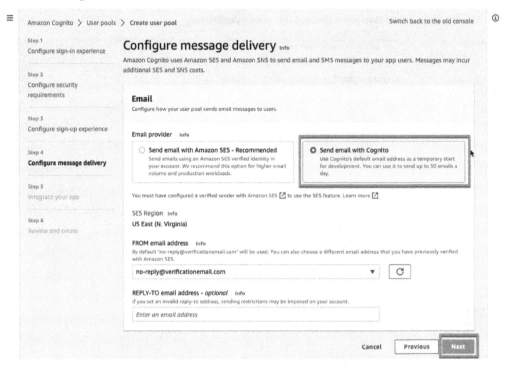

Figure 7.53 – Configure message delivery

The next step performs the integration of the app.

12. Check the **Use the Cognito Hosted UI** checkbox:

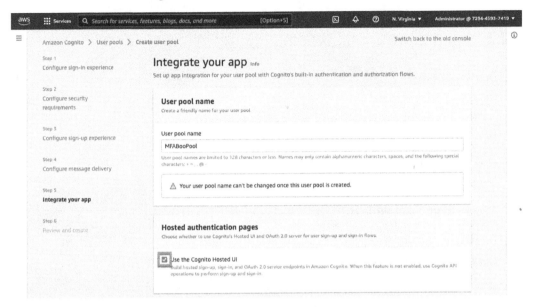

Figure 7.54 – Integrate your app

13. Select **Public client**:

Figure 7.55 – Integrate your app

14. Add a callback URL. Click **Next** to finish the configuration of the new user pool:

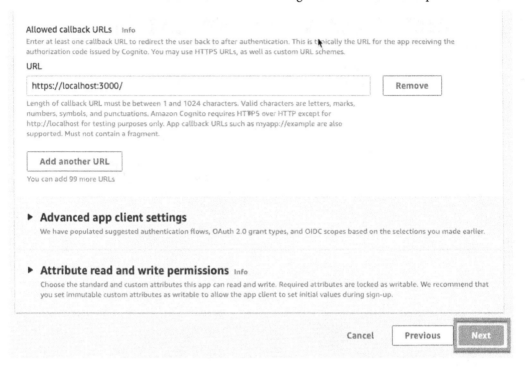

Figure 7.56 – Integrate your app

15. Review the information:

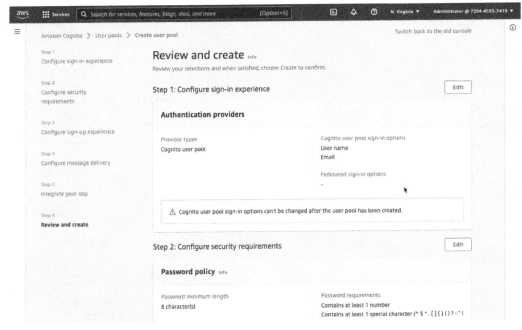

Figure 7.57 – Review and create

16. Click on **Create user pool**:

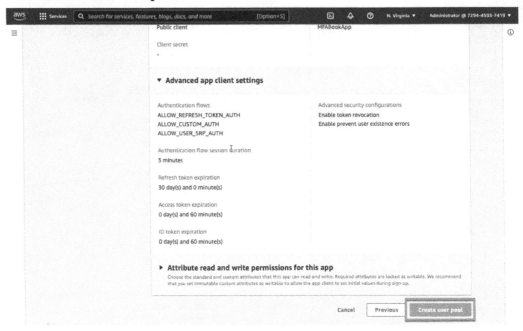

Figure 7.58 – Review and create

Now that we have a user pool, let us create an application to test it:

1. To create a sample Cognito application, clone the GitHub repository found here: `https://github.com/eugenp/tutorials/tree/master/spring-security-modules/spring-security-cognito`.

2. Edit the `src/main/resource/application.yml` file:

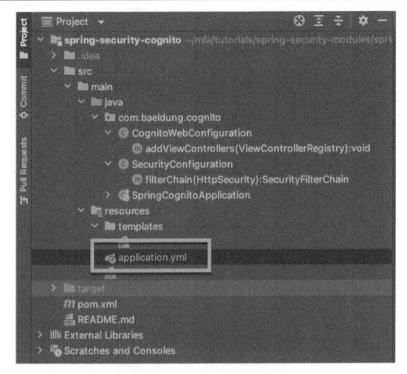

Figure 7.59 – resources directory

3. Enter `client-id`, `client-secret`, `clientName`, and `issuerUri` from the Cognito configuration. Add `port: 3000` (or another unused port) to the configuration. The port needs to match the configuration for `redirect-uri`:

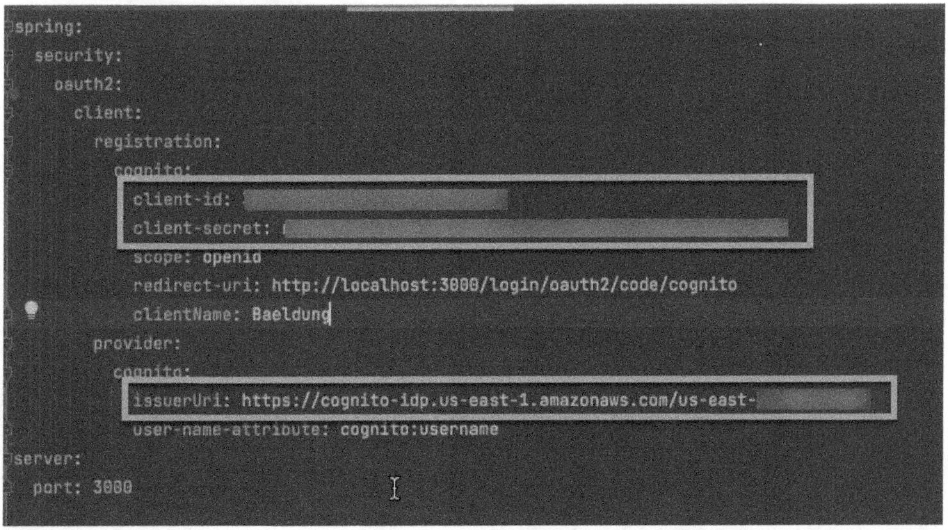

```yaml
spring:
  security:
    oauth2:
      client:
        registration:
          cognito:
            client-id:
            client-secret:
            scope: openid
            redirect-uri: http://localhost:3000/login/oauth2/code/cognito
            clientName: Baeldung
        provider:
          cognito:
            issuerUri: https://cognito-idp.us-east-1.amazonaws.com/us-east-
            user-name-attribute: cognito:username
server:
  port: 3000
```

Figure 7.60 – application.yml

4.   Run the `com.baeldung.cognito.SpringCognitoApplication` Java application.

With the application running, we can start testing Cognito using our sample application:

1.   Open `http://localhost:3000`. Click on **Log in with Amazon Cognito**:

Figure 7.61 – Demo application main page

2.   Click on **Sign up** to register a new user in Cognito:

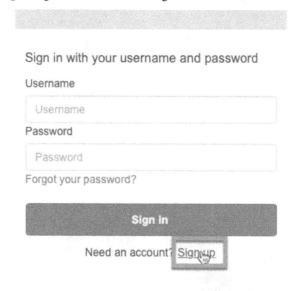

Figure 7.62 – Cognito sign-in page

3. Enter the new user information. Click on **Sign up**:

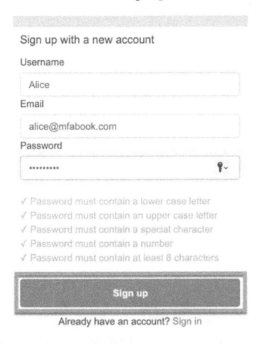

Figure 7.63 – Cognito sign-up page

4. Open `http://localhost:3000` again. Click **Log in with Amazon Cognito**, as depicted in *step 1*. Enter the sign-in information. Click **Sign in**:

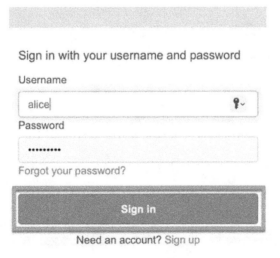

Figure 7.64 – Cognito sign-in page

5. If the user has not been confirmed automatically or manually after being created, sign-in will fail with a **User is not confirmed** message:

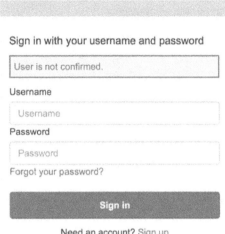

Figure 7.65 – Cognito sign-in page

6. Back in the Amazon Cognito console, select **User pools**. Select the unconfirmed user:

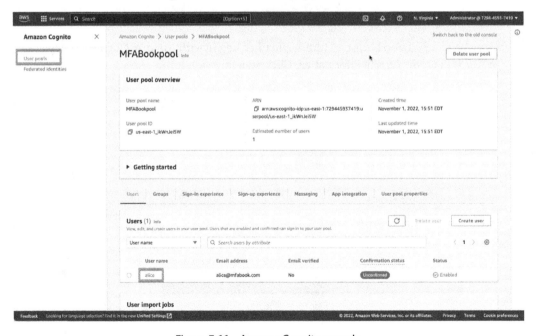

Figure 7.66 – Amazon Cognito console

7.   Select **Actions | Confirm account**:

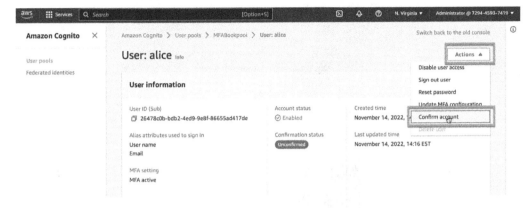

Figure 7.67 – Amazon Cognito console

8.   Click **Confirm** to confirm the user's account:

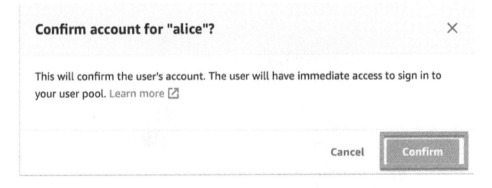

Figure 7.68 – New user confirmation

9.   After being confirmed, the user can try to sign in again. As we required MFA as a condition for login, the user is now prompted to set up an authenticator app for MFA. Click **Sign up** when done with the configuration:

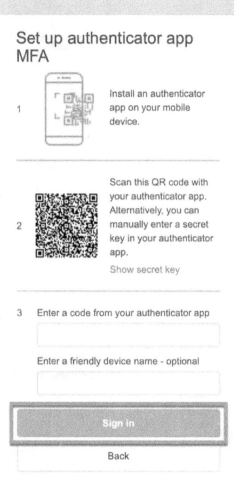

Figure 7.69 – New user MFA setup

Finally, the user is able to log in to the application:

Figure 7.70 – Login successful

This completes our section about CIAM on AWS, in which we saw how to configure and test Cognito user pools for customer-focused apps.

Summary

In this chapter, we discussed MFA on AWS for customers and Workforce users. We covered the free tools that Amazon provides to better manage Workforce users in a secure and scalable way. We also covered Amazon Cognito, a pay-as-you-use tool for end-user self-service, identity management, and SSO.

In the next chapter, we are going to explore the use of Keycloak, an open source IAM product that provides MFA and IAM services for your workforce and customers as well.

Google Cloud Platform and MFA

The *top three* cloud platform service providers are **Amazon Web Services**, **Microsoft Azure**, and **Google Cloud Platform** (**GCP**). These three providers are the dominant players in the cloud computing market and offer a wide range of cloud computing services, including **infrastructure as a service** (**IaaS**), **platform as a service** (**PaaS**), and **software as a service** (**SaaS**), which are available across multiple regions around the world.

Each of these top cloud providers offers its own unique value proposition. While Azure, which we discussed in *Chapter 3*, claims to provide a more user-friendly way for organizations to migrate to the cloud, AWS claims to be the most developer-friendly for its wide range of services offered, while Google amplifies its machine learning and data analytics capabilities. AWS being the oldest of the three providers, claims to offer over 200 services from 31 Regions and 99 Availability Zones (at the time of writing), Azure delivers over 150 services from over 60 regions and Google delivers over 100 products from over 20 regions.

Each platform has strengths and is well suited to different workloads and use cases. As a result, Acme, like many other companies, may want to employ more than one of these platforms, depending on its specific needs and requirements.

This chapter will complete the *big three* coverage with GCP.

We are going to cover the following topics:

- Google Cloud Identity
- Google Cloud Identity Platform

Technical requirements

For this chapter, you are going to need a Google Cloud account. If required, you can create a new account at `https://cloud.google.com/free`. With a new account, you can use the free credits for new customers to follow the examples in the book. In addition, if your organization controls an internet domain name, you can use it for optional protection in Google Cloud Identity.

Cloud security involves a combination of different security measures to protect hardware, software, applications, data, and users in the cloud. Both cloud providers and customers have a role to play in ensuring the security of their workloads and applications. Every cloud provider defines a shared responsibility model to clearly articulate what the responsibility of the cloud provider is versus the responsibility of the customer for different types of services (IaaS, PaaS, and SaaS). Cloud providers are responsible for securing the infrastructure on which their services run. At the same time, customers are responsible for securing their data and following the best practices and guidelines provided by the provider. The specific responsibilities of customers may vary depending on the cloud computing model they are using, such as IaaS, PaaS, or serverless. Ensuring the security of cloud systems requires collaboration and cooperation between providers and users.

As we have seen with the other cloud service providers, we will use GCP's **identity and access management** (**IAM**) services to authenticate the users that we want to allow or deny the use of GCP.

Google Cloud Identity

Google Cloud Identity is a cloud-based IAM platform provided by Google Cloud. It is designed to help organizations manage user access to cloud and on-premises resources, devices, and applications. With Cloud Identity, administrators can create and manage user accounts, set up and enforce policies for access and authentication, and monitor and audit user activity. The platform also includes features such as **single sign-on** (**SSO**) and **multifactor authentication** (**MFA**) to help ensure the security of user accounts and protect against unauthorized access. In addition, Cloud Identity integrates with other Google Cloud products, such as Google Workspace, GCP, and Google Identity Platform (formerly Google Cloud Identity for Customers and Partners), to provide a unified and comprehensive approach to IAM.

Setting up Cloud Identity

Let's start setting up Cloud Identity for Acme:

1. Go to `https://admin.google.com/u/3/ac/home`. Click **BEGIN THE SETUP**:

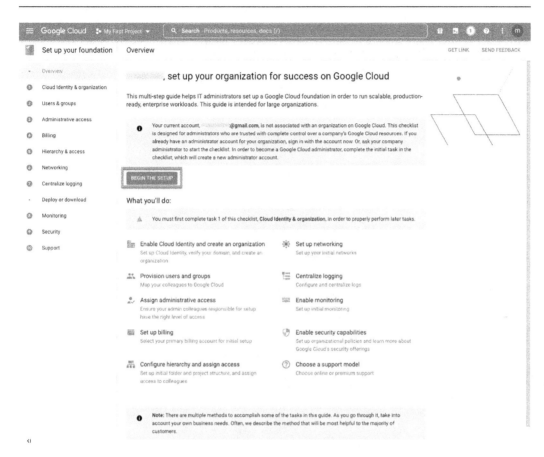

Figure 8.1 – Google Workspace Admin console

2.   On the **Cloud Identity & organization** page, scroll down to the second page:

Figure 8.2 – Cloud Identity & organization page (top)

3. Select the **I'm a new customer** option under **Select an option to view detailed steps**. Click the **SIGN UP FOR CLOUD IDENTITY** button:

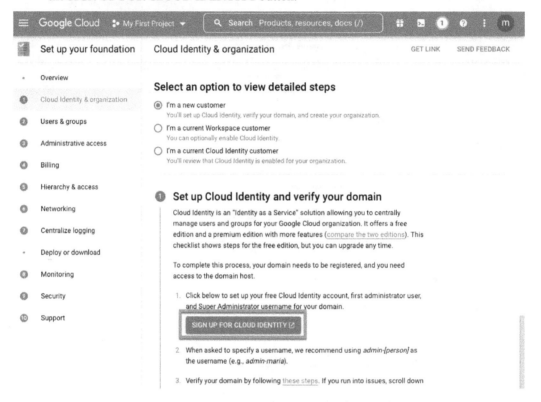

Figure 8.3 – Cloud Identity & organization page (bottom)

4. On the pop-up page, complete the organization information and click **Next**:

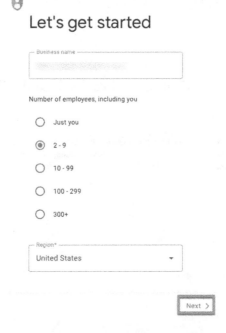

Figure 8.4 – Cloud Identity account signup

5.   Add the contact info and click **Next**:

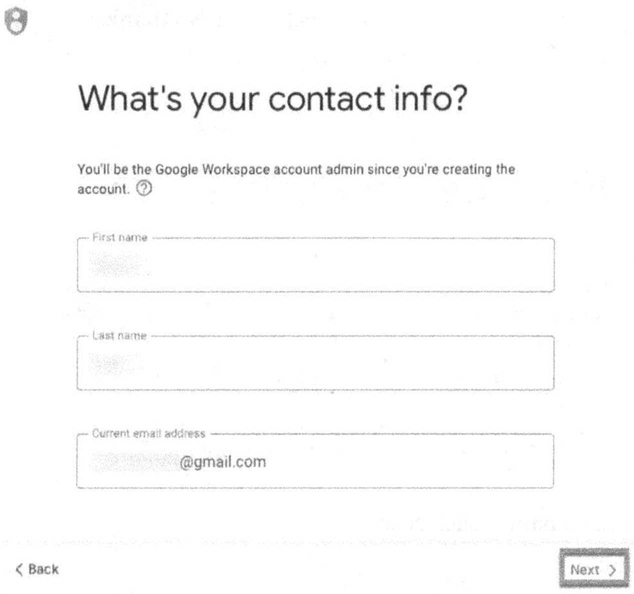

What's your contact info?

You'll be the Google Workspace account admin since you're creating the account. ⓘ

First name

Last name

Current email address

@gmail.com

< Back   Next >

Figure 8.5 – Cloud Identity account sign-up

6. Add the business domain name and click **Next**:

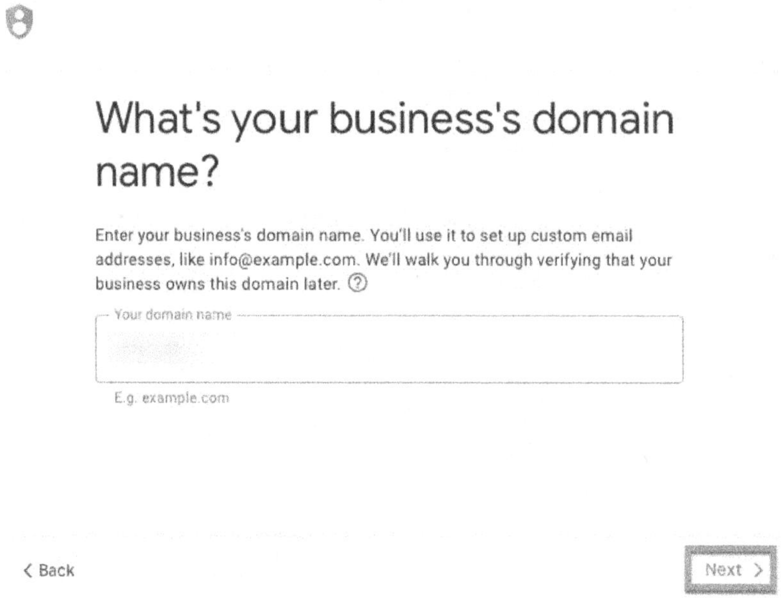

What's your business's domain name?

Enter your business's domain name. You'll use it to set up custom email addresses, like info@example.com. We'll walk you through verifying that your business owns this domain later. ⓘ

Your domain name

E.g. example.com

< Back   Next >

Figure 8.6 – Cloud Identity account signup

7. Choose whether you want your organization's users to automatically receive information about Google Workspace, new features, and tips. Click **Ok** or **No thanks**:

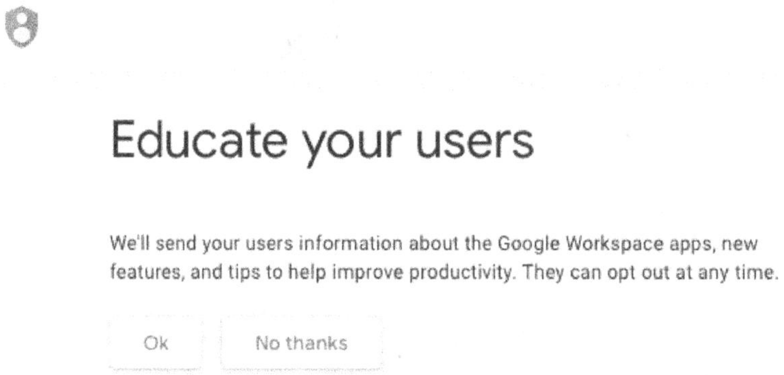

Educate your users

We'll send your users information about the Google Workspace apps, new features, and tips to help improve productivity. They can opt out at any time.

Figure 8.7 – Cloud Identity account signup

8. Confirm the domain name. Click **Next**:

Use this domain to set up the account?

Your domain

Emails sent to ____ won't be affected until you set up email with this account.

< Back

Next >

Figure 8.8 – Cloud Identity account sign-up

9. Select the username and password to sign into the Google Workspace account. Click **Agree and continue** to complete the Cloud Identity account signup steps:

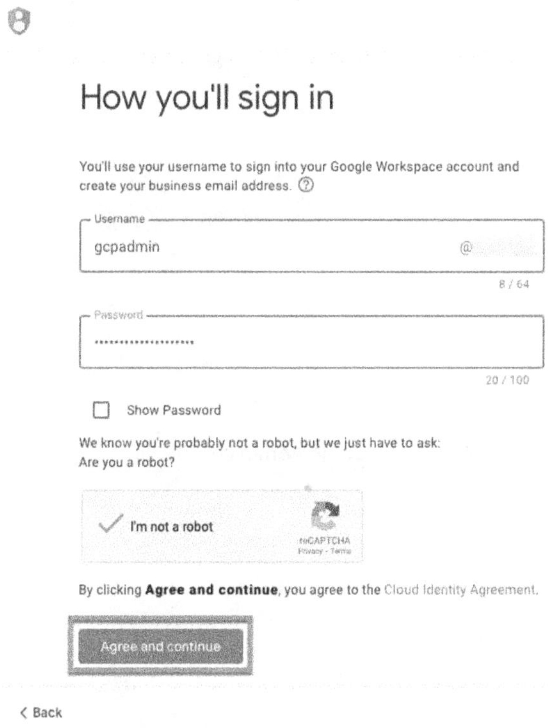

How you'll sign in

You'll use your username to sign into your Google Workspace account and create your business email address. ⑦

Username
gcpadmin   @____

8 / 64

Password
....

20 / 100

☐ Show Password

We know you're probably not a robot, but we just have to ask: Are you a robot?

✓ I'm not a robot

reCAPTCHA
Privacy - Terms

By clicking **Agree and continue**, you agree to the Cloud Identity Agreement.

Agree and continue

< Back

Figure 8.9 – Cloud Identity account signup

This completes the setup of the free Cloud Identity account, the creation of the first administrator user and setting up the super administrator username for the domain.

We can now log back to Google Cloud Identity as the administrator user:

1. Go to https://admin.google.com/u/3/ac/home. Enter the username and click **Next**:

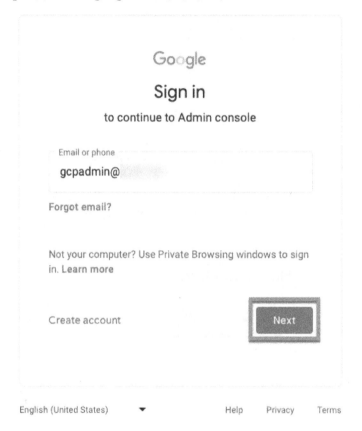

Figure 8.10 – Google Admin console sign-in

2. Enter the password and click **Next**:

Figure 8.11 – Google Admin console sign-in

3. If presented with the **Welcome to your new account** terms of service page, click **I understand**:

Welcome to your new account

Welcome to your new account: gcpadmin@   . Your   administrator decides which Google Workspace and other Google services you may access using this account.

Your organization administrator manages this account and any Google data associated with this account (as further detailed here). This means that your administrator can access and process your data, including the contents of your communications, how you interact with Google services, or the privacy settings on your account. Your administrator can also delete your account, or restrict you from accessing any data associated with this account.

If your organization provides you access to administrator-managed services, like Google Workspace, your use of those services is governed by your organization's enterprise agreement. Besides these terms, we also publish a Google Cloud Privacy Notice.

If your administrator enables you to use other Google services besides Google Workspace while logged in to this gcpadmin@   account, your use of those services will be governed by their respective terms, such as the Google Terms of Service and the Google Privacy Policy and other service-specific Google terms. If you do not agree to these terms, or do not wish Google to handle your data in this way, do not use those other Google services with this gcpadmin@   account. You may also customize your privacy settings at myaccount.google.com.

Your use of Google services with this account is also governed by your organization's internal policies.

I understand

Figure 8.12 – Google Admin console terms of service

The first step when setting up Cloud Identity is to protect your domain from unauthorized use. To protect your domain from unauthorized use, it is important to verify your ownership with Google Workspace. This will ensure that no one else can use your domain to sign up for Google Workspace without your permission. By verifying your ownership, you can help keep your domain secure and prevent any potential misuse.

4. On the **Welcome** page for Cloud Identity, click **Protect**:

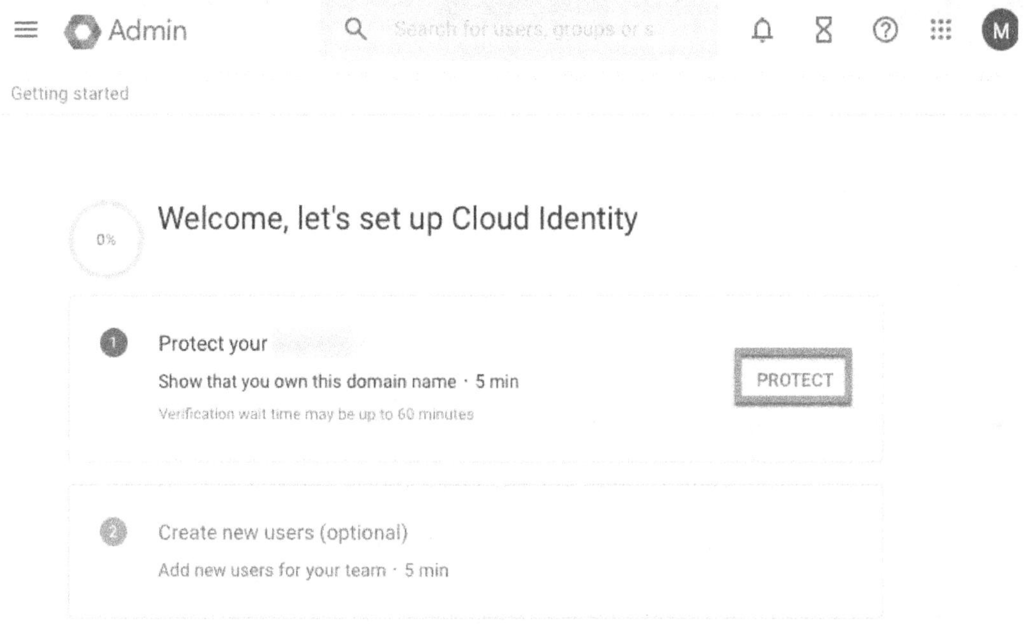

Figure 8.13 – Cloud Identity setup

Domain Name System (**DNS**) records are publicly available information that directs internet traffic to your domain's website and email servers. These records provide the necessary information for computers to locate your website and deliver emails to your domain. By modifying your DNS records, you can control how internet traffic is routed to your domain and manage your company's online presence.

5. In the **Protect your domain** page for Cloud Identity, click **I'M READY TO PROTECT MY DOMAIN**. This can be a complicated task, and the steps are different for each domain registrar where your domain records reside:

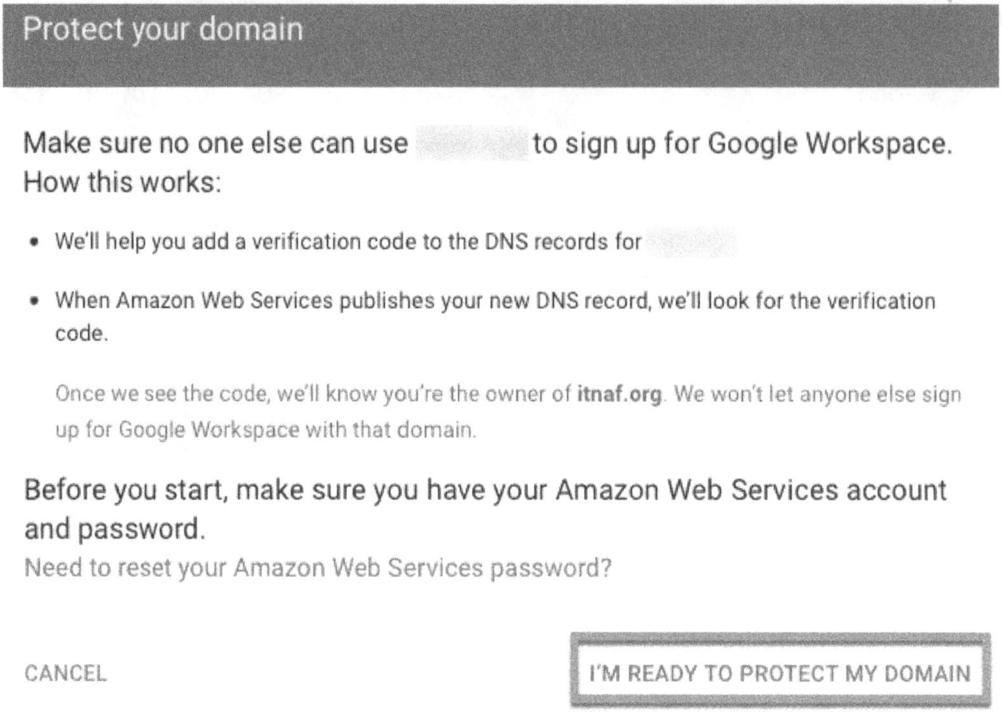

Protect your domain

Make sure no one else can use _____ to sign up for Google Workspace. How this works:

- We'll help you add a verification code to the DNS records for _____

- When Amazon Web Services publishes your new DNS record, we'll look for the verification code.

Once we see the code, we'll know you're the owner of **itnaf.org**. We won't let anyone else sign up for Google Workspace with that domain.

Before you start, make sure you have your Amazon Web Services account and password.
Need to reset your Amazon Web Services password?

CANCEL   I'M READY TO PROTECT MY DOMAIN

Figure 8.14 – Protect your domain page

6.   Follow actions (**a**) through (**d**) on a new browser tab. When finished, return to this tab, and click **NEXT: GO TO STEP 2**.

The instructions in *Figure 8.15* are automatically created by Google according to where your domain is registered. Follow the instructions provided for your domain registrar, and not the instructions in the following figure. More information on this step can be found at `https://support.google.com/cloudidentity/answer/6248925?hl=en`. Let's have a look at *Figure 8.15* now:

Find your DNS records or settings

(a)   Click to open Amazon Web Services in a new browser tab. ☑

(b)   On the AWS Management Console, under either **Recently visited services** or **Networking & Content Delivery**, click **Route 53**.

(c)   Under **DNS management**, click **Hosted zones**.

(d)   Select the domain you want to verify.

Then **come back here** and click Next: Go to Step 2 to add a verification record.

NEXT: GO TO STEP 2

Figure 8.15 – Protect your domain – step 1

7.   Perform actions **(b)** through **(d)** from *Figure 8.15* specific to your domain registrar. Remember, this page and the steps may look different for your domain registrar.

8.   Perform actions **(e)** through **(i)** in **STEP 2** (*Figure 8.16*) in the new tab opened for your domain registrar. When finished, you have to click **PROTECT DOMAIN**:

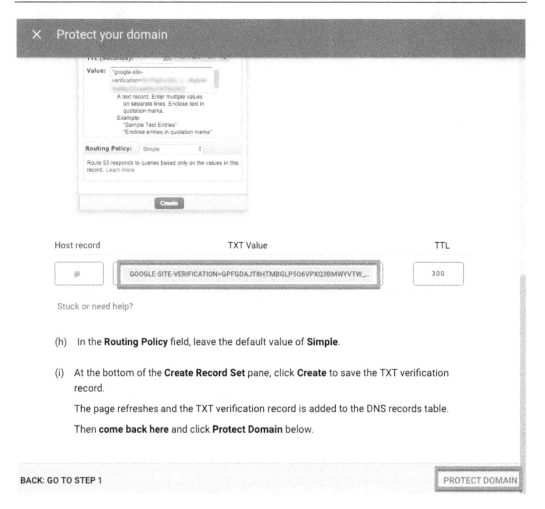

Figure 8.16 – Protect your domain – step 2

9. The instructions in *Figure 8.16* will create a new record for your domain registrar. Again, *Figure 8.17* may look different for your domain registrar. The following example shows the result on AWS:

Figure 8.17 – AWS Route 53 administration page

After the domain is protected, the next step is optional. Just like we did for Azure and AWS, let's create some users in the organization using Cloud Identity:

1.   On the **Welcome** page for Cloud Identity, click **CREATE**:

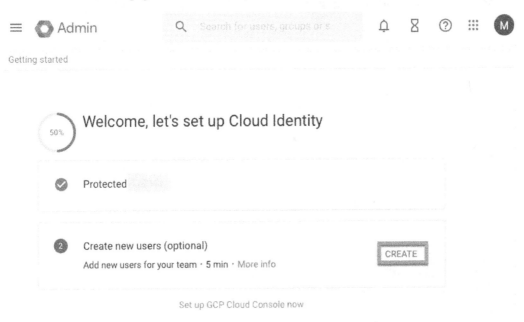

Figure 8.18 – Cloud Identity setup

2. On the **Add new user** page, enter the user info and click **ADD NEW USER**:

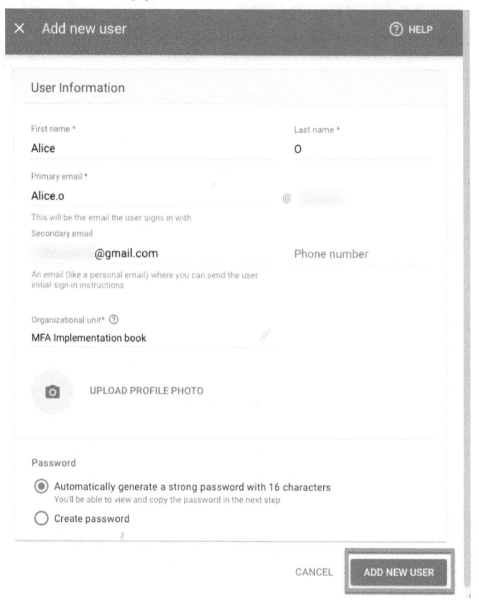

Figure 8.19 – Add new user page

3. Copy the new password, if required. Click on **DONE** or **ADD ANOTHER USER** if more users are required:

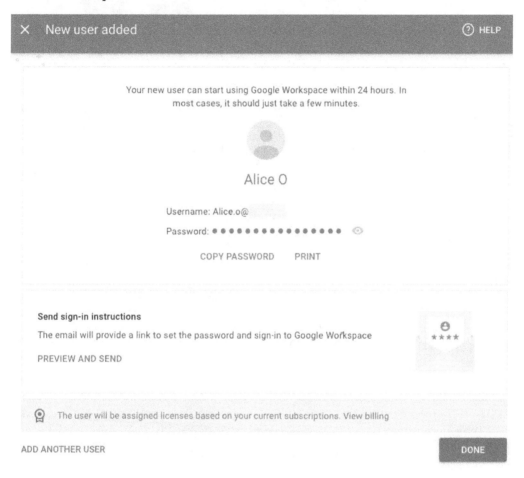

Figure 8.20 – New user confirmation page

In addition to manually creating users, the **Bulk update users** page allows you to upload users from a CSV file. If selected, that option allows the download of a CSV file template.

4. After updating the template with the required users, attach the new CSV file and click **UPLOAD**:

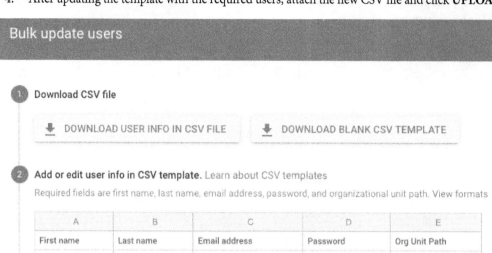

Figure 8.21 – Bulk update users page

We are almost done. The final step is to agree to the terms of service for Cloud Identity:

1.   Click on **Set up GCP Cloud Console now**:

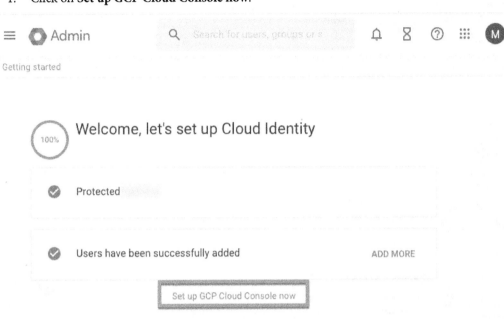

Figure 8.22 – Cloud Identity setup

2.   Click **AGREE AND CONTINUE:**

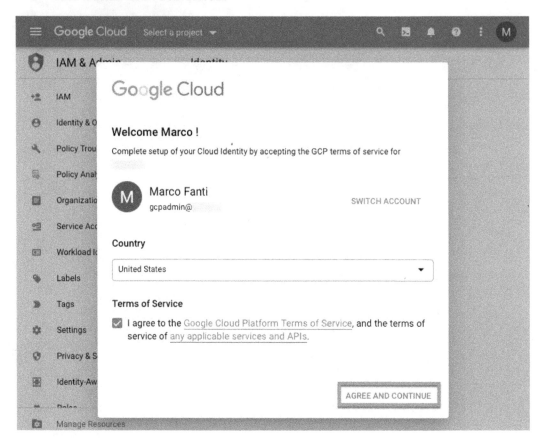

Figure 8.23 – Cloud Identity setup terms of service agreement

That's it, the basic setup for Google Cloud Identity is complete.

Managing user accounts and administrative functions

With the basic Cloud Identity setup complete for an organization, we are going to look at some of the recommended steps from Google for managing user accounts for employees of the organization while using an organizational structure and different roles for the different administrative functions as required by the organization:

1. Log in to `https://console.cloud.google.com/cloud-setup` using the administrator account. Click **GO TO THE CHECKLIST**:

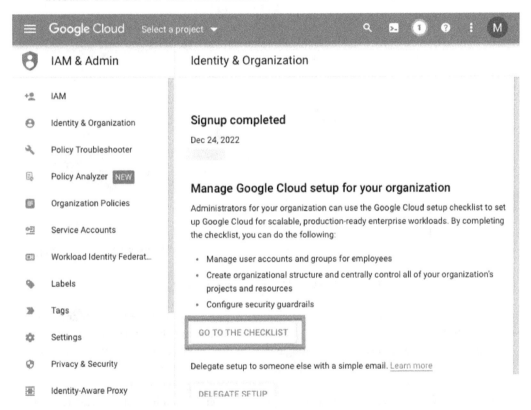

Figure 8.24 – Cloud Identity – Identity & Organization

2. The checkmark next to **Enable Cloud Identity and create an organization** indicates we completed that task already (in the previous section). **Provision users and groups** will guide you through the setup of the recommended set of administrative users and groups to administer the core functions for Acme Software.

    Click **Provision users and groups**:

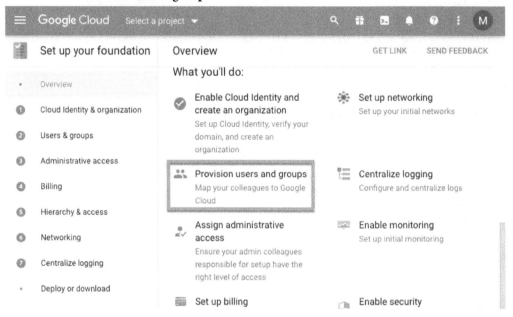

Figure 8.25 – Cloud Identity setup checklist

A Google Group is a way to organize and manage a collection of Google Accounts and service accounts. It is identified by a unique name and email address, such as `grp-gcp-billing-admins@example.com`. With a Google Group, you can easily manage access to resources and communication for a group of people, rather than adding them individually. This checklist step will create the core functions of an organization. The checklist step will also add the current admin user as a member to all the groups. When other administrators are created, it is recommended that they are replaced in the appropriate groups.

3. Click **START USERS & GROUPS**:

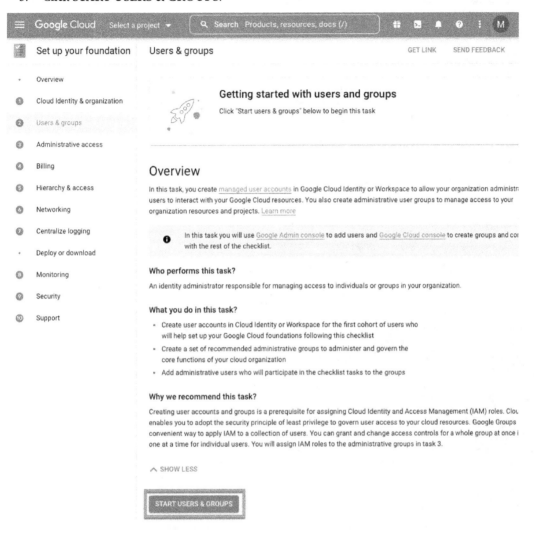

Figure 8.26 – The Getting started with users and groups page

4. Select **CREATE ALL GROUPS** and click **CONTINUE**. This will create all the Google Cloud administrative groups listed on the page:

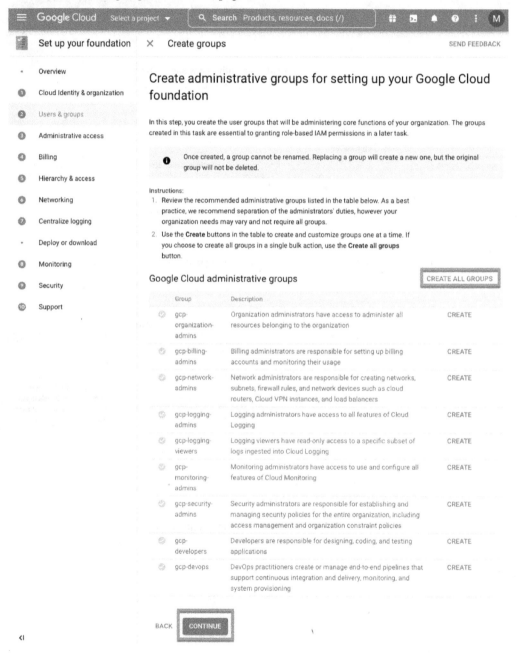

Figure 8.27 – Create groups page

With this, we have created the administrative groups and are ready to administer the organizational tasks using the appropriate roles.

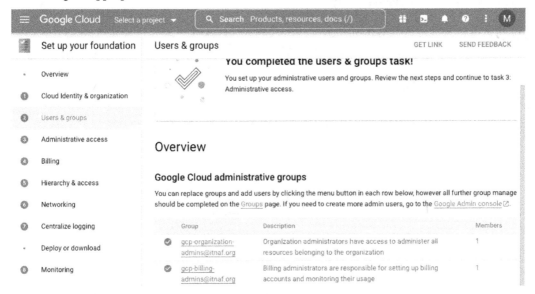

Figure 8.28 – Users & groups page

You can continue following the steps of the checklist (**Administrative access**, **Billing**, and so on), but we are going to skip the remaining steps and set up MFA for our organization.

Setting up MFA in Cloud Identity

Let's now set up MFA in Cloud Identity:

1. Go to https://admin.google.com/. Select **Security**:

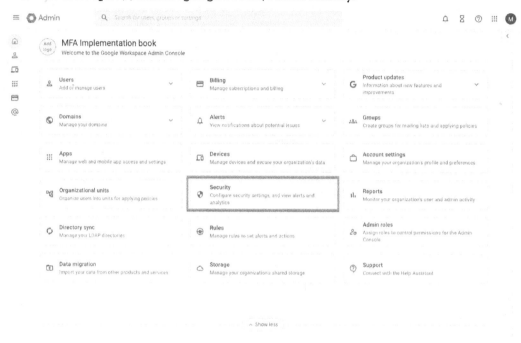

Figure 8.29 – The Admin console

The **Security** tab contains many important settings for Google Cloud Identity. **Alert center**, for example, allows administrators to view, filter, and set up alerts for different types of security issues:

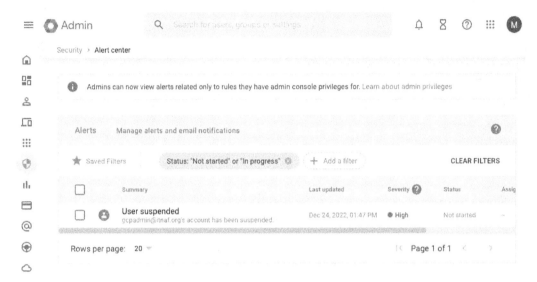

Figure 8.30 – Admin console – Security – Alert center

2.   Select **2-Step Verification**:

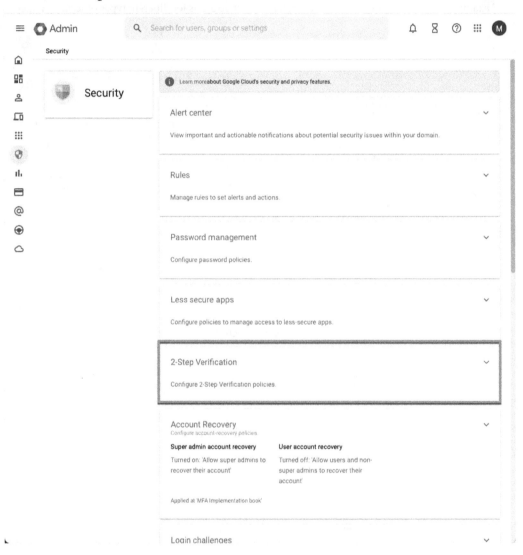

Figure 8.31 – Admin console – Security

Google allows 2-step verification as optional or required for your users, depending on your security needs. For example, we recommend enforcing 2-step verification for administrator accounts, as administrators have the most privileges and access to sensitive information. Additionally, it is a good idea to require 2-step verification for users who work with essential business data, such as financial records and employee information.

To enforce 2-step verification for an organizational unit, select that organizational unit before configuring 2-step verification. Similarly, to enforce 2-step verification only for a chosen child organizational unit or for a configuration group, select that organizational unit or configuration group before performing the configuration steps.

The example in *Figure 8.32* applies to the parent organizational unit, **MFA Implementation book**:

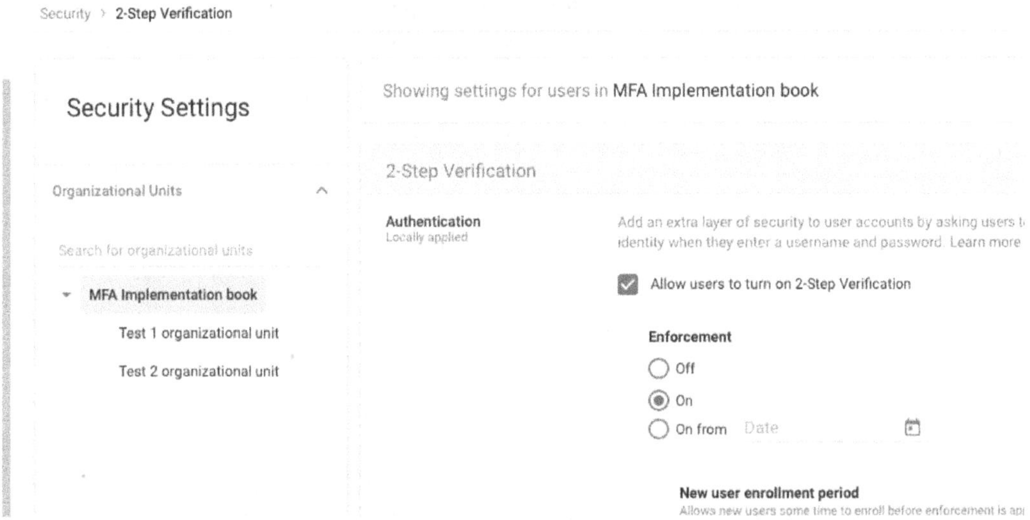

Figure 8.32 – Admin console – Security – 2-Step Verification

Selecting a specific organizational unit, however, as in *Figure 8.33*, only applies to the **Test 1 organizational unit**:

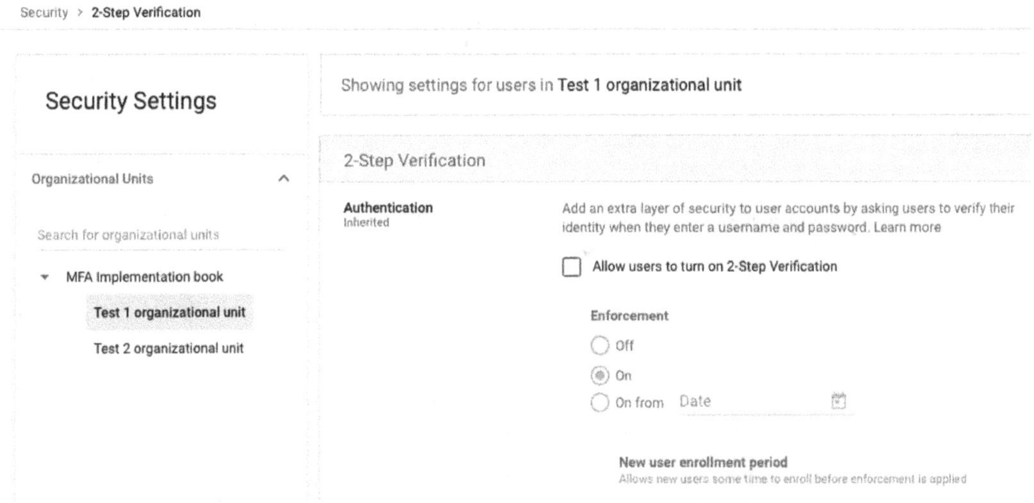

Figure 8.33 – Admin console – Security – 2-Step Verification

After selecting the organizational unit and group (if any), configure 2-step verification for users in that organizational unit and group.

The recommended way to deploy 2-step verification is to start by educating the users about what 2-step verification is and why the organization is doing it. If 2-step verification is required, give users a date by which they must turn on 2-step verification on their account. Specify a verification method if one is required or recommended. You can also deploy 2-step verification for a pilot group of users by following the same recommendation but selecting a specific group or organizational unit.

The following configuration (*Figure 8.34*) implements the recommended way of deploying 2-step verification for all users in **Test 1 organizational unit** by January 31, 2023:

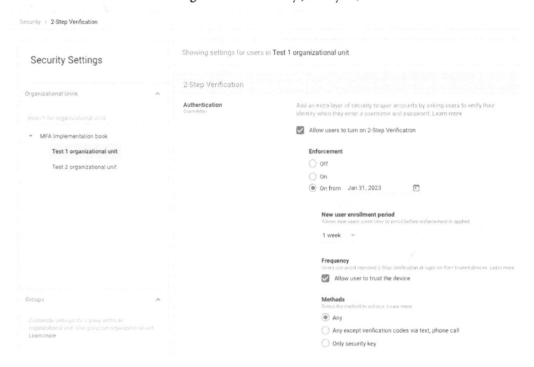

Figure 8.34 – Admin console – Security – 2-Step Verification

The configuration of *Figure 8.35* enforces the use of security keys for members of the **administrators** group:

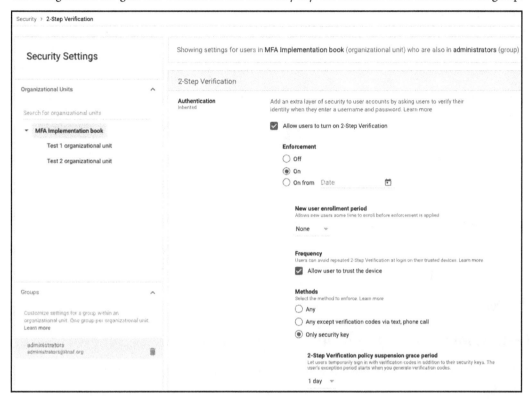

Figure 8.35 – Admin console – Security – 2-Step Verification

> **Important**
>
> When selecting **Only security key** as the allowed method for 2-step verification, make sure the users affected have security keys configured by going to **Reporting | Reports | User Reports | Security**. For more details, see https://support.google.com/a/answer/9176805.

To generate a report of users and their enrollment status and see whether 2-step verification is enforced on them or not, go to **Reporting | Reports | User Reports | Security**:

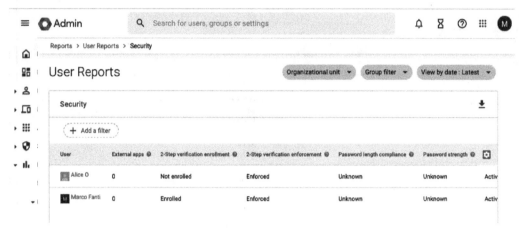

Figure 8.36 – Admin console – Reporting – Reports – User Reports – Security

That concludes our section on configuring MFA (in this case, 2FA specifically) for Google Cloud Identity.

Testing MFA enforcement in Cloud Identity

Now that we have configured 2-step verification for certain users, let's confirm it is being enforced when required.

We are going to create a new user, Bob G., and add that user to **Test 1 organizational unit**. Let's see what happens when he tries to log in to `https://accounts.google.com`.

After logging in with a username and password, Bob is shown the prompt from *Figure 8.37*. If Bob chooses **Do this later**, the registration will be skipped, and he will be redirected to the account page at Google. In this case, Bob chooses to register a security key by clicking on **REGISTER KEY**:

Google

Don't get locked out

Register your Security Key now so you don't lose access to your account.

As of Jan 31, 2023 your domain will require you to use 2-Step Verification to sign in to your account. This makes your account more secure, because you'll use a password and a Security Key to sign in, instead of just a password.

REGISTER KEY

Do this later

Google   Privacy & Terms   Help   Feedback

Figure 8.37 – REGISTER KEY prompt

Bob clicks **NEXT**:

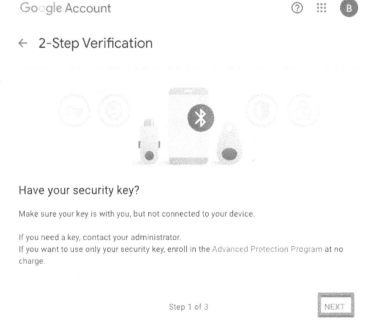

Figure 8.38 – Security key registration

Bob inserts the key on his laptop and touches it as directed:

Register your Security Key

1. Insert your Security Key in your computer's USB port or connect it with a USB cable.

2. Once connected, tap the button or gold disk if your key has one of them.

Figure 8.39 – Security key registration

Bob clicks **Allow**:

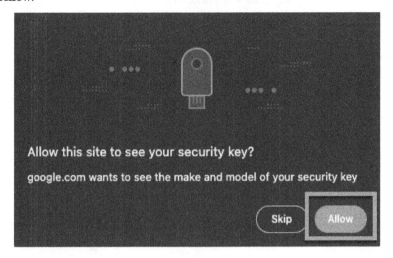

Figure 8.40 – Security key registration

And to complete the registration, Bob gives the security key a unique name and clicks **DONE**:

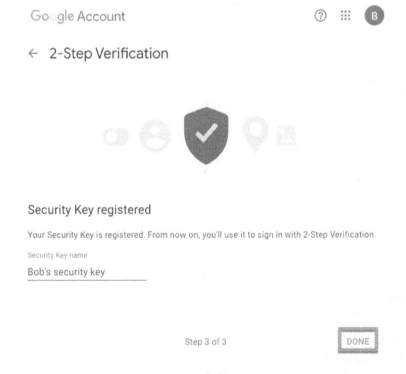

Figure 8.41 – Security key registration completed

After confirming that a security key is now available as a second step after entering the password, Bob signs out from his account:

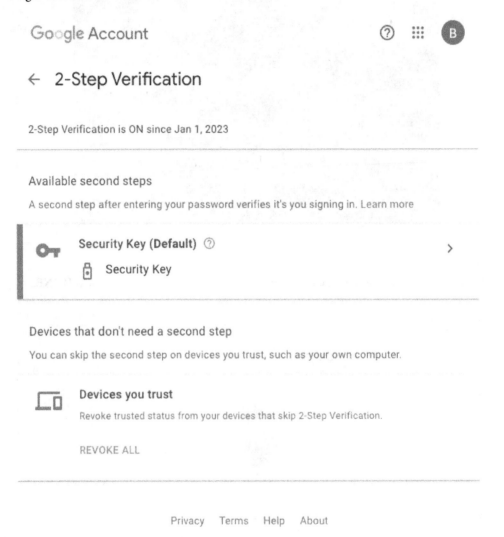

Figure 8.42 – 2-Step Verification setup

Instead of waiting until the end of the month (remember, enforcement was going to start on January 31), we are going to change the security policy and enforce it now, leaving all the other parameters the same:

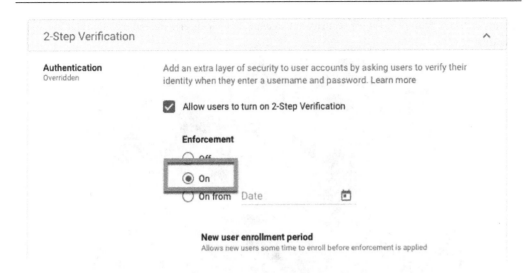

Figure 8.43 – 2-Step Verification configuration for Test 1 organizational unit

Bob signs in to `https://accounts.google.com` again. He enters the username and password, and then clicks **Next**:

Figure 8.44 – Account sign-in at Google

And now Bob gets prompted for his security key as expected. He inserts the security key and touches it to activate:

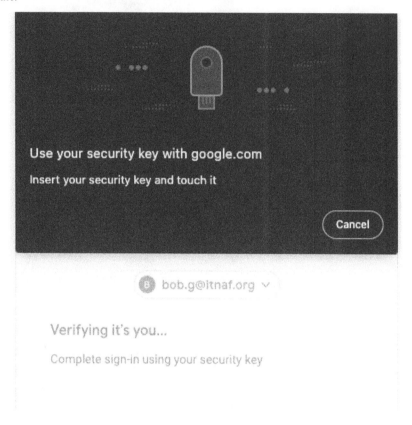

Figure 8.45 – Account sign-in at Google with 2-step verification

After selecting to trust sign-ins from the same laptop in the future without being prompted for 2-step verification, Bob clicks **Next** one more time:

2-Step Verification

To help keep your account safe, Google wants to
make sure it's really you trying to sign in

You're all set

☑   Don't ask again on this device

Next

Figure 8.46 – 2-Step Verification completed

And as expected, Bob gets to his Google account home page:

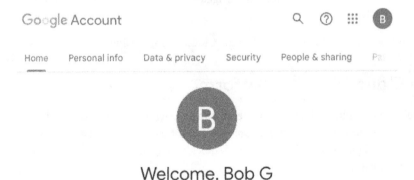

Figure 8.47 – Bob's Google account page

Google Workspace provides access to Google Keep by default. **Google Keep** is a free note-taking and to-do list app developed by Google. It allows users to create notes, lists, and reminders, and to organize and store them in one place. Notes can be text-based, or they can include images, audio recordings, and other media. Google Keep can be accessed through a web browser or through a mobile app, and it can be synced across multiple devices. Users can also set location- and time-based reminders, share notes with others, and collaborate on notes in real time.

To test SSO, Bob enters the URL for Google Keep in his browser (`https://keep.google.com`) without signing out from Google:

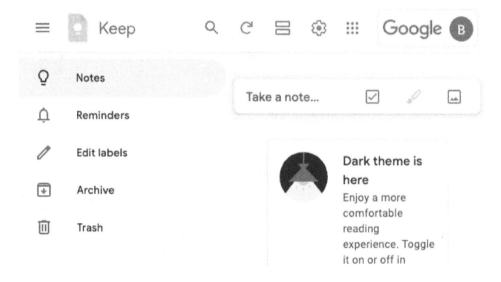

Figure 8.48 – Bob's Google Keep account page

Again, as expected, SSO worked, and Bob can navigate between his SSO-enabled applications without being prompted to sign in again.

This also completes our setup and tests of Google Cloud Identity using MFA (2-step verification).

Google Cloud Identity Platform

Google Cloud Identity Platform is a set of identity and security services provided by Google Cloud that allows organizations to securely manage user identities, access to resources, and application permissions. It is designed to enable developers to quickly build and manage applications that require user authentication and authorization.

Some of the key features of Google Cloud Identity Platform include the following:

- **SSO integration**: Allows users to use a single set of credentials to access all their applications and resources, eliminating the need to remember multiple usernames and passwords

- **User management**: Provides tools for creating, managing, and deleting user accounts, as well as setting up roles and permissions

- **MFA**: Enhances security by requiring users to provide additional verification when logging in, such as a one-time code sent to their phone or a fingerprint scan

- **IAM**: Allows administrators to control who has access to which resources and at what level of access

Google Cloud Identity Platform is built on top of Google's infrastructure and is designed to be scalable, reliable, and secure. As a result, it can be used by organizations of all sizes, from small start-ups to large enterprises.

BeyondCorp

BeyondCorp is a security model developed by Google that shifts the focus of traditional network security from a perimeter-based approach to a user- and device-centric approach. Google introduced BeyondCorp as a response to an evolving enterprise environment. There needed to be more than the traditional perimeter-based security to protect an organization's resources, especially with the rise of remote work, **Bring Your Own Device (BYOD)** policies, and cloud-based services.

BeyondCorp is grounded in the *zero-trust* security principle, which means that no user or device should be automatically trusted. Instead, resource access is granted based on user identity, device information, and other contextual factors, such as location and time.

Google Cloud Identity Platform is an essential building block in implementing the BeyondCorp model within an organization. They work together in the following ways:

- **User authentication and authorization**: Google Cloud Identity Platform offers various authentication methods, including SSO, MFA, and risk-based authentication. These methods help verify users' identities before granting access to resources, which is a fundamental requirement of the BeyondCorp model.

- **Centralized identity management**: Google Cloud Identity Platform provides a centralized platform for managing user identities and access permissions. This makes it easier for organizations to enforce consistent access controls across their applications and resources, in line with the principles of the BeyondCorp model.

- **Access control policies**: Google Cloud Identity Platform allows administrators to define access control policies based on user attributes, device information, and context. These policies are essential for implementing the granular, context-aware access controls required by the BeyondCorp model.

In summary, Google Cloud Identity Platform enables organizations to adopt the BeyondCorp security model by providing robust tools and services for managing user identities and access control. In addition, it serves as the foundation for implementing zero-trust principles and ensuring a more secure user- and device-centric approach to network security.

More information on Google Cloud Identity Platform can be found at `https://cloud.google.com/identity-platform`.

Summary

In this chapter, we discussed MFA on Google Cloud Platform for workforce users using Google Cloud Identity. We tested a few configurations and validated that enforcement is easy to configure. To test everything, we used a couple of SSO-enabled applications provided by Google Workplace.

Finally, we introduced Google Cloud Identity Platform, Google's solution for external customers of the organization.

In the next chapter, we are going to explore the use of Keycloak, an open source IAM product that provides MFA and IAM services to the workforce and customers as well.

MFA without Commercial Products – Doing it All Yourself with Keycloak

Acme Corporation has relied heavily on cloud service providers' commercial products and default services for its **identity and access management** (**IAM**) infrastructure. While these solutions have served their purpose, Acme's management is eager to explore open source alternatives that offer flexibility, cost-effectiveness, and a vital feature set for its expanding business needs.

In this chapter, we will dive into Keycloak, an open source IAM solution licensed under Apache License 2.0 (at the time of writing). Developed and maintained by the Keycloak community, this powerful tool aims to simplify the complex world of authentication and authorization for both customers and the workforce. By examining Keycloak's features, implementation process, and how it compares to commercial products, Acme hopes to make an informed decision on whether it is the right choice for the organization.

In this chapter, we will cover the following topics:

- What is a Keycloak server?
- Running Keycloak using Docker
- Running Keycloak using Java
- Keycloak administration
- Using Keycloak for **single sign-on** (**SSO**)
- Keycloak and **multifactor authentication** (**MFA**)

Technical requirements

Java **software development kit** (**SDK**) version 11 or above is required to run Keycloak on a server. Instructions on how to install Java SDK version 11 can be found in *Appendix A*.

Alternatively, Keycloak can be deployed using container technologies. To run Keycloak using Docker, **Docker Desktop** should be installed (https://www.docker.com/get-started/). If **Podman** is preferred instead of Docker, instructions for running Keycloak with Podman can be found at https://www.keycloak.org/getting-started/getting-started-podman. To run Keycloak using Kubernetes, instructions are available at https://www.keycloak.org/getting-started/getting-started-kube.

Maven is required to build Keycloak from source and also for some of the sample applications available from the Keycloak website.

What is Keycloak?

Keycloak is an open source IAM product. Keycloak provides SSO, user federation, MFA, and identity brokering (think of it as a proxy between your users and some external identity provider or providers). It supports OpenID Connect, OAuth 2.0, SAML 2.0, and social logins/identity providers such as Google, Twitter, Facebook, and so on.

The scope of Keycloak services' usage in this chapter is limited to authentication and MFA. Integration with social logins is not detailed in this chapter, and you are encouraged to refer to the Keycloak documentation for details on that topic.

Running Keycloak using Docker

Container-based deployment provides the quickest way to enable and start using Keycloak. Keycloak supports Docker, Podman, and Kubernetes as containers:

1. To run Keycloak in development mode using Docker, execute the following command: `docker run -p 8081:8080 -e KEYCLOAK_ADMIN=admin -e KEYCLOAK_ADMIN_PASSWORD=admin quay.io/keycloak/keycloak:19.0.3 start-dev`:

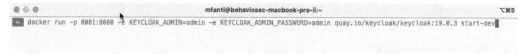

Figure 9.1 – Starting Keycloak using Docker

The developer mode Docker image of Keycloak runs on port 8080. In the preceding command, the -p 8081:8080 parameter exposes port 8081 on the local machine to port 8080 on the Keycloak server. The preceding command also initializes the administrator username and password as admin.

When the ... Listening on: http://0.0.0.0:8080 message is displayed, Keycloak is ready for use:

Figure 9.2 – Running Keycloak

With Keycloak ready and operational, we are ready to start using the Keycloak administration console. If using Docker is the only way you want to run Keycloak, feel free to skip this next section.

Running Keycloak using Java

Keycloak is a Java-based application and can be started from any compute instance as a standard Java runtime application:

1. Download the latest server distribution for Keycloak from the https://www.keycloak.org/downloads website.

> **Important note**
> Keycloak is available in two distributions. One uses **Wildfly**, which is deprecated and will be discontinued in the future. The other uses **Quarkus** and will be the one used in this book.

2. Click on the ZIP (sha1) link of the Keycloak server distribution powered by Quarkus to download the binaries:

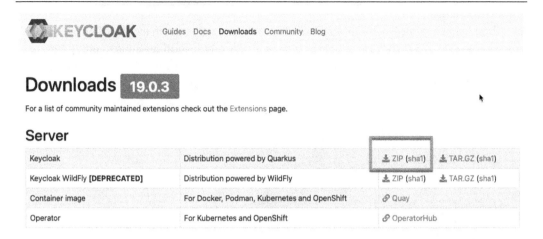

Figure 9.3 – Downloading Keycloak

3. Extract the contents of the file. Change into that directory (cd keycloak-19.0.3):

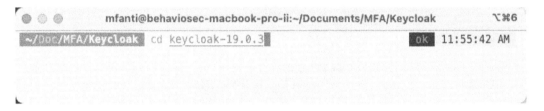

Figure 9.4 – Changing into the Keycloak directory

Keycloak will prompt you to create the initial administrator on your first visit to the welcome page. Alternatively, that can be configured using environment variables.

If the KEYCLOAK_ADMIN and KEYCLOAK_ADMIN_PASSWORD environment variables are set before the server starts, it will use them to create an initial administrator account, and you will not be prompted on the welcome page.

4. To run Keycloak in development mode, execute the following command: bin/kc.sh start-dev --http-port=8081:

Figure 9.5 – Starting Keycloak using Java

The developer mode image of Keycloak runs on port `8080`. In the preceding command, the `--http-port=8081` parameter uses port `8081` on the local machine for the Keycloak server.

When the ... `Listening on: http://0.0.0.0:8081` message is displayed, Keycloak is ready for use:

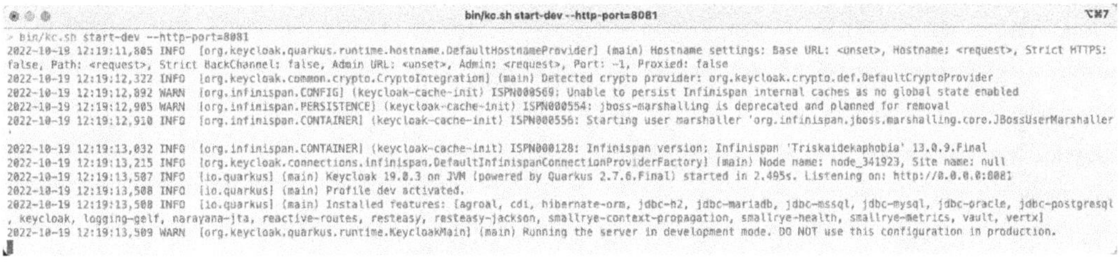

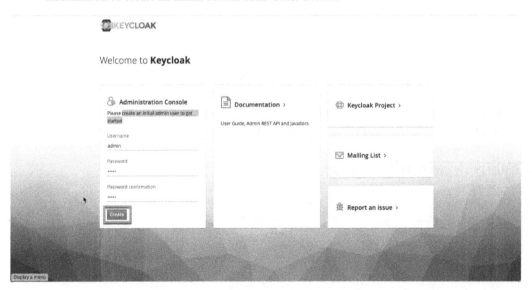

Figure 9.6 – Running Keycloak

Having successfully set up Keycloak and ensured that it's functioning correctly, we are ready to start using it.

5. Open the following URL on your browser: `http://localhost:8081/`. Enter the information to create an initial admin user. Click **Create**:

Figure 9.7 – Keycloak initial admin page

6. The **User created** message indicates that the user creation was successful:

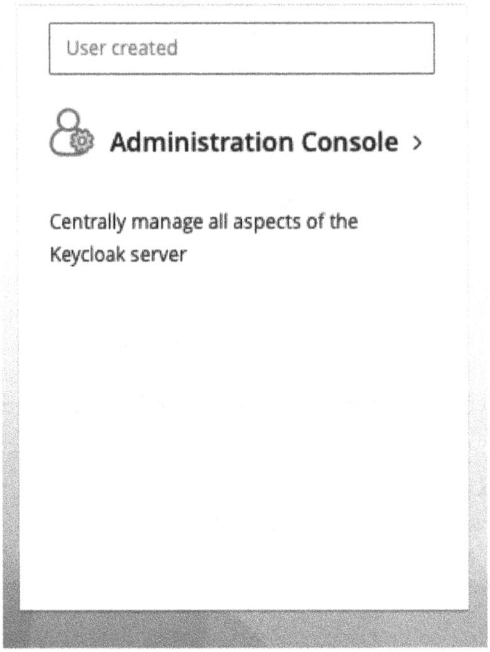

Figure 9.8 – The User created message

With Keycloak now successfully installed and operational, and the initial administrator account established, we are well-prepared to commence the utilization of Keycloak.

Keycloak administration

Keycloak administration and configuration can be done using the Keycloak command-line tool or using the browser and the Keycloak administration console:

1. To access the administration console, open the following URL on your browser, `http://localhost:8081/`, and click **Administration Console**:

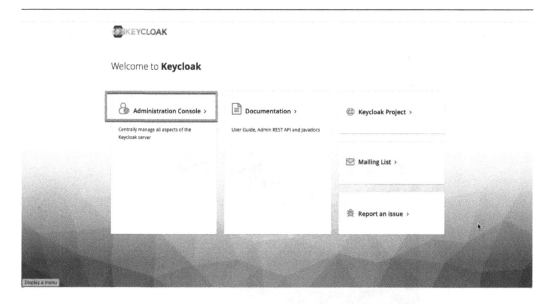

Figure 9.9 – Keycloak welcome page

2. Enter the username and password for the administrator (admin/admin in this example):

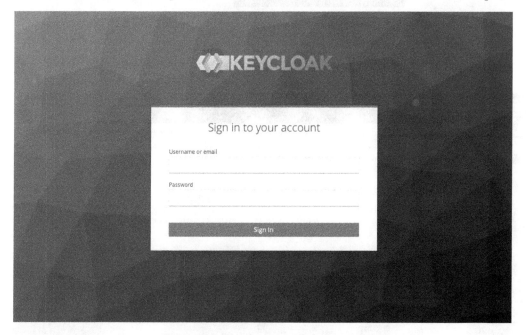

Figure 9.10 – Keycloak console login page

After deploying Keycloak for the first time, it needs to be configured before it can be used. The first recommended configuration is to define a **realm**. A realm is a separate namespace where you configure users, applications, roles, and groups. A user belongs to and logs in to a specific realm. One Keycloak deployment can define multiple realms.

3.   After login, in the top-left corner, click the **Master** realm name and then click **Create Realm**:

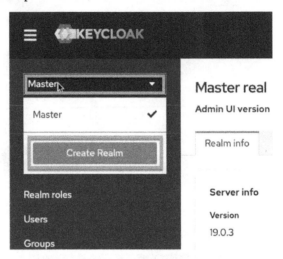

Figure 9.11 – The Create Realm button

4.   On the **Create realm** page, enter a value into the **Realm name** field, toggle **Enabled** to on, and then click **Create**:

Figure 9.12 – The Create realm page

The newly created realm will be immediately selected, and all actions performed will apply to that realm until a new one is selected:

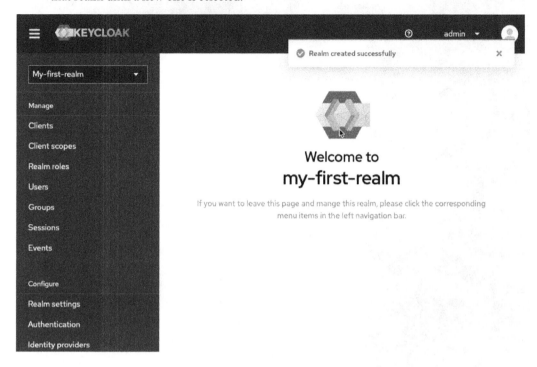

Figure 9.13 – New realm successfully created

After the new realm is created and selected, let us create a user in that realm.

5. Click **Users** to open the **Users** page, and then click **Create new user**:

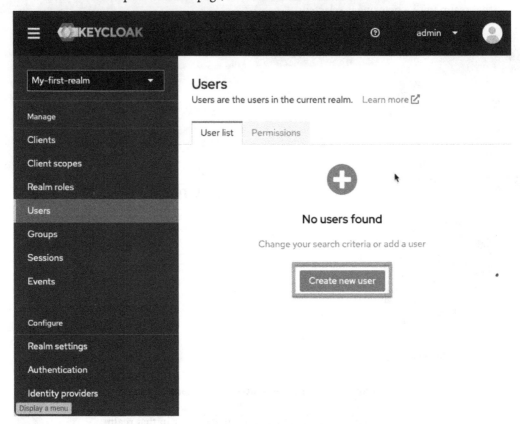

Figure 9.14 – The Users page

6. Enter a value in the **Username** field. If desired, fill out the optional details about the user and toggle **Enabled** to on. Click **Create**:

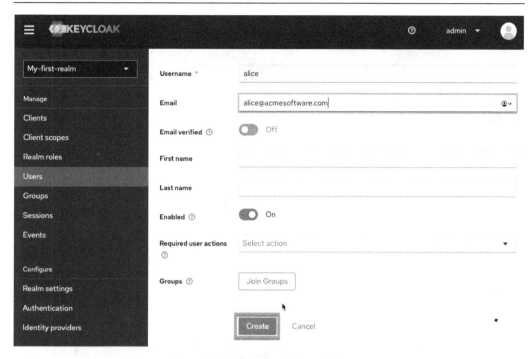

Figure 9.15 – The Create new user page

Once a user account has been successfully created, it's not immediately ready for use. There's an additional step that needs to be carried out before the user can commence with logging in—that's the configuration of user credentials.

7. After the user is created, the **User details** page is displayed. Click the **Credentials** tab to set the password for the user:

Figure 9.16 – The User details page

8. Click **Set password**:

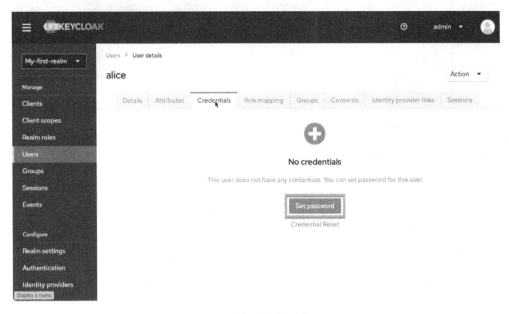

Figure 9.17 – The Credentials page

9. Enter values in the **Password** and **Password confirmation** fields. If the **Temporary** toggle button is on, the user will have to reset the password during the next login. Click **Save**:

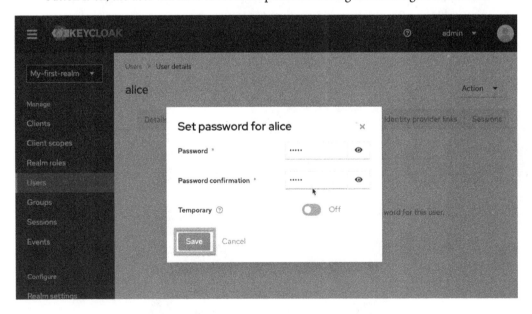

Figure 9.18 – Setting a new password for the user

There are additional configuration actions that can be done to a user account, such as adding the user to a role or group, but those are optional. Setting the password is the minimum required to activate the account and enable us to use it to test the basic functionality of Keycloak:

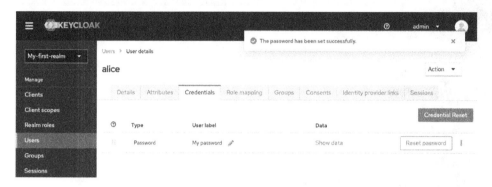

Figure 9.19 – The password is configured for the user

Now that we have created a test user (or multiple users), let us create an application to test the new user.

Using Keycloak for SSO

Applications or services that request authentication from a user are called **clients** in Keycloak. Generally, clients are applications and services that want to use Keycloak as an SSO solution:

1. Click **Clients** in the left menu, and then **Create client**:

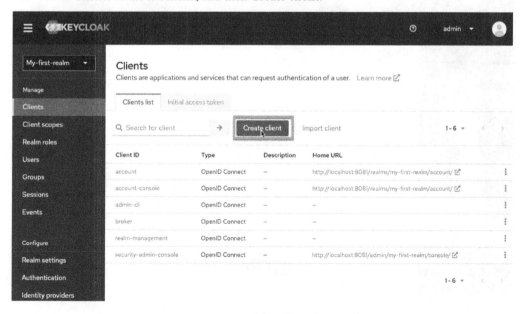

Figure 9.20 – The Clients page

2. Enter a unique **Client ID** value and click **Next**:

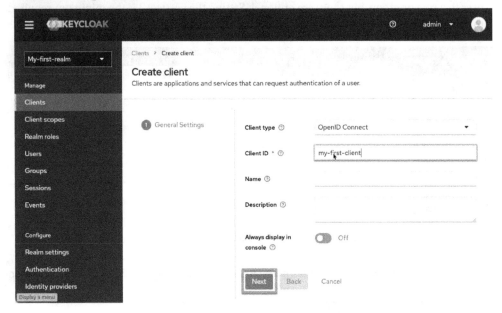

Figure 9.21 – The Create client page

3. Click **Save**:

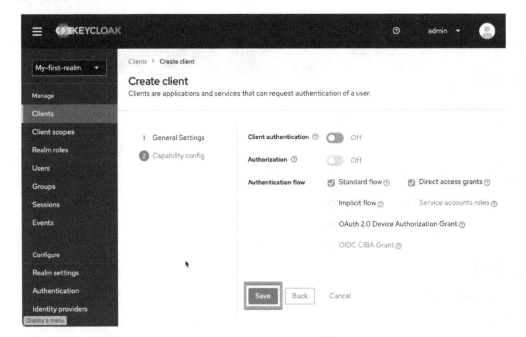

Figure 9.22 – The Create client capability page

After the client is created, it is not very useful unless it protects an application or service. Keycloak provides a test application at `https://www.keycloak.org/app` and we are going to use that application to test Keycloak.

4. Enter `https://www.keycloak.org/app/` in the **Root URL** field. Enter `https://www.keycloak.org/*` in the **Valid redirect URIs** field. Enter `https://www.keycloak.org/app/*` in the **Valid post logout redirect URIs** field. Click **Save**:

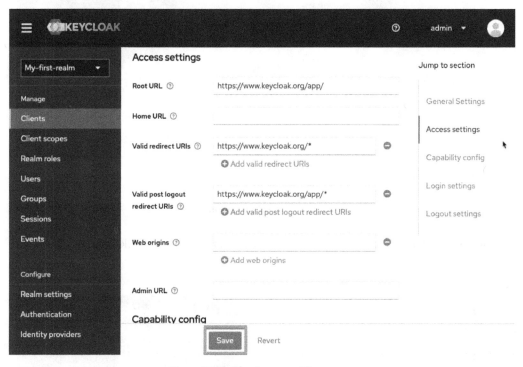

Figure 9.23 – The Access settings page

Finally, let us test the Keycloak test application:

5.   Open the browser at `https://www.keycloak.org/app`:

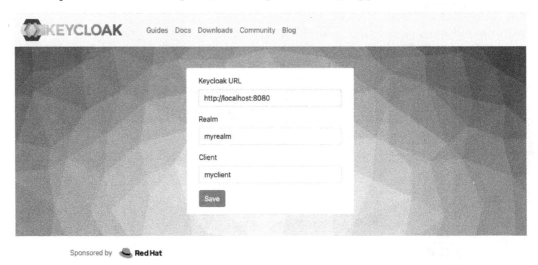

Figure 9.24 – Keycloak test app

6.   Enter the URL for Keycloak, the realm, and the client name to test. Click **Save**:

Figure 9.25 – Keycloak test app

7.   Click **Sign in**:

Figure 9.26 – Configured test app

8.   Enter the username and password for the user created previously. Click **Sign In**:

Figure 9.27 – The test app sign in page

If the username and password are correct, the login should succeed, and the test is complete:

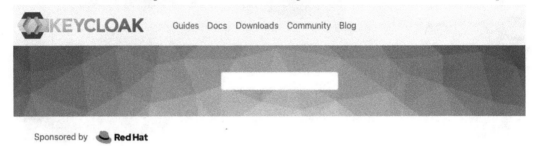

Figure 9.28 – Test app sign in page

Now that we have tested our configuration using the app provided by Keycloak, let us create and deploy our own applications.

Creating and deploying sample applications in Keycloak

Keycloak provides a number of applications and sample code. The latest QuickStart distribution can be found on the **Downloads** page. We are going to test this using a vue.js (https://vuejs.org) application. The npm software registry (https://docs.npmjs.com/about-npm) is required to build and run this example:

1. On the Keycloak console, click **Clients** in the left menu, and then **Create client**:

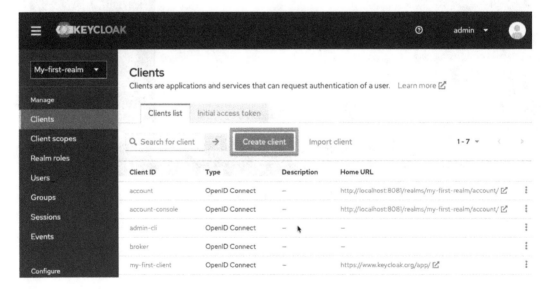

Figure 9.29 – The Clients page

2. Enter a unique value in the **Client ID** field and click **Next**:

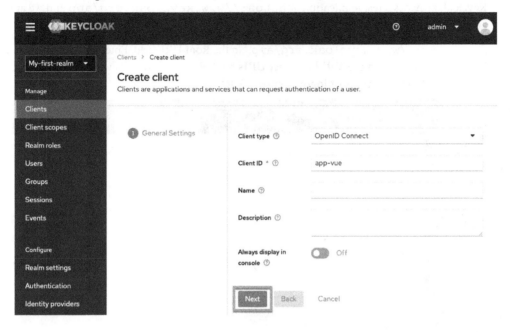

Figure 9.30 – The Create client page

3. Click **Save**:

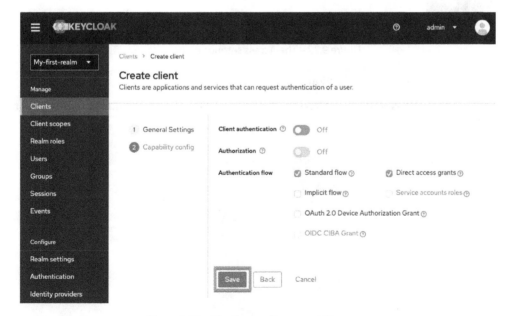

Figure 9.31 – The Create client capability page

After the client is created, it is not very useful unless it protects an application or service. Keycloak provides a test application at `https://www.keycloak.org/app` and we are going to use that application to test Keycloak.

4. Enter `https://www.keycloak.org/app/` in the **Root URL** field. Enter `https://www.keycloak.org/*` in the **Valid redirect URIs** field. Enter `https://www.keycloak.org/app/*` in the **Valid post logout redirect URIs** field. Click **Save**:

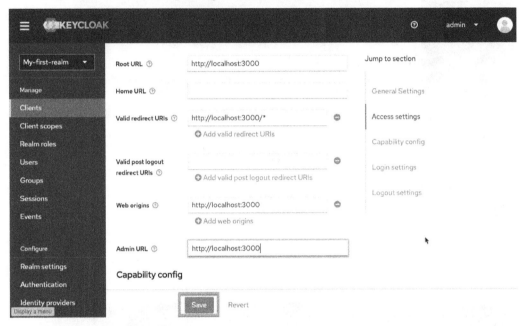

Figure 9.32 – The Access settings page

Next, let us create a test application.

5. Open `https://www.keycloak.org/downloads`. Download the **Quickstarts distribution** ZIP file:

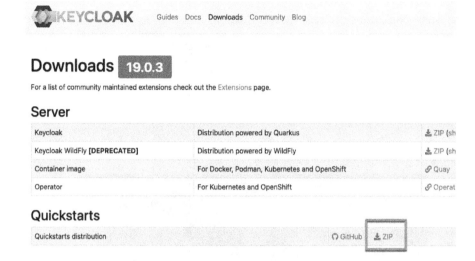

Figure 9.33 – The Keycloak Downloads page

6.   Extract the contents of the ZIP file:

Figure 9.34 – The Keycloak Downloads page

7. Change directory to `keycloak-quickstarts-latest/applications/app-vue`:

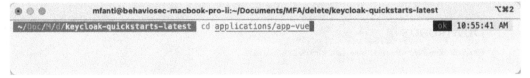

Figure 9.35 – The Quickstarts distribution folder

8. Edit the `main.js` file in the `src` directory:

Figure 9.36 – The app-vue directory

9. Change the default values in `initOptions` (the URL for Keycloak, the realm, and the client name) to match the client defined previously:

```
JS main.js    ✕

src > JS main.js > ...
  1    import Vue from 'vue'
  2    import App from './App.vue'
  3    import VueLogger from 'vuejs-logger';
  4    import * as Keycloak from 'keycloak-js';
  5
  6    Vue.use(VueLogger);
  7
  8    let initOptions = {
  9      url: 'http://127.0.0.1:8081/', realm: 'my-first-ream', clientId: 'app-vue', onLoad: 'login-required'
 10    }
 11
 12    let keycloak = Keycloak(initOptions);
 13    |                                                             I
 14    keycloak.init({ onLoad: initOptions.onLoad }).then((auth) => {
 15      if (!auth) {
 16        window.location.reload();
 17      } else {
 18        Vue.$log.info("Authenticated");
 19
 20        new Vue({
 21          el: '#app',
 22          render: h => h(App, { props: { keycloak: keycloak } })
 23        })
 24      }
```

Figure 9.37 – initOptions in main.js

After `main.js` is saved, we can build and start the test application.

10. Run `npm install`, then run `npm run dev`:

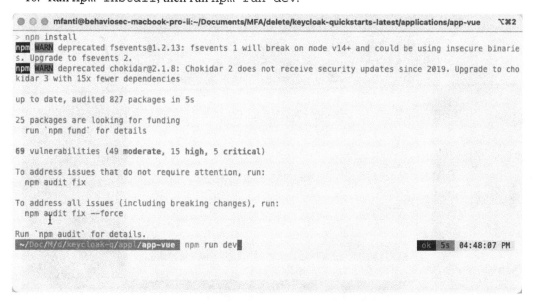

Figure 9.38 – Building and running the test app

11. Open the browser at `http://localhost:3000/` if the previous command didn't open it automatically:

Figure 9.39 – The test app sign-in page

12. Enter the username and password for the user created previously. Click **Sign In**:

Figure 9.40 – The test app sign-in page

If the username and password are correct, the login should succeed, and the test is complete:

Figure 9.41 – The Vue.js test app

We have now tested password login using the app provided by Keycloak, as well as the one we created ourselves. It's time to make access to the applications more secure using MFA.

Keycloak and MFA

The simplest way to enable MFA in Keycloak is on a user-by-user basis. When a user is created, an administrator can also configure additional steps that must be performed during the first login. Those include **one-time password** (OTP) and **Web Authentication** (**WebAuthn**) registration:

1. In the Keycloak console, go to **Users | Create User**. Enter a value in the **Username** field and select **Configure OTP** in the **Required user actions** drop-down menu:

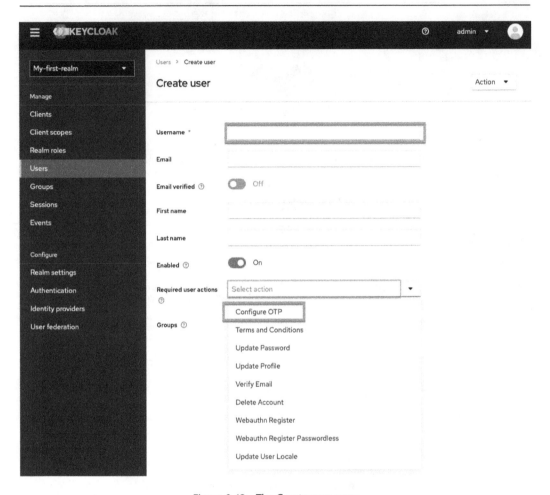

Figure 9.42 – The Create user page

2. Click **Create**:

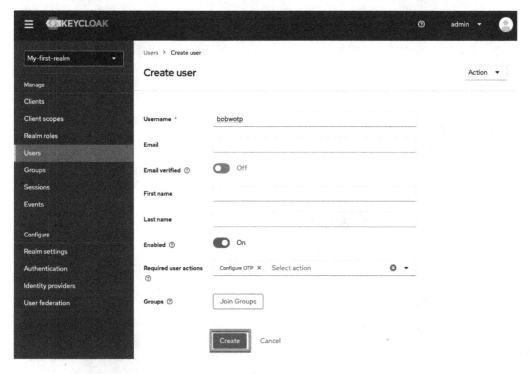

Figure 9.43 – The Create user page

After setting the credentials for the user, let's try accessing the same application we tested before.

3. Open `http://localhost:3000/`. Enter the username and password for the user we just created. After the sign-in page, the **Mobile Authenticator Setup** page opens, as this is the first time this user is logging in, and the user doesn't have an OTP authenticator configured.

4. Scan the QR code with FreeOTP or Google Authenticator:

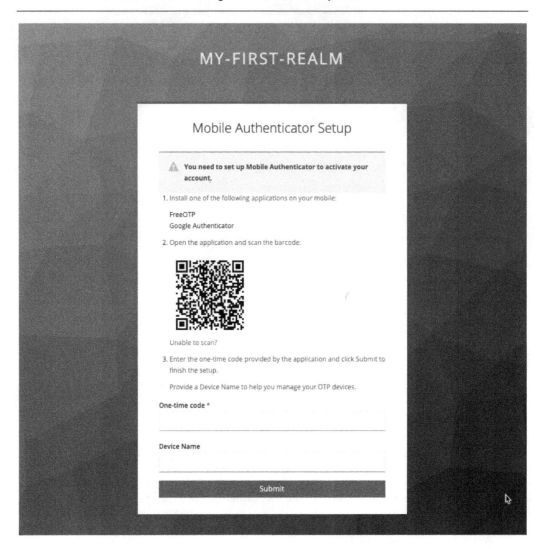

Figure 9.44 – The Mobile Authenticator Setup page

5. Enter the one-time code displayed by FreeOTP or Google Authenticator.

6. Your access to the application is granted.

    If you log out of the application and try to log in again with a user that has an OTP authenticator configured, you won't see the mobile authenticator setup after entering the username and password. The one-time code prompt is shown automatically, as this user has already registered the authenticator.

7. The username used on the sign-in page is displayed again on the second sign-in page. Enter the one-time code, and click **Sign In**:

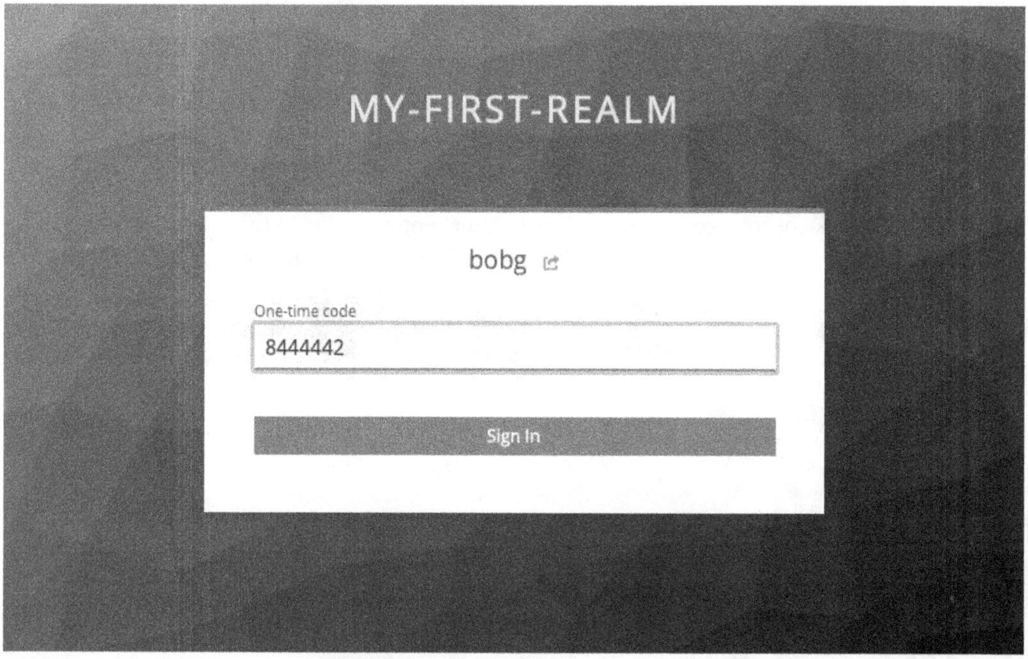

Figure 9.45 – The One-time code prompt

And again, we are allowed access to the application:

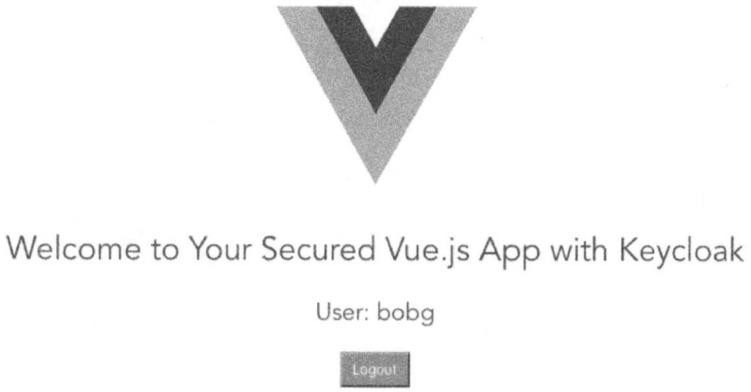

Welcome to Your Secured Vue.js App with Keycloak

User: bobg

Logout

Figure 9.46 – The Vue.js app

If you try logging in with a user that has an OTP authenticator configured, you will be presented with the one-time code prompt. If the user doesn't have an OTP authenticator configured (and it is not required for that user), then a simple username and password will suffice to sign in successfully.

Let's now create a third user, one that requires WebAuthn. WebAuthn is an open standard for secure and passwordless authentication on the web. Developed by the **World Wide Web Consortium (W3C)** and the **Fast Identity Online (FIDO)** Alliance, WebAuthn aims to provide a more secure and user-friendly method for authenticating users by leveraging public key cryptography:

1. In the Keycloak console, go to **Users | Create User**. Enter a value in the **Username** field and select **Webauthn Register** in the **Required user actions** drop-down menu:

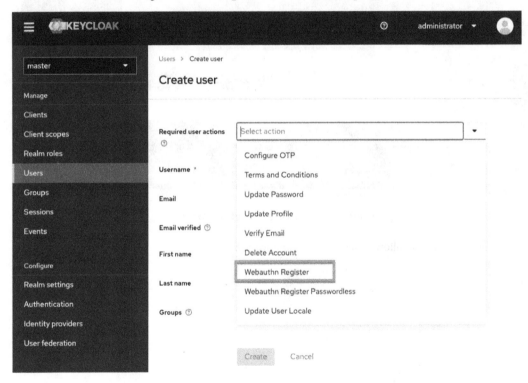

Figure 9.47 – The Create user page – Required user actions

2.   Click **Create**:

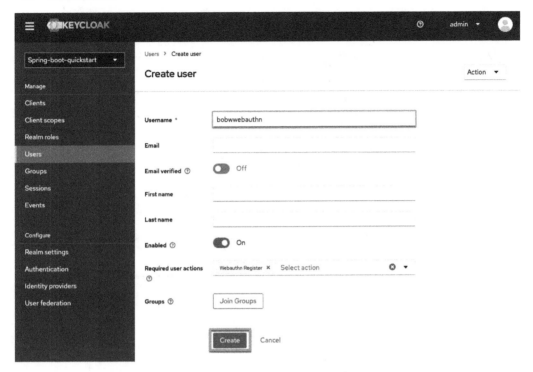

Figure 9.48 – The Create user page – Create

3.   Open http://localhost:3000/. Enter the username and password for the user we just created. After the sign-in page, the **Security Key Registration** page opens, as this is the first time this user is logging in, and the user doesn't have WebAuthn configured. Click **Register**:

Figure 9.49 – The Security Key Registration page

4. Select **USB security key** if one is available:

Verify your identity with localhost
Pick an option
Use phone with a QR code
USB security key
This device
Cancel

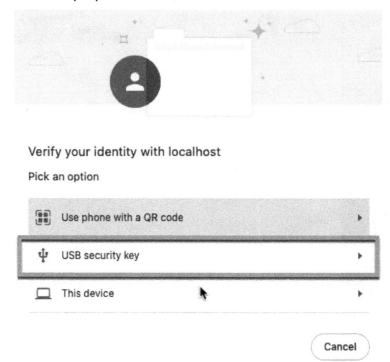

Figure 9.50 – The security key selection page

5. Click **OK**:

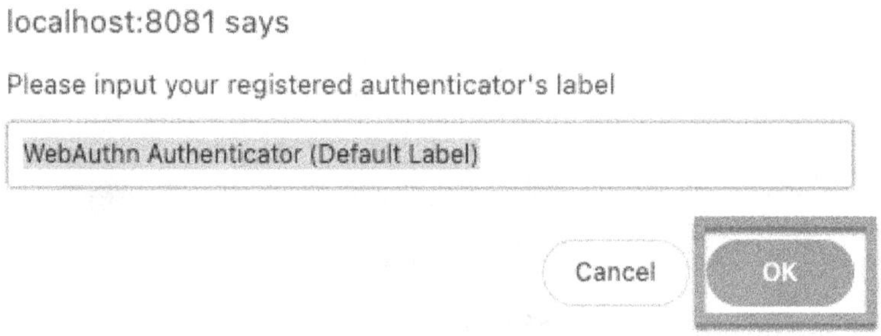

localhost:8081 says
Please input your registered authenticator's label
WebAuthn Authenticator (Default Label)
Cancel
OK

Figure 9.51 – The security key label page

When you finish configuring the security key, you will be allowed access to the application.

If you try logging in with a user that has WebAuthn configured, you will not be presented with a prompt, unlike users with an OTP authenticator configured. Why is that the case?

Keycloak realms allow for different authentication flows. One of those flows is the **browser flow**. Let's take a look at the default browser authentication flow for the **My-first-realm** realm:

1. Select the **My-first-realm** realm and **Authentication**:

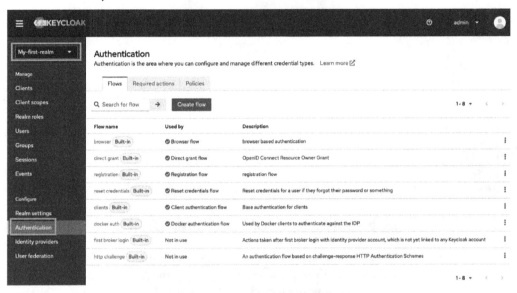

Figure 9.52 – Realm authentication configuration

2. Click on the browser link:

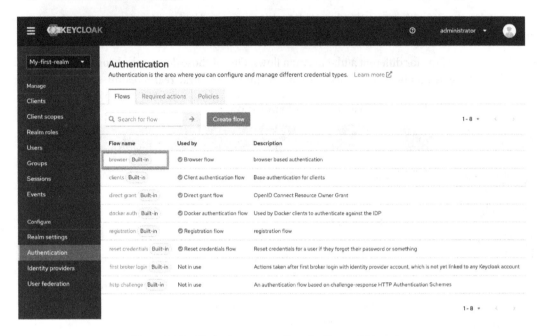

Figure 9.53 – Realm browser authentication configuration

The browser authentication flow contains several steps used for browser-based authentication. Some steps, such as **Kerberos**, are disabled and are not used. The **Alternative** steps are steps that are sufficient to allow the flow to proceed if valid. **Required** are steps that must be true for the flow to proceed.

The browser flow accepts **Cookie** as the first alternative to evaluate whether the user is authenticated. If a valid cookie is present, the user is not presented with a sign-in page.

If cookies are not present, the flow proceeds down the list. The next alternative is **Identity Provider Redirector** if a different identity provider is required.

The final alternative, **forms**, is the one we have used for sign-in. As part of the **forms** sub-flow, **Username Password Form** is required, meaning that we need to provide a valid username and password for the sub-flow to proceed. If the username and password are correct, the next sub-flow is conditional on an OTP being configured for the user. If that is the case, then the user must also present a valid OTP code:

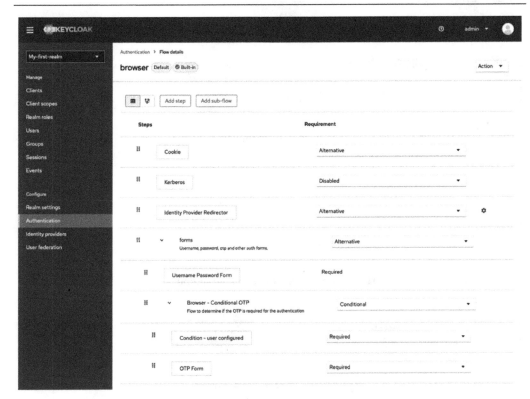

Figure 9.54 – The test app sign-in page

As seen in the **Default** browser flow, users with an OTP configured must enter an OTP code every time. Users with WebAuthn configured, however, do not need to use their security keys at all.

MFA with required OTP

To require an OTP (in addition to a password) for all users, a simple change to the browser flow is required:

1. With the browser flow open, select **Duplicate** as the action in the top-right corner of the page:

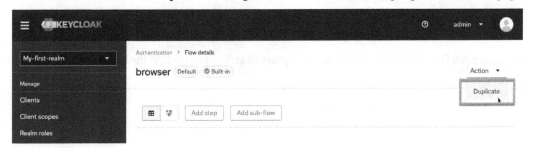

Figure 9.55 – Duplicating the browser flow

2. Give the new flow a name. Click **Duplicate**:

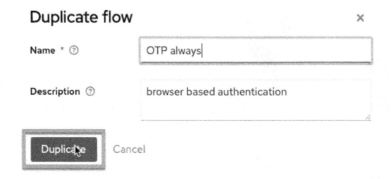

Figure 9.56 – Duplicate flow naming

3. Select **Required** in the **OTP always Browser** drop-down menu:

Figure 9.57 – Test app sign-in page

4. Click the **Action** drop-down menu in the top-right corner of the page and select **Bind flow**:

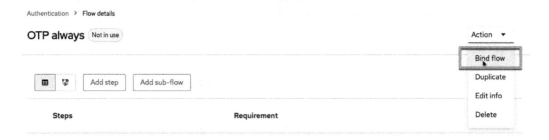

Figure 9.58 – Bind the new flow

That's it. Now the **OTP always** flow will be executed every time the users log in to the **My-first-realm** realm.

If we try to sign in to the application with a user that does not have an OTP authenticator registered, now all users will be required to use the OTP authenticator for every login. Users that do not have authenticators configured will have to configure them before completing the sign-in to the application:

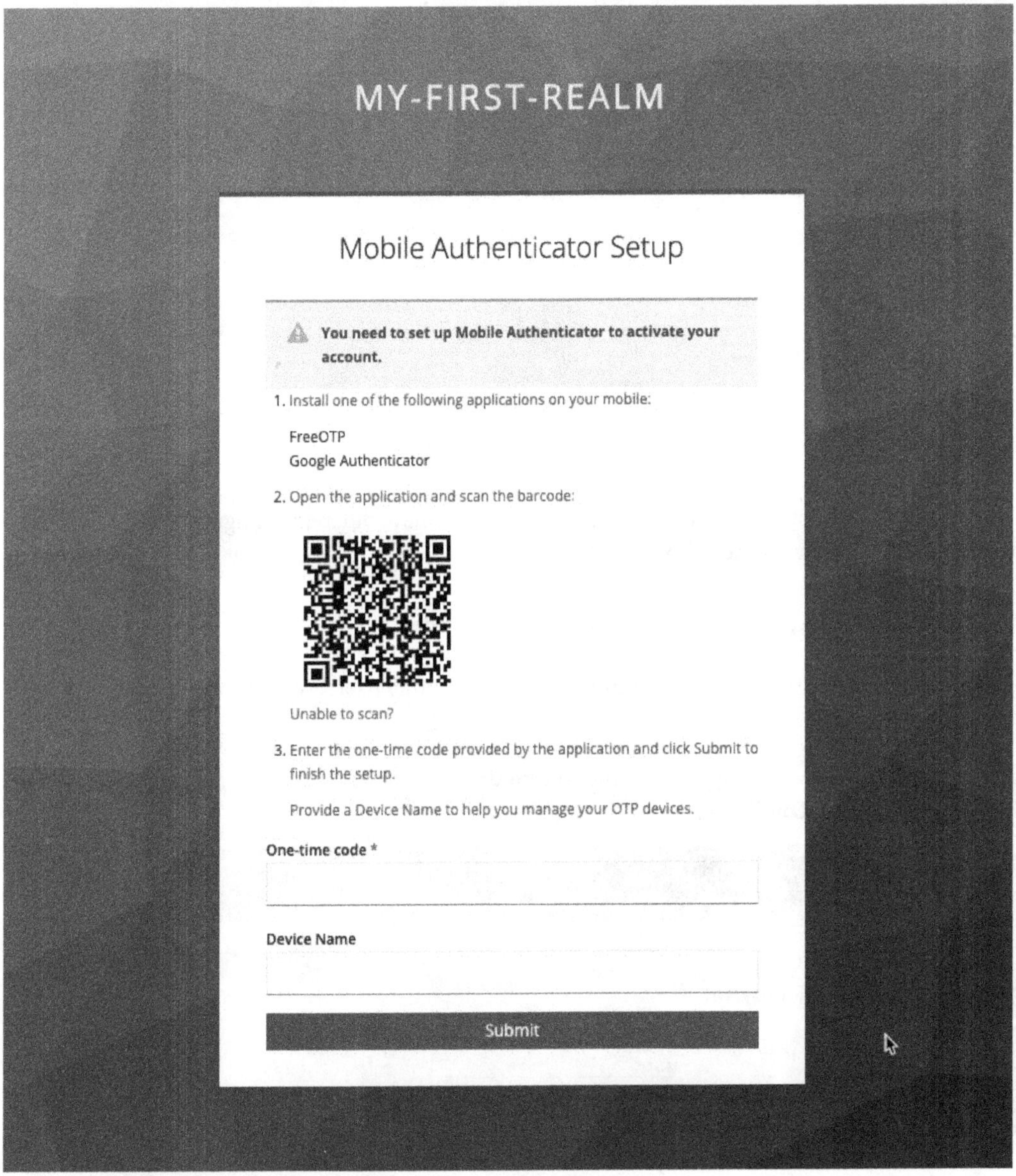

Figure 9.59 – Mandatory authenticator setup

After authenticator registration (or OTP code validation), users will be allowed access to the application:

Welcome to Your Secured Vue.js App with Keycloak

User: alice

Logout

Figure 9.60 – The Vue.js app

Modifying the browser authentication flow allows Acme to achieve different configurations for different applications, depending on the realms that they are created on.

MFA with OTP or passwordless WebAuthn

To allow users to select passwordless WebAuthn or OTP for all users, the following authentication flow can be used:

1. With the browser flow open, click the **Action** drop-down menu in the top-right corner of the page and select **Duplicate**:

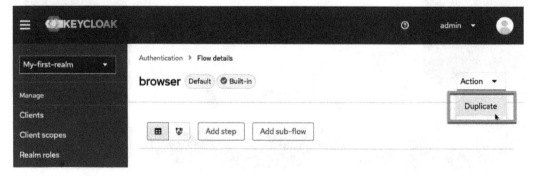

Figure 9.61 – Duplicating the browser flow

2. Give the new flow a name. Click **Duplicate**:

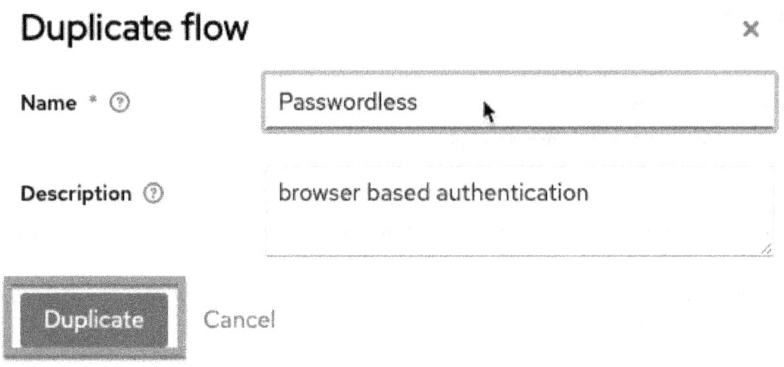

Duplicate flow   ✕

Name * ⑦   Passwordless   ▸

Description ⑦   browser based authentication

[Duplicate]   Cancel

Figure 9.62 – Duplicate flow

3. For the **Passwordless forms** step, click + and then **Add step**:

| ⠿ | Identity Provider Redirector | Alternative | ▾ | ⚙ | 🗑 |

Add

| ⠿ ⌄ | Passwordless forms
Username, password, otp and other auth forms. | Alternative | ▾ | + ▾ | ✎ | 🗑 |

Add step
Add condition
Add sub-flow

| ⠿ | Username Password Form | Required |

| ⠿ ⌄ | Passwordless Browser - Conditional OTP
Flow to determine if the OTP is required for the authentication | Conditional | ▾ | + ▾ | ✎ | 🗑 |

Figure 9.63 – Add step

4.   Select **Username Form** and then click **Add**:

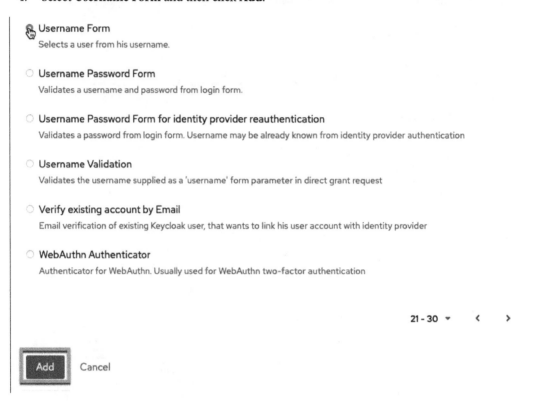

Figure 9.64 – The Username Form step

5.   Drag **Username Form** above **Username Password Form** and click the delete button on **Username Password Form**:

Figure 9.65 – Delete Username Password Form

6. Confirm the deletion of **Username Password Form** by clicking **Delete**:

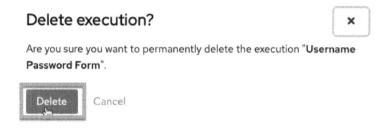

Figure 9.66 – Delete Username Password Form

7. For the **Passwordless Browser – Conditional OTP** step, click on + and then **Add step**:

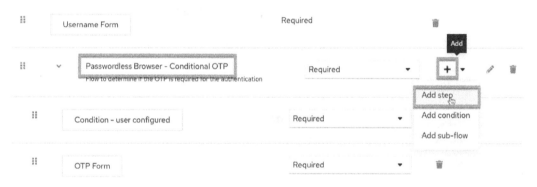

Figure 9.67 – Add step

8. Select **WebAuthn Passwordless Authenticator** and then click **Add**:

Figure 9.68 – The WebAuthn Passwordless Authenticator step

9. Click the delete button on **Condition – user configured**:

Figure 9.69 – Delete Condition – user configured

10. Confirm the deletion of **Condition – user configured** by clicking **Delete**:

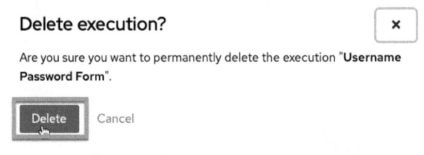

Figure 9.70 – Delete Condition – user configured

11. Bind the new flow to the browser authentication flow by clicking the **Action** drop-down menu in the top-right corner of the page and selecting **Bind flow**:

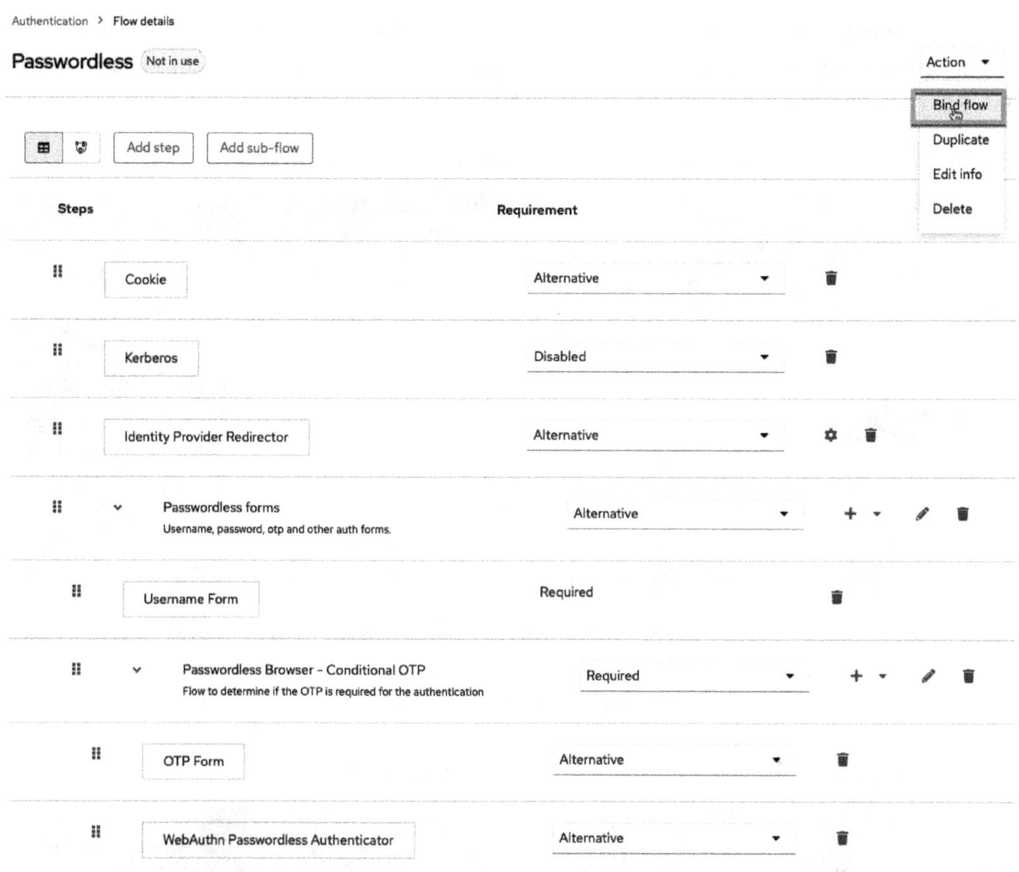

Authentication > Flow details

Passwordless Not in use

Action ▼

Bind flow

Duplicate

Edit info

Delete

Steps	Requirement	
⊞		
⊞ Cookie	Alternative ▼	🗑
⊞ Kerberos	Disabled ▼	🗑
⊞ Identity Provider Redirector	Alternative ▼	⚙ 🗑
⊞ ⌄ Passwordless forms Username, password, otp and other auth forms.	Alternative ▼	+ ▾ ✏ 🗑
⊞ Username Form	Required	🗑
⊞ ⌄ Passwordless Browser - Conditional OTP Flow to determine if the OTP is required for the authentication	Required ▼	+ ▾ ✏ 🗑
⊞ OTP Form	Alternative ▼	🗑
⊞ WebAuthn Passwordless Authenticator	Alternative ▼	🗑

Add step   Add sub-flow

Figure 9.71 – Passwordless authentication flow

That's it. Now the authentication allows the user to select to use an OTP or WebAuthn for a passwordless login in the **My-first-realm** realm.

If we try to sign in to the application with a user that does not have an OTP authenticator registered, they will have to configure it before completing the sign-in to the application. The same applies if the user has not configured a security key. Let's try the test application again:

1. Go to http://localhost:3000. Enter a username and click **Sign In**:

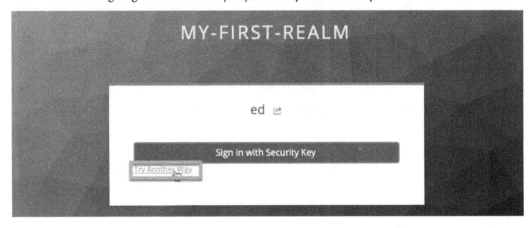

Figure 9.72 – Username prompt

2. Instead of signing in with a security key, click **Try Another Way**:

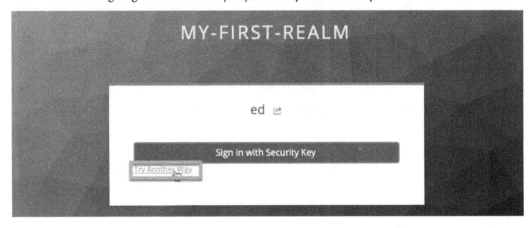

Figure 9.73 – Sign in with Security Key

3. Enter the OTP code and click **Sign In**:

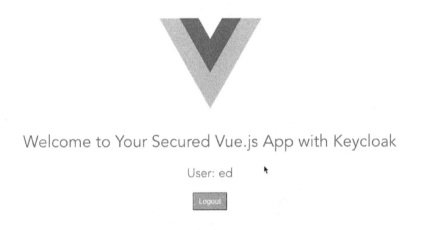

Figure 9.74 – The OTP prompt

After authenticator registration (or OTP code validation), users will be allowed access to the application:

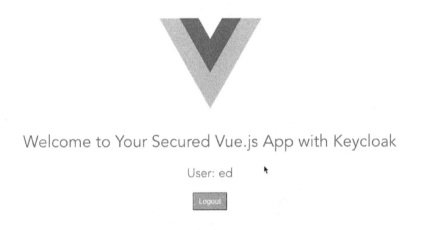

Welcome to Your Secured Vue.js App with Keycloak

User: ed

Logout

Figure 9.75 – The Vue.js app

Modifying the browser authentication flow allows Acme to achieve different configurations for different applications, depending on the realms that they are created on.

This completes our configuration and test of Keycloak using MFA.

Summary

This chapter described the installation and configuration of Keycloak, an open source product, and how Acme has used mobile OTP authenticators and passwordless WebAuthn to achieve a higher level of security for its workforce.

In the next chapter, we will explore what to do when MFA is not enough. MFA is only part of an IAM solution, but sometimes MFA (or certain types of MFA) must not be used or only used in conjunction with other products.

Part 3:
Proven Implementation Strategies and Deploying Cutting-Edge Technologies

Part 3 of the book delves into the exciting realm of cutting-edge technologies and innovative new products shaping the future of authentication security to help your company stay ahead of the curve in an ever-evolving cybersecurity landscape. It also covers effective implementation strategies with best practices and proven tactics for successfully integrating MFA solutions into your organization's security infrastructure.

This part has the following chapters:

- *Chapter 10, Implementing MFA in the Real World*
- *Chapter 11, The Future of (Multifactor) Authentication*

10

Implementing MFA in the Real World

By reading this chapter, you will gain a comprehensive understanding of cybersecurity and information security concepts, the importance of authentication, and the business impact of cybersecurity. You will also obtain a deeper understanding of the legal and ethical responsibilities of directors and officers of an organization to prioritize security and protect the assets and sensitive information of the organization. In the chapter, we also discuss the evolution of web applications and the requirement to evolve cybersecurity, as well as strategies to implement MFA successfully in the real world.

We are going to cover the following topics:

- Understanding the business side of cybersecurity
- Strengthening cybersecurity
- Strategies for implementing MFA

Technical requirements

There are no technical requirements for this chapter.

Understanding the business side of cybersecurity

By now, you have heard of cybersecurity many times in this book, so let's recap. Cybersecurity protects computer systems, networks, and data from unauthorized access, use, disclosure, disruption, modification, or destruction. It involves a combination of technologies, processes, and practices designed to secure sensitive information and prevent cyber threats.

Information security is a broader term that refers to the protection of information and information systems from unauthorized access, use, disclosure, disruption, modification, or destruction. This includes physical security measures and digital security measures to ensure the confidentiality, integrity, and availability of information, the CIA Triad.

These three objectives form the foundation of a comprehensive cybersecurity strategy and are considered to be the cornerstones of information security.

Authentication is essential for maintaining confidentiality, as it verifies the identity of users before granting access to sensitive information. By implementing secure authentication methods, such as multifactor authentication, organizations can ensure that only authorized individuals have access to sensitive information, reducing the risk of unauthorized disclosure.

Cybersecurity is no longer just a technical issue for IT departments to handle. Instead, it has become a social phenomenon, with investors, employees, customers, and governments all paying more attention to the cybersecurity practices of organizations. Customers also want to know that the businesses they deal with are secure. According to analysts, 88% of boards view cybersecurity as a business risk rather than just a technical problem, and 13% have even created cybersecurity-specific committees to address the issue.

Cybersecurity is a business issue because it affects an organization's overall health and success. If an organization's security measures are inadequate, they can lead to financial losses, damage to the company's reputation, and loss of customer trust.

Additionally, if a company fails to comply with industry regulations and standards for security, it may face legal consequences and fines. Therefore, businesses must prioritize security to protect themselves, their customers, and their assets.

Cybersecurity can also be considered a **fiduciary duty** for directors and officers of an organization. A fiduciary duty is a legal obligation to act in the best interests of the shareholders and customers. For directors and officers, this means that they must act with due care and diligence in the management of the organization, including protecting its assets and ensuring the safety and security of its information and systems.

Organizations are responsible for exercising due care in cybersecurity to protect their assets and the interests of their stakeholders. This means taking reasonable steps to safeguard their information and systems from cyber threats. The specific measures that an organization should take will depend on the size and nature of the organization and the types of threats it is likely to face. However, some common elements of due care in cybersecurity include the following:

- Developing and implementing a **cybersecurity policy** that sets out the roles, responsibilities, and practices required to secure the organization's assets and data
- Providing **employee training** on cybersecurity best practices and policies
- Regularly reviewing and updating the organization's **security measures** to ensure they are appropriate and effective
- Ensuring that appropriate **security controls** are in place for all information systems and networks, including those that third parties manage
- Establish **incident response procedures** to follow in case of a security breach or other cyber incident

By exercising due care in cybersecurity, organizations can minimize the risk of a cyber-attack or data breach and reduce the potential impact of any incidents that do occur.

For a start-up such as Acme Software, the incentives are in place for it to start thinking about cybersecurity from the beginning. However, reviewing or establishing a cybersecurity policy is also an opportunity for more mature companies. We will discuss this in detail in the following section.

Cybersecurity policy

A **cybersecurity policy** is a set of processes, guidelines, and standards that an organization uses to protect its information and information systems from cyber threats. It defines the roles, responsibilities, and practices that employees, contractors, and other stakeholders must follow to secure the organization's assets and data. A cybersecurity policy typically covers a range of areas, including network security, access control, incident response, and employee training. In addition, a cybersecurity policy aims to prevent cyber-attacks and other security incidents that could compromise the availability, integrity, and confidentiality of the organization's information and systems.

Addressing risks through effective policy

The early stage of the **World Wide Web**, known as **Web 1.0**, was a time of rapid growth and innovation in the world of the internet. However, with this growth came an increased risk to the security and privacy of users. Web 1.0, also known as the **static web**, refers to the early days of the internet when websites were primarily used to display information and were not interactive. It was a time when the internet was still in its infancy, and the landscape of online threats was constantly evolving. In Web 1.0, the focus was primarily on the development of websites and the exchange of information, with limited attention paid to security measures.

In order to understand the security risks in Web 1.0, it is important to consider the historical context of the internet and the limited security measures that were in place at the time. Understanding these risks can help organizations and individuals take steps to better protect themselves and their data in the present and future.

Cybersecurity risks in Web 1.0

In the mid-90s, the days of Web 1.0, I became fascinated with the web and cybersecurity. As I learned more about how cybercriminals could exploit vulnerabilities in networks and websites, I became more and more invested in finding ways to protect against these threats. It was a challenging and exciting time to work in the field, and it sparked a lifelong passion for everything related to web and cyber security.

Interactivity was limited in Web 1.0 due to the technological constraints of the time. Web 1.0 was primarily a one-way communication platform, with users able to consume content but not able to interact with it or contribute to it in any meaningful way. There were some exceptions to this, however. For example, some Web 1.0 sites included forms that allowed users to input data and submit it to the server for processing. Users would submit these forms to provide information or to enable them to provide feedback or comments. During this time, authentication was not a significant concern as customers could execute very few online transactions. I observed this at my first cybersecurity job at a financial company. My company, like most, had primarily static web pages, as shown in the following figure:

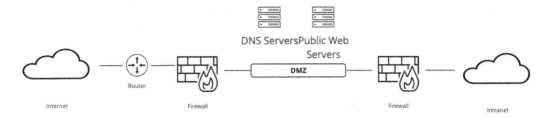

Figure 10.1 – Network diagram of static web pages

To interact with the company, customers had to go to branches and show valid forms of identification, or at least call the help desk (and provide a PIN) to get access to their account information.

To address this, a lot of my time was dedicated to planning the security of the company's network and protecting the content and availability of the web servers. Protecting the website against defacement was the most considerable risk.

The cybersecurity policy was relatively simple and consisted of the following components:

- **Access control**: Establish procedures for controlling access to the website, including the authentication and authorization of users

- **Network security**: Implement measures to protect the website from network-based threats, such as firewalls, intrusion detection/prevention systems, and secure protocols

- **Web application security**: Ensure that the website is designed and developed with security in mind, including measures to prevent injection attacks, cross-site scripting, and other common web application vulnerabilities

- **Data protection**: Implement measures to protect any sensitive data that is collected or stored on the website, including encryption, access controls, and regular backups

- **Incident response**: Establish procedures to follow in the event of a security incident, including identifying and containing the issue, reporting the incident to appropriate parties, and restoring the website to a secure state

- **Employee training**: Provide training to employees on the importance of cybersecurity and their roles and responsibilities in protecting the website

- **Regular review and updates**: Regularly review the website's security measures and update them as needed to ensure that they are effective in protecting against emerging threats

Most of the details for the policy were based on the *Guidelines on Securing Public Web Servers - Recommendations of the National Institute of Standards and Technology*, available at `https://nvlpubs.nist.gov/nistpubs/Legacy/SP/nistspecialpublication800-44ver2.pdf`. **NIST (National Institute of Standards and Technology)** is a US government agency that develops and promotes standards, guidelines, and best practices to enhance cybersecurity and information security.

In the **Europe, Middle East, and Africa (EMEA)** region, the closest equivalent to NIST is the **European Union Agency for Cybersecurity (ENISA)**. ENISA is a center of expertise for cybersecurity in Europe that works with EU Member States and private stakeholders to develop and promote cybersecurity best practices and standards.

In the **Asia-Pacific (APAC)** region, there are several organizations that serve as equivalents to NIST. One example is the **Cyber Security Agency of Singapore (CSA)**, which is responsible for overseeing and coordinating Singapore's cybersecurity strategy and implementation. Another example is the **Japan Cybersecurity Alliance**, a public-private partnership that aims to enhance cybersecurity in Japan by promoting best practices, developing human resources, and providing technical support.

> **Note**
>
> To have a good idea of what my job was when I started, one only needs to read the description of the book *The Complete Guide to Internet Security* (`https://www.amazon.com/Complete-Guide-Internet-Security/dp/081447070X`), published in 2000, which discusses how *hardly a week goes by without a report of some hacker, disgruntled employee, or techno-thief breaking into a computer system–vandalizing web sites, stealing confidential data, compromising trade secrets, or worse.*

As the web grew in popularity, developers started using client-side technologies such as JavaScript to create more interactive and dynamic web pages. In the late 1990s, server-side technologies such as **Common Gateway Interface (CGI)** and **Active Server Pages (ASP)** were introduced, allowing developers to create web applications that could generate dynamic content and interact with databases. These technologies allowed the creation of more complex web applications, such as online stores and forums.

One of the projects I worked on at the time was supporting the implementation of a system that allowed customers to view their account statements online. It was a simple but innovative feature at the time, and I was excited to be a part of bringing it to life. Customers could register for online access by entering their account number and choosing a username and password. From there, they could log in anytime to see their account information and stay up-to-date on their financial health. It was the first time I had to write an **Request for Proposal (RFP)**, select a vendor for an **Access Management** product, and follow the deployment process flow: planning, development, testing, deploying in production, and monitoring.

Server-side scripts, such as CGI, introduced more security risks because they allowed the server to execute code and generate dynamic content in response to user requests. If a server-side script is not implemented or configured correctly, it can be exploited by malicious users to gain unauthorized access to the server or to execute code with the permissions of the user running the script.

With the addition of server-side code, cyber security responsibilities now included adopting best practices for developing and deploying server-side scripts. This included using secure coding practices, properly validating user input to prevent injection attacks, and configuring the server to run scripts with the least privileges necessary. It was also essential to keep the server and scripts up to date with the latest security patches.

Attacks against individual accounts were not a considerable risk, because there was no big money to be made by stealing someone's account statements. Instead, attacking the network and stealing the company's confidential data was much more profitable. The cybersecurity policies had to be extended to allow for the additional risk.

In addition to static content, we now had Access Manager components providing authentication and authorization services for the web server, with the help of a directory of users hosted inside the internal network. We also had a content management application running on the internal network that the scripts on the web server would call to generate dynamic content specific to each user. The number of processes grew, as well as roles required to ensure the separation of duties and an additional level of security in case of a compromised account. These are illustrated in the following figure:

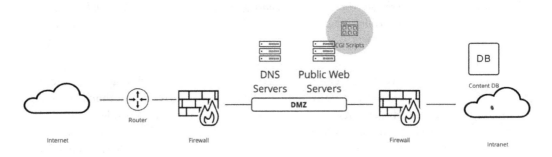

Figure 10.2 – Network diagram for access management and content management systems

Some consider the transition from static to dynamic web pages the beginning of **Web 2.0**. Also known as the **dynamic web**, Web 2.0 refers to the evolution of the internet into a platform for interactive and collaborative communication.

From static to dynamic: the evolution of web applications and the importance of authentication in Web 2.0

The rise of Java and the J2EE platform led to the development of servlets and application servers, further extending web applications' capabilities. In recent years, there has been a shift toward using web frameworks, such as Ruby on Rails, Django, and ASP.NET, to build web applications. These frameworks provide a structure and set of tools for developing web applications more efficiently. Additionally, there has been an increase in the use of **single-page applications** (**SPAs**). These are web applications that only load a single HTML page and dynamically change the content of the page as the user interacts with the application. SPAs are often built using JavaScript libraries such as React, Angular, and Vue.js.

This also marked a significant shift in my career. In the process of due diligence, choosing an access management product, and deploying it in production, I decided to join my first start-up, now as a software engineer. I started working for the company that created the access management product one week after we deployed it in production.

The evolution of web applications has involved the development of increasingly powerful technologies and frameworks for building dynamic and interactive web-based applications. With the increase in online transactions and the need to protect sensitive information, authentication has become a critical concern in Web 2.0.

Strengthening cybersecurity

One major difference in authentication requirements between Web 1.0 and Web 2.0 is the use of passwords. In Web 1.0, passwords were not widely needed as a means of authentication. However, with the increasing number of online transactions and the need to protect sensitive information in Web 2.0, passwords became an important security measure used to protect access to our online accounts, including bank accounts, medical records, and social media profiles. They helped to ensure that only authorized individuals could access these accounts and view or modify the information contained within them.

Because of the increased need for passwords, data encryption also increased with the help of the **Secure Sockets Layer** (**SSL**). SSL is a security protocol that establishes an encrypted connection between a web server and a client, such as a web browser. The link is established using an SSL handshake, which involves the exchange of several messages between the client and the server. Once the connection is established, all data transmitted between the client and the server is encrypted, ensuring that it cannot be intercepted or read by third parties.

Even though passwords have been a common form of security for a long time, they are not easy for humans to use. We have to remember many different passwords, rotate them regularly, and make them longer and more complicated in order to make them more secure. This can be inconvenient and annoying, especially when we have to type them on a computer. Additionally, the need to remember complex passwords can lead people to engage in unsafe practices, such as reusing passwords or writing them down on a Post-it note. Passwords rely on something that the user knows, which works well in a physical space, but in the digital world, machines can impersonate humans and try to crack passwords. With the rapid advancement of technology, the number of password-cracking attempts has increased exponentially in recent years.

Passwords become vulnerable when they are shared or stored on other people's devices. If an attacker is able to access a database that contains passwords, they can potentially use those passwords to gain access to other accounts. Security experts have worked to improve the safety of passwords by developing methods such as **hashing** and **salting**. Hashing involves encrypting passwords using an intricate formula, while salting involves adding dummy characters to the password before encryption. However, the increasing prevalence of **malware** and **phishing attacks** means that anyone can potentially become a target, from individuals to large corporations.

Just like corporations and individuals, cybercriminals often target specific individuals or organizations based on the value of the information they have. This can include personal information, financial data, intellectual property, and other types of sensitive or valuable information. By targeting these individuals or organizations, cybercriminals may be able to gain access to valuable information or resources, or they may be able to monetize the information in some way, such as by selling it on the black market. Overall, the increasing number of online accounts has made it easier for cybercriminals to find targets, and they are constantly developing new tactics to try to gain access to these accounts.

The rapid growth of malware and phishing attacks means that nearly anyone can potentially become a target, regardless of their level of technical expertise or the size of their organization. **Brute-force attacks** are another tactic that cybercriminals use. In a brute-force attack, the cybercriminal uses a computer program to try to guess a user's login credentials by trying every possible combination of characters. While these attacks can be time-consuming, they can be successful if the victim is using a weak password.

The increase in **password dumps** is also a concerning trend. In a password dump, a large database of stolen passwords is shared among attackers, often on the dark web. These passwords can then be used by the attackers to try to gain access to other accounts, either by using them directly or by using them in combination with other tactics, such as phishing or brute-force attacks. The fact that these password dumps are often bundled and sold in a convenient format makes it easier for attackers to use them, and it increases the risk of individuals and organizations having their accounts compromised.

Password-based authentication was not the only change during the evolution of Web 1.0 to Web 2.0.

Cybersecurity is a never-ending process

In addition to the need for more powerful technologies, Web 2.0 also brought a trend toward migrating web applications from on-premises servers or private clouds to **public clouds**, such as Amazon Web Services, Microsoft Azure, and Google Cloud Platform. This has allowed organizations to benefit from the scalability, reliability, and security of the public cloud while also reducing the cost and complexity of maintaining their infrastructure.

As web applications have migrated to the cloud, there has been an increase in the number of roles and responsibilities related to cybersecurity. One of the main benefits of using a public cloud provider is that the provider is responsible for the security of the underlying infrastructure and takes care of many of the security tasks that organizations would have to handle themselves if they were running their applications on-premises. However, organizations still need to ensure the security of their applications and data in the cloud, which requires a range of cybersecurity roles and responsibilities.

The shift toward **Software as a Service (SaaS)** has been driven by the increasing demand for applications to be available online. Just as customers expect to be able to access their applications from any device and location, employees, contractors, and business partners also want the convenience and flexibility of being able to access their day-to-day applications as a service. This trend has had a significant impact on the way that businesses and organizations deliver software and IT services.

The COVID-19 pandemic has certainly had a major impact on the way that companies deliver their services, including a shift toward online delivery. Many businesses and organizations were forced to rapidly adapt to remote work and online operations in order to continue serving their customers and stakeholders. This has led to an increase in the use of digital tools and platforms for communication, collaboration, and service delivery.

For some companies, the shift toward online service delivery has been a challenge, as they had to quickly adapt to new technologies and ways of working. However, for others, the move to online delivery has been an opportunity to innovate and find new ways of serving their customers and stakeholders. The pandemic has also highlighted the importance of digital infrastructure and the need for businesses to be able to adapt and thrive in an increasingly digital world.

It is also true that the COVID-19 pandemic has led to an increase in cybersecurity threats, including fraud. In addition, as more people have started working from home and using digital tools to communicate and access sensitive information, the value and, consequently, the risk of cyber-attacks has increased. Passwords have long been considered the first defense against unauthorized access to accounts and systems, but they are no longer sufficient to protect against advanced threats.

Are password managers a solution for password risks?

As we've seen before, users are bad at creating unique passwords and remembering them on each website. As a result, **password managers** can be an important tool for helping to protect against cyber-attacks, including phishing attacks. Certain organizations mandate the use of password managers by their employees and anybody else with access to their applications. A password manager software application allows users to generate, store, and securely manage their passwords. In addition, some password managers can automatically fill in user login forms, saving them the hassle of remembering and typing in their passwords manually.

There are several benefits to using a password manager:

- They help users create and use strong, unique passwords for all their accounts:

    - Strong, unique passwords are harder for hackers to guess or crack

    - Users are less likely to be compromised in a brute-force attack

- They reduce the risk from password dumps by generating and storing strong, unique passwords for each online account

- They protect against phishing attacks:

    - Password managers include features to detect and warn users about phishing websites and fake login forms

- They increase confidence in protecting accounts and personal information from cyber threats

However, independent of the use of password managers, passwords have an inherent vulnerability in that they have to be stored by the entity (e.g., a website, an online service, or an organization) that verifies them. When a user creates an account with a website or online service and chooses a password, that password is typically stored in the company's database. This is necessary in order for the company to be able to verify the user's identity and grant them access to their account whenever they log in. The problem with this approach is that if the company's database is hacked, the hackers could potentially gain access to all of the passwords stored in the database. This is why it is so important for companies to implement strong security measures to protect their databases from cyber-attacks.

> **Note**
>
> The vulnerability of passwords also applies to password managers. A password manager typically stores passwords in an encrypted form, using a robust encryption algorithm to protect the data. The password manager software includes a **vault** where users can store their passwords and other sensitive information, such as account numbers and security questions. The password manager typically uses a master password to encrypt and decrypt the data in the vault. In the case of a data breach such as the one suffered by **LastPass** (`https://blog.lastpass.com/2022/12/notice-of-recent-security-incident/`), if the attackers can break a user's master password, they will have access to all usernames and passwords for that user. Even without attempting to break a user's master password, attackers can still use the unencrypted data stored in LastPass, including URLs. Such data could give hackers insight into which websites the user has accounts with. When combined with phishing or other types of attacks, this information could be powerful for hackers targeting specific users.

While password managers can help users create, store, and manage their passwords securely, they do not eliminate the need for users to remember multiple passwords. Additionally, if a user's password manager is compromised, the attacker can gain access to all of the user's accounts.

In light of these risks, alternative authentication methods have been developed to provide stronger and more secure authentication than passwords.

Identifying alternatives to passwords

The adoption of **WebAuthn** as a web standard in 2019 was a significant milestone in developing the FIDO Alliance's **FIDO2** specifications, as it means that web applications could use **FIDO-based authentication** methods in a standardized way.

FIDO uses public key cryptography to create a unique set of keys for each user and device. When a user wants to authenticate to a service, they use their device (such as a smartphone or security key) to sign a message with their private key. Then, the service can verify the message's authenticity using the user's public key.

To register with an online service, their authenticator creates a new cryptographic key pair, securely storing the private key locally and registering the public key with the service. The online service may accept different authenticators, allowing the user to select which one to use. The user unlocks the authenticator using a PIN, fingerprint reader, or face ID.

The registration process is illustrated in the following figure:

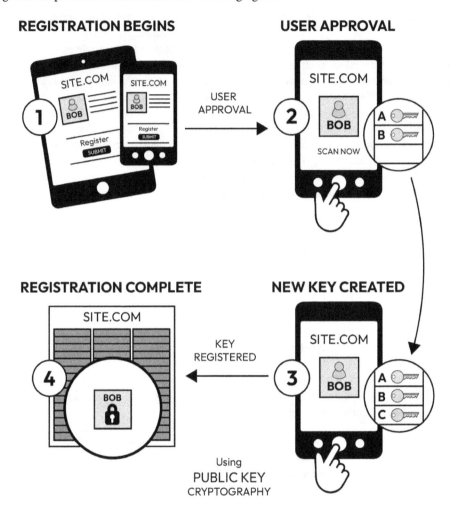

Figure 10.3 – FIDO registration

To authenticate to the service using a registered device, the user will use a client such as a browser or app to access the service. The service presents a login challenge, including an account identifier, which the client passes to the authenticator. It prompts the user to unlock the authenticator and uses the account identifier in the challenge to select the correct private key and sign the challenge. The user's client sends the signed challenge to the service, which uses the public key it stored during registration to verify the signature and authenticate the user.

In this way, FIDO enables secure and convenient authentication to online services without the need for passwords, using unique keys and local authentication methods:

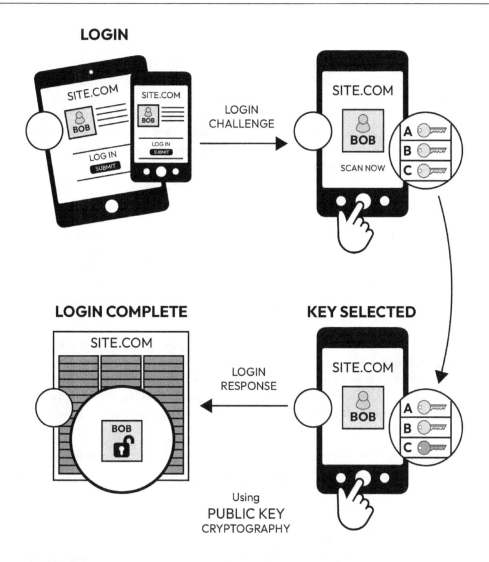

Figure 10.4 – FIDO login

In summary, FIDO works by using a device to sign a message with a private key, which can be verified by the service using the corresponding public key. This allows users to authenticate to online services without the need for passwords (and also without storing any secrets on the server).

Even though FIDO allows passwordless authentication, most online services still use passwords as the first factor of authentication and FIDO as a second factor or part of MFA. The main reason for passwords being present in most authentication processes is familiarity. Users are used to registering a new account using a username (usually an email) and password, especially on customer-facing websites.

Strategies for implementing MFA

The list of strategies in this section will provide suggestions that can help in a successful MFA implementation. As we discussed, each strategy needs to be considered according to business needs and as a balance between usability and security, as well as the costs involved. Also, security is a business issue, and a major initiative such as the implementation of an MFA system can only be successful with the support from high-level sponsors in the organization. It will be very difficult to succeed if it is just an IT or security team-imposed solution. In the following sections, I will also identify when the strategies and tips apply only to the workforce or only to customer MFA.

Eliminating passwords should be the goal

Even for small companies like Acme, it may be impossible to completely eliminate the use of passwords for all users accessing the organization's internal applications. This can be due to a number of factors, such as the lack of support for alternative authentication methods in certain applications, cost and resource constraints, and resistance to change among users. Additionally, external partners, contractors, and suppliers may not have access to the same types of authentication factors that internal users may have, such as **security keys** or **biometric authentication**.

While eliminating passwords for all applications and users may be an ideal goal, it is important to recognize that it may not be feasible in certain scenarios. Therefore, it is important to take other steps to address the scenarios where password-based authentication is still necessary. This may include implementing additional security measures such as regular password audits, and user education and training on best practices for password management. Additionally, organizations can explore password managers for those who are still required to use them.

In summary, even though it may not be possible to eliminate all passwords for all users and all applications, it is important to strive for that goal and take other steps to address the scenarios where it is not achievable.

Get the right people

Most security-related projects, including the implementation of MFA and the related processes, are complex in nature and require a wide range of skills and expertise to successfully execute. These types of projects involve many different roles, each with its own unique responsibilities and tasks. Some of the essential roles that are often required for security projects include a project manager, a security architect, security and network engineers, and an identity and access management engineer.

In addition to these conventional team roles associated with most security projects, Ian Glazer, during his Identiverse keynote (`https://www.youtube.com/watch?v=ZK9QOrEmpKw`), had the following recommendations:

> *IAM project teams bring on a technical writer who can clarify complex ideas for those who are not well-versed in technology. This writer should be capable of crafting the right message so that users understand the secure login process is in their own best interest.*

> *You hire the best tech writer you can … Why? Because in the second and third acts, we're going to need to communicate a ton of things to people that don't normally do identity stuff. And you're going to need time for them to come up to speed.*

Deploying MFA and eliminating passwords can improve security by reducing risks from password-only access, but it may be seen as an inconvenience to some users. To ensure buy-in and participation from the entire organization, it's important to clearly communicate to non-identity, non-technical people about how things are going to change and why it's better. And for this, you need a good writer.

The buy-in and the success of the project require that all users involved (which should be most of the organization) understand what is happening and why. The writer will also be responsible for major updates, sending out emails with advance notice of the changes, documentation, guides, and FAQs. It is also important to clearly communicate all tools available for support and ways of providing feedback. One good example of the communication deliverables is the FAQ created by Salesforce for their customer MFA rollout: `https://help.salesforce.com/s/articleView?id=000388806&type=1`.

Glazer also recommended hiring a data analyst who can analyze and communicate IAM customer usage patterns.

> *You'll want to analyze your data and understand things like, 'This region in this industry is not getting good adoption,' 'We need to improve,' or 'Let's give credit to this cloud for reaching their numbers.'*

Glazer further explained that this individual will be responsible for understanding and analyzing the various data points related to the project, including user behavior, system performance, risk levels, implementation velocity, and adoption rates. This person will also be responsible for working with the project manager and the writer to explain the data to the rest of the team, so that everyone is aware of the progress of the project, and where things are going according to plan, as well as where they may be falling behind.

A data person can also help evaluate alternatives and choose the best option when roadblocks or issues arise. They can also provide insights on how to optimize the system, by looking at user adoption rates and other **key performance indicators** (**KPIs**). This can be especially important after the system is in production, as it will be crucial to monitor the system and make any necessary adjustments to ensure that it is meeting the needs of the organization.

Good metrics on user adoption and other use cases will also help in understanding how the system is being used and what kind of impact it has on the organization. This data can be used to make informed decisions about how to scale the system and allocate resources to support it.

Another unconventional suggestion from Ian Glazer is to hire *non-identity, non-technical people* to work with the customers, and operate the MFA rollout program without subject matter experts. According to Ian:

> *I was deeply, deeply concerned about this because I thought that my team, as the subject matter experts, were going to spend all their time explaining what identity is and not doing the work.*

Although Glazer was right that the identity team would need to spend considerable time educating the programmers, it turned out to be the perfect dry run for communicating the same concepts to the customer base later on.

Focus on three use cases

The three main use cases to consider when implementing MFA are **registration**, **login**, and **account recovery**.

Registration (or sign up)

The registration process will vary greatly between workforce registration and end user (customers) registration. It will also vary depending on the maturity of the system in place.

Let's start with **workforce registration**. For workforce users, registration is usually performed by an administrator, or by an identity management system. Depending on the level of automation, information will come from Human Resources in the form of an email, phone call, or via an automated process, and a user will be provisioned in the system. This same process should also perform additional steps depending on the type of user being onboarded in the organization. If a user is being hired for a position where they will have access to confidential information, or as an administrator that requires a higher security assurance, the company may ship a security token automatically to the user's permanent address, for example.

For established organizations that are starting to implement MFA, a **phased approach** is recommended, starting with the riskiest users and going all the way to end users. Even within user groups, it is important to deploy users in phases so that processes can be evaluated and refined as needed. It will also prevent locking out all administrators out of the system, in case of a bad policy, for example.

Other things to consider are the types of factors allowed by an organization. Requiring security keys from all employees may not work well for remote workers that may need to be onboarded immediately or organizations with large turnovers. Similarly, allowing only SMS or voice authenticators may not work for places where the user may have internet service but not very good or nonexistent cell service (my house, for example).

Customer registration, contrary to workforce registration, is mostly done via self-service. For existing systems, implementation of MFA requires education of the user, especially if there is a target date for the users to be using MFA. Users should be nudged to register different factors of authentication. At the same time, they need to understand the benefits and reasons for change, and they must have access to resources that will help them in case of need. As is the case with workforce users, a phased implementation helps resolve issues and improve the process with a small number of users before deploying to all.

Companies can increase sign-up conversion (users that do not abandon the sign-up process) by eliminating a lot of friction to establish a new account. Instead of asking for all information upfront (unless required for other reasons, such as identity verification or industry regulations) companies can use **passwordless progressive profiling** to increase onboarding success. Instead of asking for emails, phone numbers, and other information and expecting users to enter an acceptable password twice, companies can ask for a phone number or username and a biometric login. Especially now, with passkey support from Microsoft, Apple, and Google, users can register their credentials once, the credentials are automatically backed up, and they can be used on other devices the user has, even on different platforms. It also solves the issue of account recovery greatly reducing one of the biggest costs associated with MFA implementation.

Here is one example of passwordless registration using passkeys:

1. The user enters an email address and clicks **Continue**:

Figure 10.5 – Passwordless registration

You will see that in this example, the only input from the user is a valid email address. If the user continues with registration, the email can also be validated using **Email Intelligence (EI)**. EI is a type of technology that helps to detect and prevent fraud by analyzing email addresses. This technology works by offering a real-time email risk assessment that uses machine learning to evaluate historical data and various attributes of an email address, such as the domain, the format, and the IP address, to determine the likelihood that the email address is associated with fraudulent activity, indicated in a score. Several commercial products are available in the market, and each uses proprietary technology when generating an email risk score.

2. The email is not registered in the system. The user selects **Sign Up** to create the account:

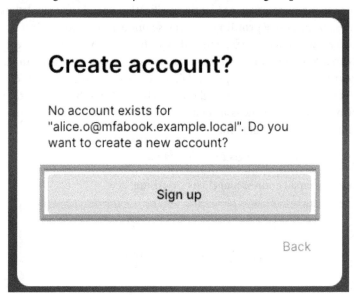

Figure 10.6 – Account verification

An additional step may be to confirm the user owns the account by sending a **One-Time Passcode (OTP)** that the user needs to enter to continue registration.

3. A passcode is received by the user, which is used to create a new account:

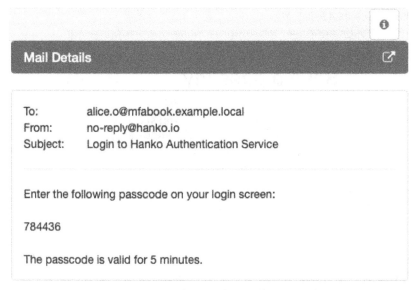

Figure 10.7 – Email ownership verification code

4. The user enters the passcode received in the email and clicks **Sign In**:

Figure 10.8 – Passcode verification

After registration, the system asks the user if the user wants to sign in using a passkey.

5. The user clicks **Save a passkey**:

Figure 10.9 – Passkey sign-in selection

If they agree, the system creates a passkey and prompts the user to save the passkey locally.

6. If the user saves the passkey locally by clicking **Continue**, they can only use the passkey to log in from that device. If the user chooses **Try another way**, they are allowed to save their passkey in the cloud and share the passkey among devices:

Figure 10.10 – Local passkey creation

That's it. The user has registered with the service and has a passkey to use when logging into the account later.

Login (or sign in)

The **login** process is not different for workforce or customer users. The biggest difference, in this case, is between organizations that have existing access management systems that the internal or external users already use and to educate their users on the new processes for login and when they are not able to login.

As discussed earlier in this chapter, users are used to entering usernames and passwords when logging in online. **Decoupling** the screen where the user enters the identification and the screen(s) where the user will enter or select the credentials has many benefits. The first one is that it starts to eliminate the mandatory link between a username and password. Also, it allows for the discovery of what authentication method is available for this user on the device they are using.

The first example of login screen decoupling is from Microsoft. On the first page, the user will select an email or phone number to sign in to Microsoft Azure:

1. The user enters their email, phone number, or Skype, and clicks **Next**:

Figure 10.11 – Microsoft Azure sign-in page

The system detects the user has registered for passwordless authentication and tells the user how to approve the sign-in. It also tells the user how to log in if the Authenticator app is not available (**Use your password instead**).

2. The user is instructed to open the authenticator app and enter the number shown:

Figure 10.12 – Microsoft Azure passwordless sign-in approval

3. The user confirms the login by entering the number displayed on the login screen to the authenticator app. The application the user is trying to log in to and the user's location are also configured to prevent phishing attempts:

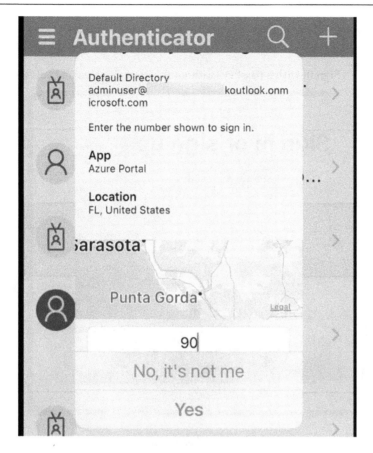

Figure 10.13 – Microsoft Authenticator sign-in confirmation with number matching

4. Upon approval, the user is logged in to the application without being prompted for a password:

Figure 10.14 – Microsoft Azure successful passwordless sign-in

Next, we show another login example in a scenario where the user registered using passkeys. In this case, the user does not even need to enter the email address registered with the application:

1. The user clicks **Sign in with a passkey** (without the need for an email address).

Figure 10.15 – Passkey sign-in

The system recognizes that the user already registered a passkey for that domain and allows the user to select the passkey directly. It also allows the user to click **Try another way** in case the passkey was created on a different device and saved to the cloud.

2. The user selects the previously registered passkey and clicks **Continue**:

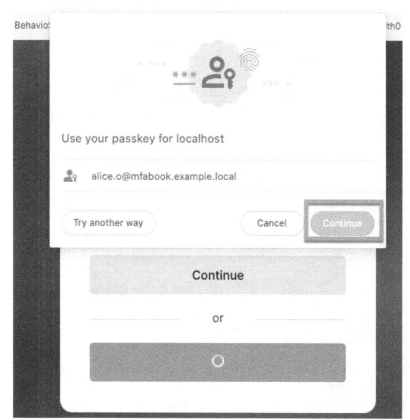

Figure 10.16 – Passkey selection

That's it. The user has logged into the application with two simple clicks.

Other considerations for requiring MFA

While you analyze the use cases, consider the types of applications and users using the site and the support cost. If you have a website that users visit once a year (such as TurboTax), you may want to avoid mandating security keys, for example. If your users change devices every year, they will have to register the security keys again on the new device the next time they go to the website. Passkeys may be a good solution in this case, but they may only be available on specific platforms, or your requirements prohibit sharing the credentials across devices.

You may also decide that SMS is a factor that you want to support. But many users complain that it can take 30 seconds to get an SMS, and they call customer support. So, you have to take into account how you maintain it and how you support it.

There are additional security measures that you can implement to enhance security during login:

- **Fraud countermeasures**: In addition to MFA, we can use additional forms of authentication to help prevent fraud when the use requirements mandate that we support an authentication scenario that we cannot avoid. For example, we may be required to support password-based authentication. **Passive authentication** (when I can get some information about this session and this user without any interaction from the user) is normally used during the login flow.

  At the user level, one common countermeasure is **breached password detection**. Breached password detection is a technology that helps identify if a user's password has been compromised in a data breach by checking the password against a database of known breached passwords.

- **Impossible travel detection**: Impossible travel detection is a security feature that uses geolocation data to detect if a user is attempting to log in from a location that is physically impossible to travel to in the time frame between the previous login and the current one.

- **Rate-limiting**: Rate-limiting is a security measure that limits the number of login attempts or other actions that can be performed within a specific time frame. This helps prevent brute-force attacks and other types of automated attacks.

- **Contextual MFA**: Contextual MFA is a security feature that uses contextual information about the user's device, location, and behavior to determine if an additional form of authentication is needed. This can help prevent account takeover attempts by blocking login attempts from suspicious devices or locations.

- **Bot detection**: Bot detection is a security feature that uses machine learning algorithms to detect and block automated bots from accessing a website or application. This can help prevent many attacks, including scraping, account takeover attempts, and DDoS attacks.

- **Remote access detection**: Remote access detection is a security feature that detects and alerts when a user attempts to access a network or device using a remote access tool. Remote access detection can help prevent coaching attacks.

  The difference between workforce and customer login is also essential. Workforce users usually work from the same location most of the time and may be required to use company-managed devices. Those users will stop seeing the first screen if the passive authentication signals confirm it is the same user device coming back and be allowed to log in using only a password without a second factor because all the risk signals indicate it is the legitimate user login in. External customers and remote users may require a lower threshold before having to use MFA when the risk signals are triggered.

- **Logging and alerting**: Unsuccessful second-factor attempts must be collected and analyzed. If multiple failed attempts occur, the system will notify the user or administrator of this suspicious behavior and prompt the user to enroll a new authentication factor.

Account recovery

MFA can only be as secure as its account recovery process. Recent incidents have revealed that attackers can take advantage of weaknesses in the recovery process to gain control of an account. For example, if a web application employs MFA using a security token and provides the option for the user to enroll their phone number as a backup second factor for recovery in case they cannot access their security token, the security of the application now depends on the security of the telecom provider's authentication process for the customer and the delivery of SMS messages.

An attacker may be able to impersonate the user and manipulate a customer service representative to perform a SIM swap or redirect SMS messages to a number they control. It is crucial to consider how to develop secure recovery processes for every second factor, as each must have a backup method.

The recovery process for the secondary authentication factor must be independent of the recovery process for the primary factor. For instance, if an attacker gains access to the primary authentication factor, the secondary factor becomes useless if it can be reset using only the primary factor. Additionally, the recovery process for the secondary factor must be completely separate from the recovery process for the primary factor. For example, if an email message is a method for recovering the primary authentication factor, ensure that the secondary factor is recovered through a completely different channel.

Enforce (or encourage) a backup secondary factor. Especially in workforce implementations, encouraging the users to enroll more than one secondary factor can simplify the recovery of the alternate secondary factor, and, in most cases, eliminate the use of a help desk or administrator. Backup secondary factors can be something inexpensive as a list of security codes that can be used only once during the recovery process.

During the design of the user experience for the three main use cases, it is important to hear from all members of the team. While the SMEs explain the implementation details, the writer can document the information, writing guides, and FAQs. Those documents can be validated by the non-technical members to verify their understanding from the point of view of the end user.

Summary

In this chapter, we discussed cybersecurity's significant impact on businesses. Moreover, we learned about the crucial role that directors and officers play in ensuring the security of an organization's assets and sensitive information, both legally and ethically. The chapter also explored the evolution of web applications and the need for continuous improvement in cybersecurity. Finally, we discovered practical strategies for implementing **multifactor authentication** (**MFA**) in a real-world setting, how to choose the appropriate authentication method for different users and scenarios, and how to roll out the system to minimize disruption and ensure user acceptance.

In the next chapter, we will discuss several emerging technologies in MFA.

The first topic discussed in the next chapter is Web 3.0, which refers to the next generation of the internet. The Web 3.0 ecosystem involves a range of technologies, such as blockchain, that enable the creation of decentralized applications and services, allowing users to control their data and identities.

The chapter further discusses passkeys and why they are nearly perfect phishing-resistant MFA methods.

The chapter also covers continuous authentication, a security technology that monitors a user's behavior and activity in real-time to detect potential threats or unauthorized access. This method uses machine learning algorithms to analyze a user's keystrokes, mouse movements, and other behaviors during login and all other user transactions throughout the whole journey using the app. Continuous authentication can help prevent data breaches and other security incidents by detecting and stopping attacks before they can cause damage.

Finally, the chapter discusses **Continuous Access Evaluation Protocol (CAEP)**, a security standard that continuously monitors a user's access to resources and revokes access if necessary. Microsoft's implementation of CAEP is designed to work with Azure Active Directory to provide real-time access control for cloud-based resources.

11

The Future of (Multifactor) Authentication

In this chapter, the final one, we take a deep dive into the world of digital identity as it evolves in the context of the groundbreaking **Web 3.0 (Web3)** paradigm. This chapter provides a comprehensive understanding of the changes Web3 will bring to digital identity and authentication, discussing the implications of these transformations for security, privacy, and user experience. Digital identity must also evolve to support human and machine interaction in this new world. As a result, **Personal Identifiable Information (PII)** data (referenced in *Chapter 1*) will become even more relevant and unique. In addition, Web3 will transform digital identity creation, verification, and use. This will undoubtedly impact how we know and use digital identity and authentication. This chapter previews such transformations using concepts of Web3, which will help improve security, increase privacy, and provide a better and more personalized user experience.

As the foundational ideas of Web3 emerge, they are poised to significantly reshape the **identity and access management (IAM)** landscape as we know it. This transformation necessitates a thorough reexamination of the underlying architectures of existing products and services. By embracing and integrating Web3 principles, organizations can revolutionize the delivery of IAM features and capabilities, unlocking new possibilities in security, privacy, and user experience. Navigating these challenges involves the following:

- Adapting to a decentralized landscape
- Embracing innovative authentication mechanisms powered by blockchain and smart contracts
- Enhancing the overall security posture
- Integrating cutting-edge technologies that form the bedrock of Web3

In this chapter, we are going to cover the following topics:

- Introducing the Web3 ecosystem
- Product trends

    - Verifiable Credentials and Microsoft Entra Verified ID
    - Identity management convergence in ForgeRock Identity Cloud
    - Are passkeys (almost) the perfect phishing-resistant **multifactor authentication (MFA)**?
    - Passkey management
    - Continuous authentication
    - **Continuous access evaluation (CAE)**—Microsoft's version of the **Continuous Access and Evaluation Protocol (CAEP)**
    - What lies ahead

In summary, this chapter provides readers with an understanding of Web3 and its impact on digital identity and authentication. This chapter provides a starting point for readers to explore more since the Web3 world is evolving rapidly, and by the time this book is published, new changes will have occurred, and it also encourages readers to learn more about these critical topics and stay up to date with the latest developments in this field.

Technical requirements

In this chapter, we will explore the practical implications of Web3 on digital IAM by examining two distinct products as examples. For Verified Credentials, this chapter requires a working Azure AD tenant with at least an Azure AD **Premium P1** or trial license enabled. The account used for this chapter must belong to one of the following roles: **Cloud Application Administrator**, **Global Administrator**, or **Application Administrator**. The passkey management example using Okta requires either Okta Classic or Okta Identity Engine with an administrator account.

Introducing the Web3 ecosystem

Web3 is the next evolution of the internet, characterized by decentralized technologies such as **blockchain** and **smart contracts**. It aims to give users more control over their data and online identity. In Web3, individuals can manage their digital identity through these decentralized systems instead of relying on one system or an application to manage and control user data. This could lead to a more secure and private internet experience and new online commerce and collaboration opportunities.

Among the key benefits of Web3, we have the following:

- **Improved security**: With Web3, users will have more control over their digital identity, enhancing the security of their online presence. Decentralized authentication mechanisms and cryptographic protocols will help protect against hacking, phishing, and other attacks.

- **Greater privacy**: By decentralizing data storage and management, Web3 will enable users to keep their personal information secure. With Web3, users can control their data and choose who can access it.

- **Better user experience**: Web3 will usher in new ways for users to interact with the web. For instance, users can create decentralized applications that provide better, more personalized experiences than traditional web applications.

Transitioning from Web2 to Web3 will occur differently for every organization. Therefore, existing Web2 applications and services must support current users and customers simultaneously as they support new Web3 technology.

An impact on existing IAM products is expected in the following areas:

- **Decentralized architecture**: Web3 is built on a decentralized architecture, meaning IAM systems will also need to work with **decentralized identities** (**DCIs**). The shift to a decentralized model will require IAM systems to be re-engineered to function in a hybrid Web2/Web3 environment.

- **New authentication mechanisms**: Web3 will introduce new authentication mechanisms, such as blockchain and smart contracts. IAM products must support these new mechanisms to provide secure and reliable authentication.

- **Changing fraudster behavior**: As IAM products become more advanced, fraudsters will develop new ways to circumvent them. IAM products must stay ahead of fraudsters by adapting to new forms of fraud and protecting against new types of attacks.

- **New technology**: Web3 is built on new technologies, such as blockchain and smart contracts. IAM products must be updated to support these new technologies to provide secure access to decentralized applications.

As we've seen with Web2, more and more consumers conduct their daily activities online. The COVID-19 pandemic has only accelerated this shift to digital dealings, and customers have massively adopted new behaviors—for example, students attending remote university classes and taking exams online, families booking a vacation rental online, or making a loan application using a mobile phone app. Therefore, identity verification is of great significance.

Let's first look at digital identity verification in the Web3 ecosystem.

Exploring digital identity in Web3

To access new apps and services, we must go through the registration process with a distinct provider every time we use a new app or service. When services were primarily face-to-face, presenting a driver's license or another form of identification was simple. But in the digital world, answering numerous questions and submitting documents to verify identity is tiresome and prone to fraud. In addition, as we saw in the previous chapter, the proliferation of online identities requires us to have as many accounts and passwords as the number of services we registered with.

In addition to the authentication fraud schemes we saw in *Chapter 2*, another type of identity fraud, and one of the fastest-growing financial crimes in the US and throughout the world, is **synthetic identity fraud**.

Synthetic identity fraud is a type of financial fraud that involves creating a fake identity using a combination of real and fake information. For example, the fraudster will often use a real person's Social Security number and combine it with a made-up name, address, and other personal information. This type of fraud is challenging to detect because the information used is often legitimate. These synthetic identities are used to open bank accounts, apply for credit cards and loans, and make purchases. The fraudster can then use these accounts to make purchases and rack up debt, which may go unnoticed for a long time. In addition, in some cases, the fraudster may establish a good credit history with the synthetic identity, making it even more challenging to detect the fraud.

In the Web3 ecosystem, identity is centered on user-controlled access to personal information through a digital wallet. This wallet, similar to a physical wallet, serves as a repository for various forms of personal identification, such as driver's licenses, credit cards, and other identity verification documents. The critical difference between this digital wallet and traditional forms of identification is that the user maintains ownership and control over their personal data and can selectively share it with others through their wallet.

This shift toward a user-centric identity management model is a significant departure from the current centralized systems where companies such as Facebook own and control user data. With a digital wallet, personal information can move seamlessly across the web and be enhanced with verified data from the real world. As we move toward a Web3 format, digital wallets will become increasingly important to manage and verify identity. This will involve leveraging traditional forms of identification such as driver's licenses, passports, and social security numbers to build and verify the digital wallet, enabling users to move and interact with the web with greater trust and security.

Adopting DCI use cases, including those for citizen identification, employee, health, and student validation, will require investment in a hybrid form of authentication.

Understanding login mechanisms

Web3 login mechanisms are expected to differ significantly from current methods, with a shift toward decentralized and user-controlled systems. One of the fundamental changes is the elimination of traditional usernames and passwords and using digital wallets for authentication and verification.

These efforts aim to transfer identity ownership to individuals and disrupt invasive social media models that compromise user privacy.

In this new model, identities are verified remotely, using a combination of traditional forms of identification such as government-issued documents, phone numbers, and other personal information. These forms of identification are checked against government databases or other trusted sources to ensure their validity.

Additionally, users will be required to provide **biometric data**, such as a facial image or fingerprint, which will be used to confirm their identity every time they log in. This serves as an additional layer of security and ensures that the person accessing the account is the person who owns it. Furthermore, users will have greater control over their personal data and the ability to share it through their digital wallets selectively. They will also have the ability to revoke access to their data at any time.

The Web3 login model is built on several established standards and protocols, including ID issuers, ID holders, and relying parties, all of which play a role in the authentication process. For example, the method includes **Know-Your-Customer** (**KYC**) proofing, which is used to build a signed credential that is then matched against what the ID relying party has on file to verify the user.

Overall, the Web3 login model is expected to provide a more secure and user-controlled method for managing and verifying identities online. Users can use their digital wallets with private keys to prove their identities using distributed identifiers and biometric authentication methods.

As part of upgrading onboarding processes, assess the feasibility of issuing and accepting DCI assertions and verifiable claims. It is important to recognize that standalone identity proofing, centralized identity schemes, and DCI will coexist for the foreseeable future, and leveraging each of them will be crucial to ensuring accessibility for a diverse range of users. As we saw in *Chapter 6*, users using 1Kosmos' DCI solution, **BlockID**, have to coexist with centralized identity schemes for the foreseeable future, and leveraging each of them will be crucial to ensuring accessibility for a diverse range of users.

Implementing decentralized solutions

DCI solutions provide a new way of managing and verifying identity by using a blockchain-based distributed ledger as the source of truth. In this approach, users can receive credentials that verify their identity from multiple issuers, such as government agencies or employers, and store these credentials in a digital wallet. **Service providers** (**SPs**) can then verify the data on the distributed ledger by requesting the minimum amount of information necessary to fulfill their request. This feature is known as **verifiable claims**.

For example, an SP can receive a simple *yes/no* response to a request to verify if a user is over 18 years old, rather than receiving the user's date of birth. This approach allows the SP to receive only the information they need while still maintaining the user's privacy. There has been significant investment in DCI solutions by vendors such as Ping Identity and IBM, as well as infrastructure providers such as Sovrin (`https://sovrin.org`).

Other companies are also interested in this space. As Sarah Clark, the SVP of digital identity at Mastercard, presented at *Authenticate 2022*, Mastercard is currently building a new Mastercard network, the **identity network** or **ID network**. This entirely new network is based on Mastercard's experience and success in building a payment network at scale. However, the ID network is not dependent on the payment network and is not directly related to it from a consumer's perspective.

The ID network focuses on digital identity and aims to create a globally interoperable, reusable digital identity network. At its core, a digital identity is a way to answer the fundamental question of *Are you who you claim to be?* The ID network is doing this by using evidence, such as government-issued IDs, that can be verified and used as a trusted root. This evidence is augmented with other signals and attributes, such as device and location information.

Mastercard's vision is for individuals to be able to present their evidence once and then quickly and seamlessly share their digital ID anywhere in the world they choose to do business. The ID network supports digital interactions across multiple channels, such as mobile, web, chat, telesales, customer support, and in-person interactions. Individuals should be able to use their reusable digital identity the same way they use their Mastercard to make payments anywhere in the world they choose to do business that requires trust. The ID network's goal is to make it easy for individuals to do business with a bank or company without going through the same onboarding process multiple times.

The different solutions for DCI are all based on the idea of creating a system where individuals can control their own digital identities and credentials. This contrasts with the current system, where individuals' identities and credentials are stored by third-party organizations, such as banks and government agencies.

DCI solutions offer a number of advantages over the current system, including the following:

- **Increased privacy and security**: Individuals will have more control over their own data and credentials, and they will be able to choose who they share them with. This will help to protect individuals from identity theft and fraud.

- **Improved efficiency**: DCI solutions will allow individuals to easily verify their identity and credentials, which will speed up transactions and make it easier to access services.

- **Reduced costs**: DCI solutions are likely to be cheaper to implement and maintain than the current system.

However, there are also some challenges that need to be addressed in order to make DCI solutions work effectively. One of the biggest challenges is interoperability. For DCI solutions to be successful, they need to be able to work together seamlessly. This means that they need to be able to share data and credentials, and they need to be able to communicate with each other.

Despite the challenges, DCI solutions offer a number of advantages over the current system, and they are likely to become increasingly popular in the future. As a security professional, it is crucial to understand the importance of interoperability in a distributed environment. To ensure the longevity and success of your solution, it is essential to stay informed about industry standards, such as those developed by the **World Wide Web Consortium** (**W3C**). Keeping up to date with these standards will help ensure that your solution is interoperable with other systems in the long term. These standards include the **W3C Decentralized Identifiers (DIDs)** (`https://www.w3.org/TR/did-core/`) and **W3C Verifiable Credentials** (`https://www.w3.org/TR/vc-data-model/`) standards.

It is also important to explore the existing options for digital credential issuing and management platforms in the marketplace before developing a new solution. By researching already established platforms that are tailored to your industry and having existing participants, you may be able to implement a solution and integrate it with your existing solutions much faster than building a new platform from scratch. Looking at implementations in government as well as vertical markets can also give you new ideas. Examples include the **European Digital Identity** wallet (`https://digital-strategy.ec.europa.eu/en/news/european-digital-identity-online-consultation-platform-european-digital-identity-wallets`), Estonians' **e-Identity** card (`https://e-estonia.com/solutions/e-identity/id-card/`), and the UK National Health Service's **Digital Staff Passport** (`https://beta.staffpassports.nhs.uk/whats-next.htm`). This approach can save time and resources while still meeting the needs of your workforce and customers alike.

Access management is constantly evolving, with new technologies and threats always emerging. DCI and verifiable credentials are just two of the recently introduced exciting developments. The products discussed in *Chapters 3* through *9* are packed with cutting-edge features and capabilities designed to keep up with the latest technological advances, fend off competitors, and combat new types of fraud.

But what exactly are these new capabilities, and how do they work? The following section will look at some essential additions to the access management landscape.

Product trends

To keep up with competitors and, more importantly, with attackers, products must constantly evolve. For instance, products have introduced advanced fraud detection capabilities, such as **machine learning** (**ML**) algorithms that can identify patterns in user behavior and flag suspicious activity.

In addition to new security features, many access management products offer enhanced integration capabilities, making integrating with a wide range of enterprise applications and systems easier. Some products also include robust analytics and reporting tools, which provide valuable insights into user behavior and access patterns.

These new capabilities have transformed the access management space, making it more secure, efficient, and user-friendly than ever before.

Verifiable Credentials and Microsoft Entra Verified ID

One company focused on developing standards for issuing, sharing, and verifying digital credentials is the **Decentralized Identity Foundation (DIF)**. DIF (`https://identity.foundation`) is a collaborative organization comprising various technology companies, start-ups, and experts working together to establish open source, decentralized standards for **digital identity management (DIM)**. DIF's primary goal is to create an ecosystem where individuals and entities can own, control, and share their digital identities without relying on centralized authorities or intermediaries. Microsoft is an active member of the DIF, working alongside other industry leaders to develop open standards and protocols for DCI management. By participating in the development of specifications such as DIDs and Verifiable Credentials Data Model, Microsoft is helping to shape the future of digital identity in the Web3 era.

Verifiable credentials are transforming how individuals share and manage their digital identities, opening up a world of possibilities for secure, private, and streamlined online interactions. Verifiable credentials are digital attestations allowing individuals to prove various attributes or claims about themselves without revealing unnecessary personal information. Trusted entities issue these credentials, which can be cryptographically verified, ensuring their authenticity and integrity.

Imagine a future where sharing your proof of age, educational qualifications, or employment history is as simple as presenting a verifiable credential from your digital wallet. No more carrying physical documents, disclosing excessive information, and relying on centralized databases that could be vulnerable to breaches. Instead, you'll have a seamless, privacy-preserving method for online-proving your identity and credentials.

Microsoft Entra Verified ID

Microsoft recently renamed **Azure Active Directory Verifiable Credentials** as **Microsoft Entra Verified ID**, and it is now part of the **Microsoft Entra** family of products.

Microsoft Entra Verified ID is an innovative DCI solution that helps safeguard organizations by enabling the issuance and verification of secure digital credentials. This powerful platform allows organizations to act as issuers, creating custom verifiable credentials that can be used to verify the identity and qualifications of individuals. Verifiers, in turn, can request and accept these credentials using the platform's REST API, ensuring that only authorized individuals gain access to sensitive data and applications.

Microsoft's Verifiable Credentials is designed using **Verifiable Credentials Data Model 1.0**, a cutting-edge framework that outlines the most secure and privacy-respecting approach to handling these types of scenarios in a digital context. This innovative approach uses a schema-based system to organize and verify the claims made by issuers, ensuring that every credential is backed up by a unique DID that confirms the authenticity of claims made in the credential.

To take advantage of Microsoft Entra Verified ID's advanced capabilities, organizations will need to set up an Azure AD tenant that can be used for both issuing their own credentials and verifying those issued by other organizations. In the following section, we are going to configure Microsoft Verified ID.

Setting up Microsoft Entra Verified ID

Configure your tenant to start using Microsoft Verified ID based on the following instruction:

1. Go to `https://entra.microsoft.com/` to sign in. The user must have the global administrator or the authentication policy administrator permission for the directory:

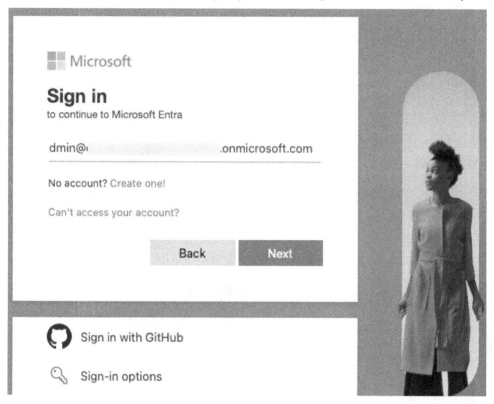

Figure 11.1 – Microsoft Entra sign-in page

2.   To set up Microsoft Verified ID for an organization for the first time, click on **Go to Verified ID** from the **Microsoft Entra admin center** dashboard:

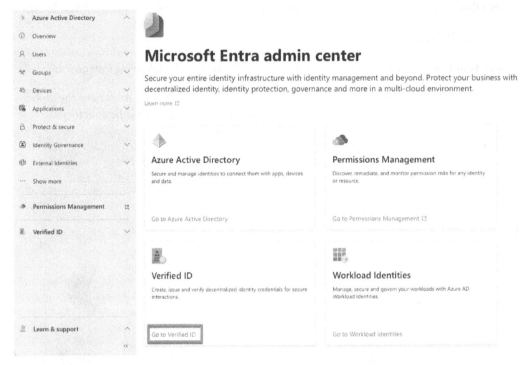

Figure 11.2 – Microsoft Entra admin center

3.   Before setting up an organization for using Verified ID, a key vault must be configured. Click on **Select keys** to select or, if necessary, create a new key vault:

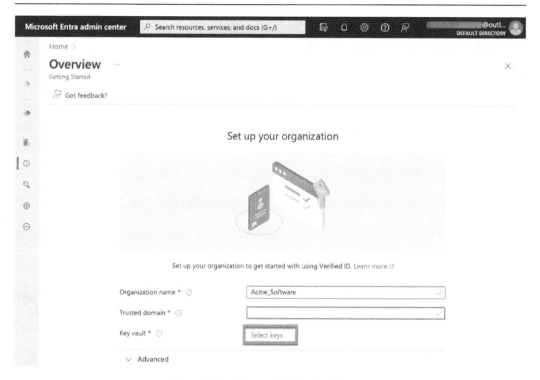

Figure 11.3 – Microsoft Entra sign-in page

4. Click on **Create new key vault** or select an existing one:

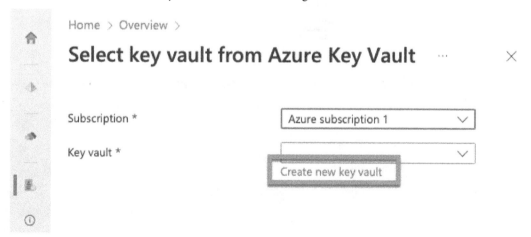

Figure 11.4 – Selecting a key vault

5.   A key vault needs a valid subscription and resource group. Select an active subscription name. Click on **Create new** or select an existing resource group:

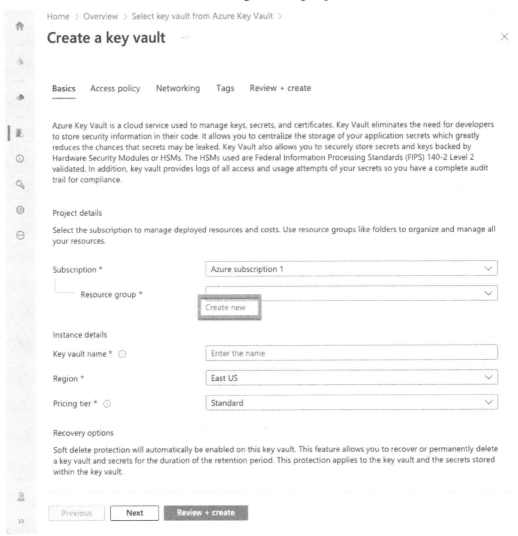

Figure 11.5 – Create a key vault page

6. Enter a unique resource group name and click **OK**:

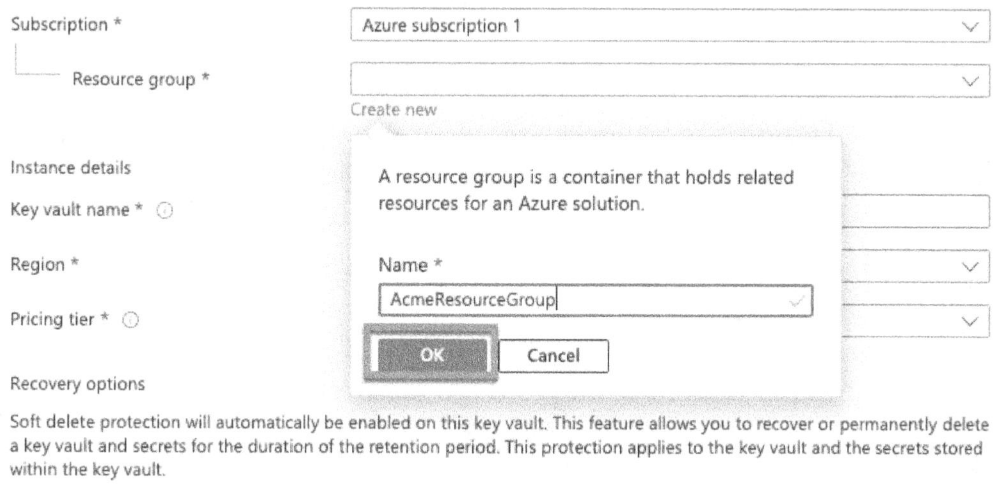

Subscription *   Azure subscription 1   ⌄

   └──── Resource group *   ⌄

                                  Create new

Instance details   A resource group is a container that holds related
                                  resources for an Azure solution.
Key vault name * ⓘ

Region *   Name *   ⌄

Pricing tier * ⓘ   AcmeResourceGroup   ✓   ⌄

                                  ┌──────────┐
                                  │   OK   │   Cancel
                                  └──────────┘
Recovery options

Soft delete protection will automatically be enabled on this key vault. This feature allows you to recover or permanently delete
a key vault and secrets for the duration of the retention period. This protection applies to the key vault and the secrets stored
within the key vault.

Figure 11.6 – Create new resource group popup

7. Enter a unique key vault name and click **Review + create**:

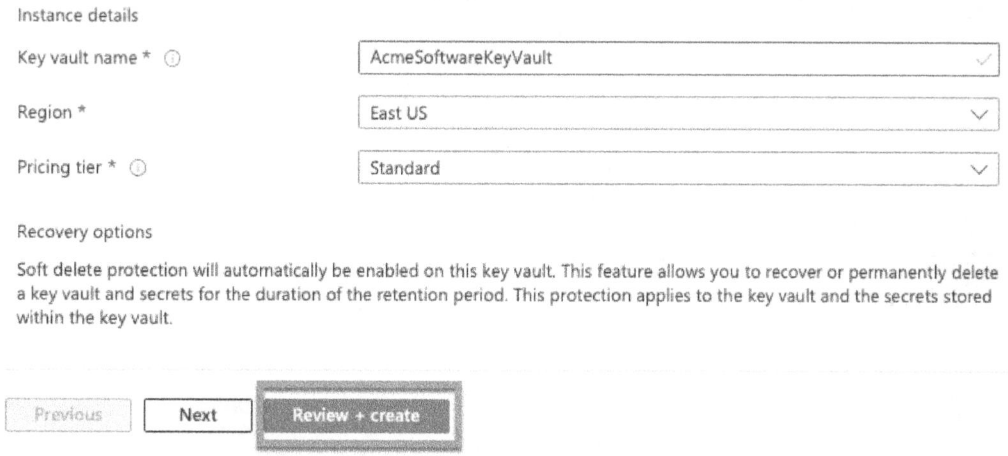

Instance details

Key vault name * ⓘ   AcmeSoftwareKeyVault   ✓

Region *   East US   ⌄

Pricing tier * ⓘ   Standard   ⌄

Recovery options

Soft delete protection will automatically be enabled on this key vault. This feature allows you to recover or permanently delete
a key vault and secrets for the duration of the retention period. This protection applies to the key vault and the secrets stored
within the key vault.

┌─────────────┐ ┌──────────┐ ┌──────────────────┐
│ Previous   │ │   Next   │ │ Review + create │
└─────────────┘ └──────────┘ └──────────────────┘

Figure 11.7 – Create a key vault page

8. Review the information and click **Create**:

Azure Virtual Machines for deployment	Disabled
Azure Resource Manager for template deployment	Disabled
Azure Disk Encryption for volume encryption	Disabled
Permission model	Vault access policy
Access policies	1

Networking

Connectivity method	Public endpoint (all networks)

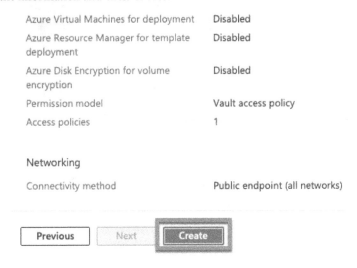

Figure 11.8 – Create a key vault page

9. After the key vault is created, select **Access policies**. Make sure the user has **Sign** under **Cryptographic Operations** selected, and click **Next** and then **Save**:

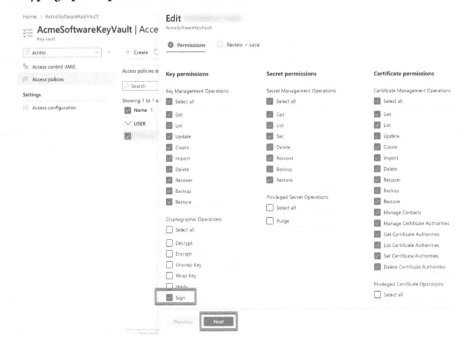

Figure 11.9 – Key vault permissions

10. Select a **Trusted domain** type and click on **Save and get started** to complete the setup of Verified ID for your organization:

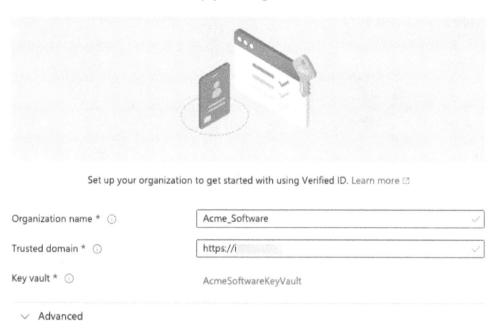

Set up your organization

Set up your organization to get started with using Verified ID. Learn more ☑

Organization name * ⓘ   Acme_Software

Trusted domain * ⓘ   https://i

Key vault * ⓘ   AcmeSoftwareKeyVault

∨ Advanced

Save and get started

Figure 11.10 – Verified ID organization setup page

This completes the setup of Verified ID to use verified credentials in the organization. Now, we need to create credentials in order to be able to issue them. Here's how we do that:

1.   Select the **Quickstart** flow for creating credentials by selecting **Start**:

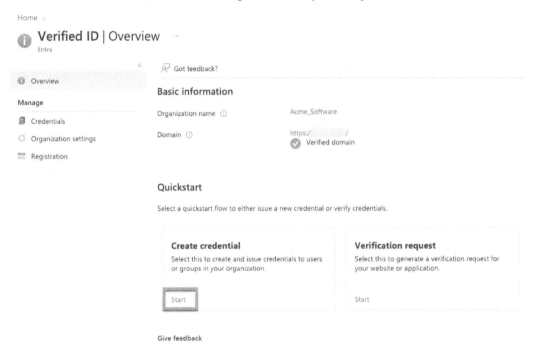

Figure 11.11 – Verified ID/Quickstart

2. Select **Verified employee**, then click **Next**:

Create credential

Linked domain ⓘ

https:// _____ /
✓ Verified domain

Select a credential type

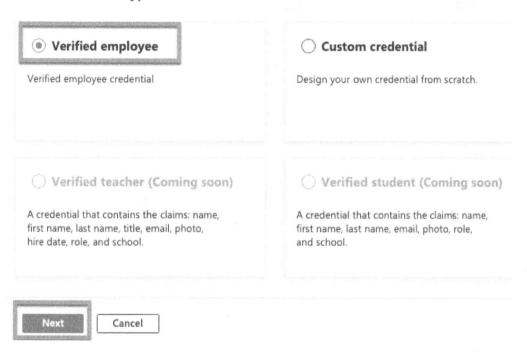

◉ **Verified employee**

Verified employee credential

○ **Custom credential**

Design your own credential from scratch.

○ Verified teacher (Coming soon)

A credential that contains the claims: name, first name, last name, title, email, photo, hire date, role, and school.

○ Verified student (Coming soon)

A credential that contains the claims: name, first name, last name, email, photo, role, and school.

Next   Cancel

Figure 11.12 – Credential type selection

3.   Click **Create**:

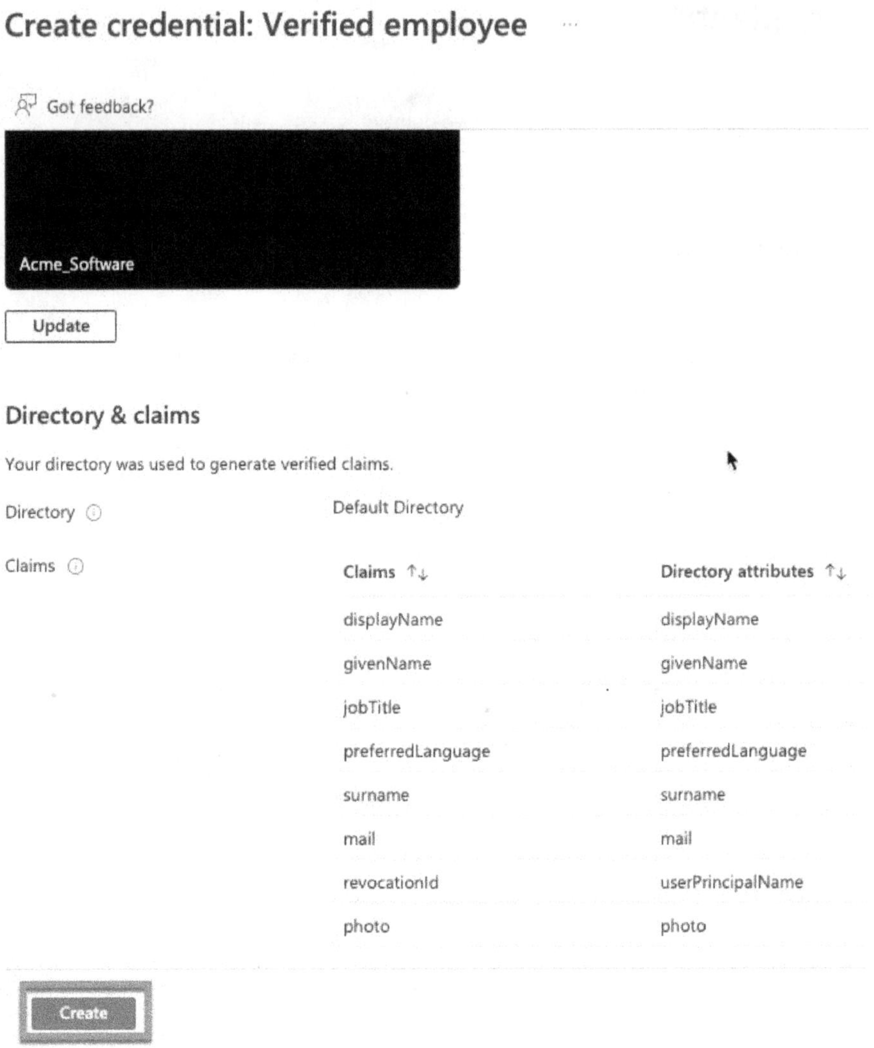

Create credential: Verified employee ...

Got feedback?

Acme_Software

Update

Directory & claims

Your directory was used to generate verified claims.

Directory ⓘ   Default Directory

Claims ⓘ

Claims ↑↓	Directory attributes ↑↓
displayName	displayName
givenName	givenName
jobTitle	jobTitle
preferredLanguage	preferredLanguage
surname	surname
mail	mail
revocationId	userPrincipalName
photo	photo

Create

Figure 11.13 – Credential type creation

Now that the Verified Employee credential has been created, applications can issue verified credentials, and users can present and share selected information with companies that request verified credentials for access. One such application can be implemented according to the instructions available at `https://learn.microsoft.com/en-us/azure/active-directory/verifiable-credentials/verifiable-credentials-configure-verifier`.

This application can be run as shown in the following instructions:

1. The user selects **GET CREDENTIAL** on the **Verifier Credential Expert Issuance and Verifier Sample** main page:

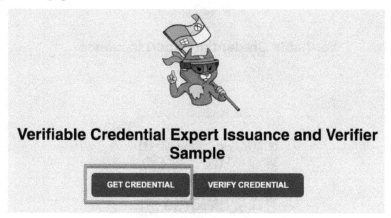

Figure 11.14 – Main page of the application

2. The user selects **GET CREDENTIAL** again:

Figure 11.15 – Getting credentials

3. The user scans the generated QR code in the Microsoft Authenticator app. The user also has to make a note of the pin code that is displayed in the following QR code (pin code 0273 in the following screenshot):

Figure 11.16 – QR code for credential creation

4. When the credential has been created and stored on Microsoft Authenticator, the application confirms that the credential was issued successfully:

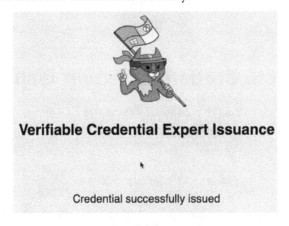

Figure 11.17 – Confirmation page

Now, the user can present the credential when challenged by an application. We are going to use the same application to verify credentials, as follows:

1. Back on the **Verifier Credential Expert Issuance and Verifier Sample** main page, the user selects **VERIFY CREDENTIAL**:

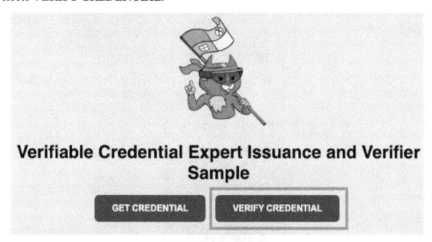

2. Then, the user selects **VERIFY CREDENTIAL** again:

Figure 11.18 – Verifying credential selection

3. The user scans the generated QR code in Microsoft Authenticator:

Figure 11.19 – Verification through QR code

4. Upon selecting to share the information from a verified credential, the application confirms the data is valid:

Figure 11.20 – Verification confirmation page

Identity management convergence is another emerging trend that is reshaping the landscape of IAM products, as it seeks to unify various identity management domains into a more cohesive, efficient, and user-friendly system. This trend has arisen in response to the growing complexity of managing digital identities across multiple platforms, devices, and applications, as well as the need for improved security and user experience.

Identity management convergence in ForgeRock Identity Cloud

In the world of business, managing identities has become an increasingly complex and vital task. It is essential to ensure the security and integrity of a company's data and systems while also providing an accessible and streamlined experience for users. This is where **converged identity management** comes in.

ForgeRock's vision is to provide one platform that can solve all of a company's identity challenges. This includes customer IAM, workforce identity management, and governance.

The benefits of this approach are many. It allows for increased agility and better value for customers. By having a pre-integrated platform, costs are reduced, and the user experience is improved. This is particularly important for companies adopting cloud technology and undergoing digital transformations.

Managing user lifecycles, providing access to apps, and governing that access when done on one common platform makes it simpler and more efficient for the customer.

ForgeRock's platform also uses AI to make better decisions and to make it simple for organizations to delegate out the administration of applications, including onboarding, assigning applications, and improving security over time.

ForgeRock Intelligent Authentication, which we used in *Chapter 5*, has been renamed **Intelligent Access**. One of the additions to Intelligent Access is **ForgeRock Go**, which is ForgeRock's implementation of the FIDO2 WebAuthn standard enabling users to register and authenticate without usernames and passwords.

ForgeRock Intelligent Access's key feature remains its drag-and-drop visual interface, which is used to create customizable user journeys called *trees* that are modular and orchestrated. The capability of ForgeRock trees has been enhanced with a larger selection of preconfigured options. Intelligent Access has introduced **device nodes,** which are contextually aware and can identify the devices that a user logs in from. Device nodes also store device profiles for authentication purposes without requiring passwords, match a device to a known location, set geo-fencing around a device, and alert the system if a device has been compromised. The device nodes available in ForgeRock version 7.2 can be seen in *Figure 11.21*:

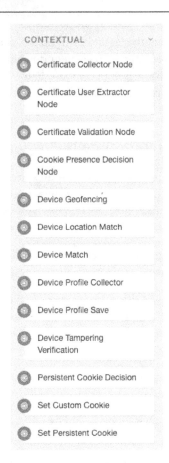

Figure 11.21 – ForgeRock's device nodes

Another enhancement, adding inner trees, allows for creating specific trees for certain actions that can be reused on other trees, similar to a function in programming languages.

As an example, the journey in *Figure 11.22* uses the inner tree in *Figure 11.23*:

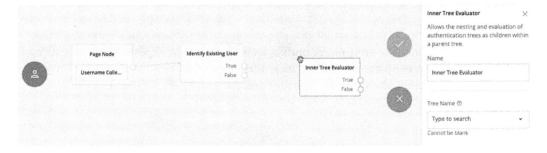

Figure 11.22 – Verification confirmation page

Here's the tree:

Figure 11.23 – ForgeRock's inner tree

More examples of the new capabilities of ForgeRock's Intelligent Access can be seen at https://www.forgerock.com/platform/access-management/intelligent-access?wvideo=qnk1616gdt.

Are passkeys (almost) the perfect phishing-resistant MFA?

Multi-device **Fast Identity Online** (**FIDO**) credentials, also known as passkeys, were developed to overcome the limitations of classic FIDO credentials and single-device credentials. This is achieved by enabling FIDO credentials to be used across multiple devices, which addresses both the recovery problem (by allowing credentials to be backed up and survive the loss of the original device) and the issue of requiring multiple enrollments (as there is no need to enroll on each device).

Despite the advantages of passkeys over passwords, which use public key cryptography to make them phishing resistant, their use has some nuances. While they improve traditional passwords, passkeys sacrifice some security properties if the starting point is classic FIDO2 platform authenticators. Additionally, some administrators prefer credentials to be bound to a specific device they control via **mobile device management** (**MDM**) so that they can be confident they meet security policies.

Passkey management

Implementing passkeys as platform authenticators can create confusion, particularly as passkey adoption grows. For example, for workforce authentication, administrators may not allow critical credentials belonging to high-level users to be backed up to devices outside of their knowledge and control.

Okta passkey management

Okta's **Passkeys Management** feature includes a **Block Passkeys for FIDO2 (WebAuthn) Authenticators** option that enables administrators to block new enrollments for passkeys across an organization.

Enabling this feature flag allows administrators to prevent users from enrolling with a multi-device FIDO credential such as passkeys. By doing so, potential risks associated with unmanaged and insecure devices accessing sensitive applications are preemptively mitigated.

To enable the **Block Passkeys for FIDO2 (WebAuthn) Authenticators** feature, select **Settings**, **Features**, and then switch **Block passkeys for FIDO2 (WebAuthn) Authenticators** to 1 (on):

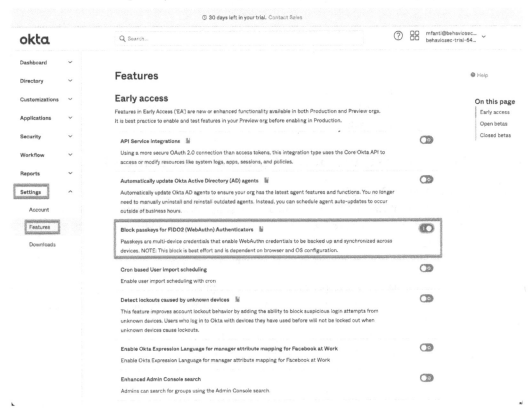

Figure 11.24 – Okta's Block passkeys for FIDO2 (WebAuthn) Authenticators feature

As can be seen in Okta's passkey blocking feature, such a process is *best effort* on the current implementation. Therefore, following the evolution of passkey support among the different authentication products is essential. As such, organizations should carefully evaluate the pros and cons of immediate passkey adoption and consider alternative authentication methods that meet their specific needs and requirements.

Continuous authentication

At the beginning of 2019, Atul Tulshibagwale, a Google software engineer, wrote a blog titled *Re-thinking federated identity with the Continuous Access Evaluation Protocol* (`https://cloud.google.com/blog/products/identity-security/re-thinking-federated-identity-with-the-continuous-access-evaluation-protocol`).

The problem described in the blog was the following:

In the current mobile and cloud computing era, enterprises often access multiple applications hosted on cloud and enterprise platforms using federation protocols. These login sessions can last for extended periods, especially on mobile devices but also on widely used applications such as Slack. However, determining whether to authorize a user session now requires dynamic data such as the user's device location, IP location, device and app health, and user privileges.

For instance, consider a user logged in to a cloud-based **customer relationship management** (**CRM**) application on their phone who then travels to China. The CRM provider must detect the user's new location and adjust their access accordingly.

Dynamic authorization decisions can benefit various scenarios, such as when a device is connected to a corporate **virtual private network** (**VPN**) but needs to be disconnected due to the presence of a malicious app. Similarly, if a file-sharing app detects that the user's IP address has changed, it needs to re-evaluate the user's access privilege based on their new IP location. Additionally, if a user is removed from a role, the app needs to be informed of the change to deny the user access to resources that are no longer allowed.

Dynamic authorization can be achieved using a variety of methods, including the following:

- **Geolocation**: This involves determining the user's location based on their IP address or mobile device's GPS coordinates
- **Device health**: This involves monitoring the device's health and security posture, such as whether it is jailbroken or rooted
- **User privileges**: This involves determining the user's privileges, such as their roles and permissions

Dynamic authorization can provide a number of benefits, including the following:

- **Increased security**: Dynamic authorization can help to prevent unauthorized access to sensitive data and applications

- **Improved user experience**: Dynamic authorization can make it easier for users to access the resources they need while still ensuring their security

- **Reduced costs**: Dynamic authorization can help to reduce the costs associated with manual authorization processes

CAEP is a new approach to cybersecurity that leverages a user's entire online footprint. It facilitates the exchange of events and signals between multiple parties to provide continuous monitoring and evaluation of a user's access privileges. This helps to identify potential risks and threats early on and to take steps to mitigate them.

Initially spearheaded by Google, CAEP has gained support from several key players in the identity and security industry, including Microsoft, Cisco, SailPoint, Amazon, Ping, and Thales. The protocol allows real-time communication between the **identity provider** (**IdP**) and relying parties, enabling better detection and response to security threats.

CAEP is based on the principles of continuous monitoring and dynamic authorization, which involves monitoring a user's activities in real time and adjusting access privileges based on the user's behavior and other contextual factors, such as the device being used, the user's location, and the time of day. This approach provides an additional layer of security beyond traditional static authorization, which is based on predefined rules and policies.

The exchange of events and signals in CAEP involves the continuous evaluation of a user's online footprint, including login history, device health, location, and other behavioral factors. This information is used to create a dynamic risk score for each access request, which can be used to make real-time authorization decisions. This approach ensures that users are granted access only when necessary and appropriate, reducing the risk of unauthorized access or account compromise.

Overall, CAEP represents a significant step forward in cybersecurity by providing a more dynamic and responsive approach to access management. By leveraging the collective expertise and resources of multiple industry leaders, CAEP is helping to establish a new standard for CAE and user protection.

CAE – Microsoft's version of CAEP

Microsoft customers have raised concerns regarding the delay in policy enforcement when conditions change for a user, such as the ones raised in the CAEP blog. In response, Azure AD has experimented with reducing token lifetimes, but it found that it could lead to poor user experiences and decreased reliability without eliminating risks. To address this, Azure AD and the relying party (enlightened app) needed a two-way conversation to respond to policy violations or security issues in a timely manner. This allows the relying party to observe changes in properties, such as network location, and inform Azure AD. It also provides Azure AD with a means of instructing the relying party to stop recognizing tokens for a specific user due to account compromise, disablement, or other issues. CAE enables this conversation. The goal is to achieve near real-time response for critical event evaluation, although latency of up to 15 minutes may be observed due to event propagation time. Nevertheless, IP location policy enforcement is instantaneous.

What lies ahead

Exciting times are ahead for the field of IAM! While we've already seen impressive trends and enhancements in this area, we can expect further evolution and advancements that will provide even better support against the latest attacks.

One of the most thrilling aspects of this evolution is the shift away from traditional password-based authentication toward phishing-resistant authentication mechanisms. IAM products are already paving the way toward more secure, modern authentication methods, and we can only expect this trend to continue.

We expect more companies to shift toward these modern authentication technologies, as traditional authentication methods are becoming increasingly vulnerable to attacks. This is especially true as phishing scams and other cyber-attacks become more sophisticated.

Summary

In this book, you learned about the journey of a fictitious software company, Acme Software, and how it chose authentication products and mechanisms to support its customers, workers, and partners while balancing security, risks, costs, and friction. In addition, the book provided step-by-step explanations of essential concepts, practical examples, and detailed hands-on implementations to help you understand the concepts and technologies of MFA.

In the book's middle section, you learned about implementing IAM products, with examples showing how to configure authorization for multiple applications using **single sign-on** (**SSO**). In addition, each chapter included a list of commercial products representative of the examples used to test the work done in each chapter.

The book demonstrated how Acme selected different products to use with its internal users, partners, and customers. Finally, the book provided instructions on where to obtain free trial versions, if available, of the products used in each example.

Toward the end of the book, you learned about different authentication methods and how to reduce the possibility of a bad actor using an account. The book guided you through selecting, deploying, and maintaining an MFA solution to reduce the risk of successful malicious attacks during user authentication.

The field of IAM is constantly evolving, and we can expect more exciting changes in the future. The shift toward more innovative, phishing-resistant authentication mechanisms is only the beginning of what promises to be an era of even greater security and convenience for users.

Appendix A
Installing the Java Software Development Kit

Before proceeding with some of the examples provided in this book, you must have Java installed on your computer. Java is a versatile, high-level programming language that allows developers to create applications that can run on almost any computer, from desktops to mobile devices. Java comes in several editions, with the **Java Standard Edition (Java SE)** being the most commonly used one. For this book, we will install the OpenJDK 11 version of the Java **Software Development Kit (SDK)**, which includes the **Java Runtime Environment (JRE)** necessary to run Java applications and the compilers and debuggers needed for development.

In the next sections, we will describe the process of installing the SDK on Windows and Mac manually, as well as using the Windows installer or Mac **Homebrew**. For Linux users, the process should be similar to installing manually on a Mac.

In this appendix, we will cover the following topics:

- Installing the Java SDK (and JRE) on Windows
- Installing the Java SDK (and JRE) on a Mac
- Testing the install of the Java SDK (with JRE)

Installing the Java SDK on Windows

There are many ways to install the Java SDK on Windows, as well as many places where the software can be downloaded from. We will use Microsoft's build of the OpenJDK software for the Windows installs (manual and using the installer).

Using the installer on Windows

The easiest option on Windows is to use the installer.

To download the software, perform the following steps:

1. Go to the download page for the Microsoft build of OpenJDK (`https://docs.microsoft.com/en-us/java/openjdk/download`) and select the OpenJDK 11, Windows x64, MSI version. Click on it to start downloading:

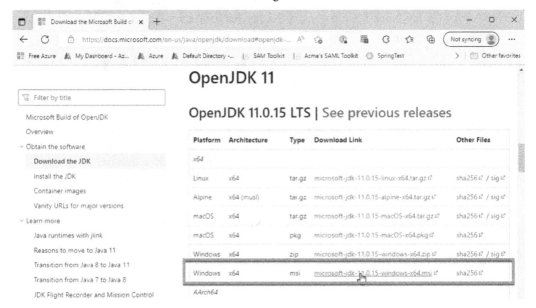

Figure A.1 – Download OpenJDK 11 installer for Windows

After the installer is downloaded, the next step is installing the software.

2. Click on **Open file**:

Figure A.2 – Open the installer

3. On the **Welcome** page, confirm the version of the install and click **Next**:

Figure A.3 – The Welcome page

4.   Accept the license agreement and click **Next**:

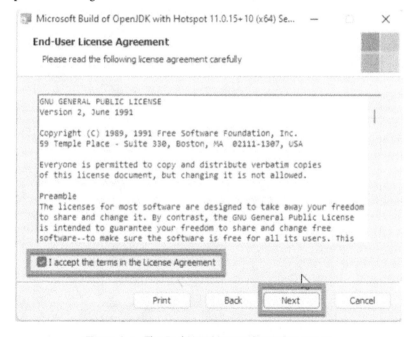

Figure A.4 – The End-User License Agreement page

5.   Make sure all options are selected and click **Next**:

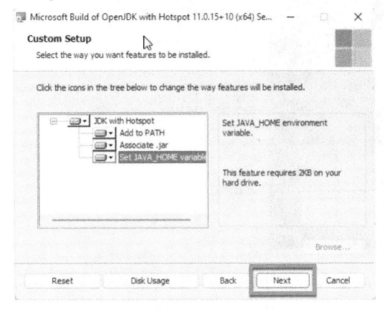

Figure A.5 – Custom setup page

6.   Click **Install**:

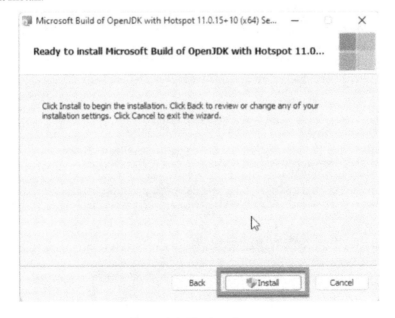

Figure A.6 – The install page

7. Click **Finish**:

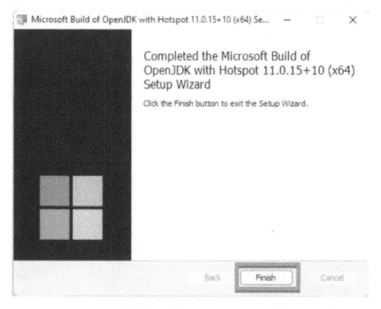

Figure A.7 – Installation complete

8. After the JDK is installed, delete the installation file (click on the trash can):

Figure A.8 – Deleting the installer file

This completes the installation. Go to the end of this appendix to test whether the installation completed successfully.

Installing the Java SDK manually on Windows

If you are encountering problems using the installer on Windows, or if your preferred method is to install manually, here are the steps to follow for a manual installation:

1. Go to the download page for the Microsoft build of the OpenJDK (`https://docs.microsoft.com/en-us/java/openjdk/download`) and select OpenJDK 11, Windows x64, the ZIP version:

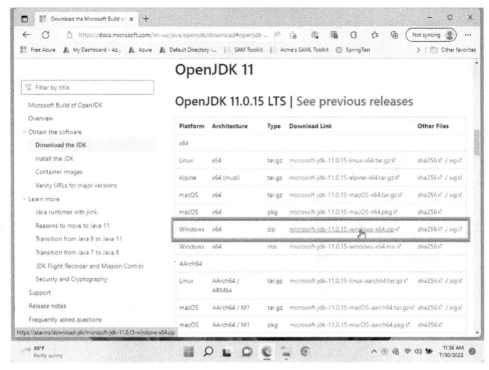

Figure A.9 – Download OpenJDK 11 ZIP file for Windows

After the installer is downloaded, the next step is installing and configuring the software.

2. In the `Downloads` folder, find the downloaded ZIP file and right-click to select the **Extract All…** option:

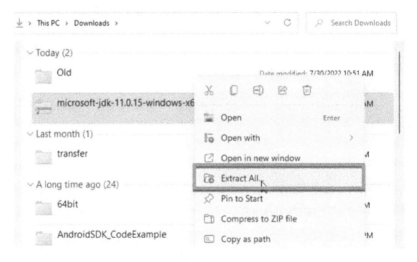

Figure A.10 – Extract All… files from the ZIP file

3. Select any directory to extract the files into. In this example, the files will be extracted to the
    `Downloads` folder:

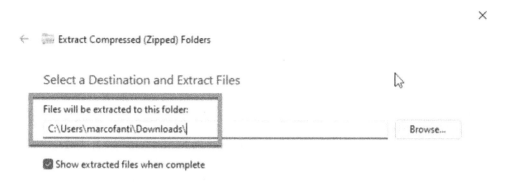

Figure A.11 – Select the destination for the extracted files

4. After the files are extracted, create a new subdirectory under `Program Files`. Select `Program
    Files`, right-click on it and select **+ New**, and then **Folder**. Name the new folder `OpenJDK`:

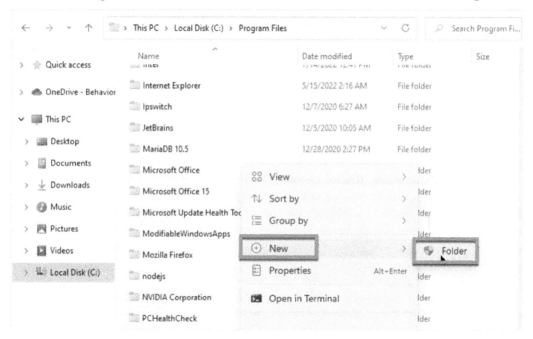

Figure A.12 – Create a new folder

5. In the `Downloads` folder, select the extracted `OpenJDK 11` folder, right-click, and select
    the **Cut** option to cut the folder:

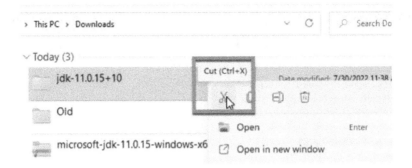

Figure A.13 – Cut the OpenJDK folder

6.   Select the OpenJDK folder created in *step 3*, right-click, and use the **Paste** option to paste the files into it:

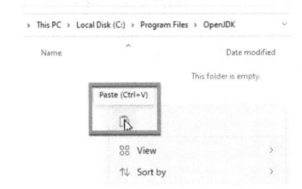

Figure A.14 – Paste the files into the OpenJDK folder

7.   Copy the full path of the extracted directory, including the \bin directory:

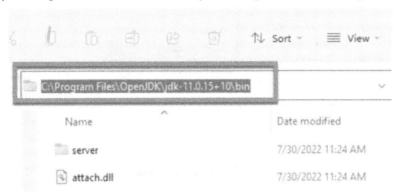

Figure A.15 – Copy the full path

8.   Using the search tool, type env and select **Edit the system environment variables**:

Figure A.16 – The Windows search tool

9. Once the **System Properties** panel opens, click on **Environment Variables**:

Figure A.17 – The System Properties panel

10. The **Environment Variables** panel will open. Select **Path** in the **System variables** sub-panel and click **Edit…**:

Figure A.18 – The Path system variable

11. To add a new entry to Path, click **New**:

Figure A.19 – The Path environment variables

12. Paste the value copied on *step 6* in the new entry and click **OK**:

Figure A.20 – The OpenJDK 11 bin directory added to Path

13. Click **New…** in the **System variables** sub-panel to create a new environment variable:

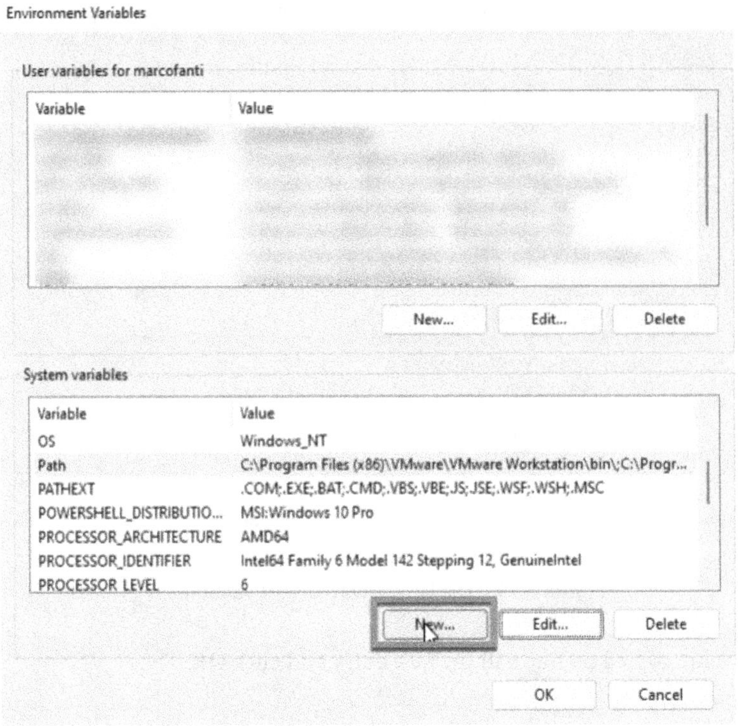

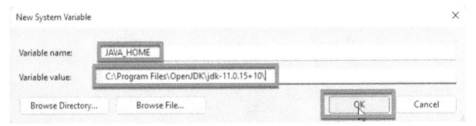

Figure A.21 – Deleting installer file

14. Enter JAVA_HOME as the **Variable name** value and paste the path from *step 6* again without the bin directory at the end as the variable value. Click **OK**:

Figure A.22 – The JAVA_HOME environment variable

This completes the manual installation and configuration of the SDK. Go to the end of this chapter to test whether all the values were configured correctly.

The next two sections describe how to install the Java SDK on a Mac. You can skip those sections if installing Java on a Mac is not required.

Installing the Java SDK on a Mac

There are also many ways to install the Java SDK on a Mac. One of the easiest ways is to use Homebrew. Homebrew is an open source software package management system that simplifies software for installation for Mac users.

Using Homebrew to install OpenJDK 11 on a Mac

OpenJDK 11 is available as a binary package and is supported for both Intel and Apple Silicon Macs.

To check the Homebrew formulae and install the JDK, perform the following steps:

1. Validate the version and supported platforms for OpenJDK 11 from the Homebrew page: `https://formulae.brew.sh/formula/openjdk@11`. Copy the **Install command** value from that page as well:

Figure A.23 – OpenJDK@11 Homebrew formulae

2. Run `brew install openjdk@11` on a Mac terminal:

Figure A.24 – Installing OpenJDK 11 using homebrew

3. Check whether the install completed successfully:

```
> brew install openjdk@11
==> Downloading https://ghcr.io/v2/homebrew/core/openjdk/11/manifests/11.0.16
################################################################## 100.0%
==> Downloading https://ghcr.io/v2/homebrew/core/openjdk/11/blobs/sha256:b0bbc489ad3d90cdb4
==> Downloading from https://pkg-containers.githubusercontent.com/ghcr1/blobs/sha256:b0bbc4
################################################################## 100.0%
==> Pouring openjdk@11--11.0.16.arm64_monterey.bottle.tar.gz              I
==> Caveats
For the system Java wrappers to find this JDK, symlink it with
  sudo ln -sfn /opt/homebrew/opt/openjdk@11/libexec/openjdk.jdk /Library/Java/JavaVirtualMa
chines/openjdk-11.jdk

openjdk@11 is keg-only, which means it was not symlinked into /opt/homebrew,
because this is an alternate version of another formula.

If you need to have openjdk@11 first in your PATH, run:
  echo 'export PATH="/opt/homebrew/opt/openjdk@11/bin:$PATH"' >> ~/.zshrc

For compilers to find openjdk@11 you may need to set:
  export CPPFLAGS="-I/opt/homebrew/opt/openjdk@11/include"

==> Summary
🍺  /opt/homebrew/Cellar/openjdk@11/11.0.16: 672 files, 299.5MB
==> Running `brew cleanup openjdk@11`...
Disable this behaviour by setting HOMEBREW_NO_INSTALL_CLEANUP.
Hide these hints with HOMEBREW_NO_ENV_HINTS (see `man brew`).
~ ▮                                                    ok  36s  02:34:46 PM
```

Figure A.25 – Homebrew post install commands

4. Homebrew will display a list of commands that need to be executed after the install. Follow the instructions to have openjdk@11 first in PATH:

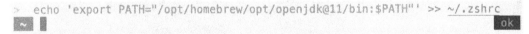

Figure A.26 – Adding OpenJDK to PATH

5. Use the same path to create the JAVA_HOME environment variable without /bin at the end:

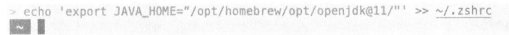

Figure A.27 – Adding JAVA_HOME

This completes the installation using Homebrew. Go to the end of this appendix to test whether the installation completed successfully.

Installing the Java SDK manually on a Mac (or Linux)

If encountering problems using Homebrew, or if the preferred method is to install manually, here are the steps to follow for a manual installation on a Mac (or Linux server):

1. Go to the download page for the Microsoft build of the OpenJDK (`https://docs.microsoft.com/en-us/java/openjdk/download`) and select the OpenJDK 11, macOS x64 (or M1), tar.gz version:

OpenJDK 11.0.15 LTS | See previous releases

Platform	Architecture	Type	Download Link	Other Files
x64				
Linux	x64	tar.gz	microsoft-jdk-11.0.15-linux-x64.tar.gz⬠	sha256⬠/ sig ⬠
Alpine	x64 (musl)	tar.gz	microsoft-jdk-11.0.15-alpine-x64.tar.gz ⬠	sha256⬠/ sig ⬠
macOS	x64	tar.gz	microsoft-jdk-11.0.15-macOS-x64.tar.gz ⬠	sha256⬠/ sig ⬠
macOS	x64	pkg	microsoft-jdk-11.0.15-macOS-x64.pkg ⬠	sha256⬠
Windows	x64	zip	microsoft-jdk-11.0.15-windows-x64.zip ⬠	sha256⬠/ sig ⬠
Windows	x64	msi	microsoft-jdk-11.0.15-windows-x64.msi ⬠	sha256⬠
AArch64				
Linux	AArch64 / ARM64	tar.gz	microsoft-jdk-11.0.15-linux-aarch64.tar.gz⬠	sha256⬠/ sig ⬠
macOS	AArch64 / M1	tar.gz	microsoft-jdk-11.0.15-macOS-aarch64.tar.gz ⬠	sha256⬠/ sig ⬠
macOS	AArch64 / M1	pkg	microsoft-jdk-11.0.15-macOS-aarch64.pkg ⬠	sha256⬠

Figure A.28 – Download the OpenJDK 11 ZIP file for Mac

After the installer is downloaded, the next step is installing and configuring the software.

2. In the `Downloads` folder, find the downloaded `tar.gz` file and extract its contents.

3. Run the `cd ~/Downloads` command.

4. Then, run the `tar xf microsoft-jdk-11.0.15-macos-*.tar.gz` command:

```
> cd ~/Downloads
> ls -ltr microsoft-jdk-11.0.15-*
-rw-r--r--@ 1 mfanti  staff  184482606 Jul 30 15:08 microsoft-jdk-11.0.15-macos-aarch64.tar.gz
> tar xf microsoft-jdk-11.0.15-macos-*.tar.gz
~/Downloads
```

Figure A.29 – Extract all files from the tar.gz file

5.   Move the JDK to the `/Library/Java/JavaVirtualMachines/` directory.

6.   Run `sudo mv jdk-11* /Library/Java/JavaVirtualMachines/:`

```
> sudo mv jdk-11* /Library/Java/JavaVirtualMachines/
Password:
~/Downloads
```

Figure A.30 – Move the JDK from the Downloads directory

7.   Check to see whether the version is installed on the system by running `/usr/libexec/java_home -V`:

```
> /usr/libexec/java_home -V
Matching Java Virtual Machines (2):
    11.0.16 (arm64) "Homebrew" - "OpenJDK 11.0.16" /opt/homebrew/Cellar/openjdk@11/11.0.16/libexec/openjdk.jdk/Contents/Home
    11.0.15 (arm64) "Microsoft" - "OpenJDK 11.0.15" /Library/Java/JavaVirtualMachines/jdk-11.0.15+10/Contents/Home
/opt/homebrew/Cellar/openjdk@11/11.0.16/libexec/openjdk.jdk/Contents/Home
~/Downloads
```

Figure A.31 – List the JDK versions installed on a Mac

8.   Copy the path for the Microsoft version of OpenJDK 11. Add that path to the beginning of your shell `PATH` environment variable, adding `/bin` to the end.

9.   Add a new environment variable to your shell named `JAVA_HOME` with this path:

```
> /usr/libexec/java_home -V
Matching Java Virtual Machines (2):
    11.0.16 (arm64) "Homebrew" - "OpenJDK 11.0.16"  /opt/homebrew/Cellar/openjdk@11/11.0.16/libexec/openjdk.jdk/Contents/Home
    11.0.15 (arm64) "Microsoft" - "OpenJDK 11.0.15"  /Library/Java/JavaVirtualMachines/jdk-11.0.15+10/Contents/Home
/opt/homebrew/Cellar/openjdk@11/11.0.16/libexec/ope......
> echo 'export PATH="/Library/Java/JavaVirtualMachines/jdk-11.0.15+10/Contents/Home/bin:$PATH"' >> ~/.zshrc
> echo 'export JAVA_HOME="/Library/Java/JavaVirtualMachines/jdk-11.0.15+10/Contents/Home/"' >> ~/.zshrc
~/Downloads
```

Figure A.32 – Configuring the new OpenJDK as the default

This completes the manual installation and configuration of the SDK on a Mac. Go to the end of this appendix to test whether all the values were configured correctly.

Testing the install of the Java JDK

To test whether the correct version is installed, try to run the Java executable from a command line using the version option. This will confirm Java is in the system path and also the version that is installed.

To test the install on Windows, perform the following step:

1. Open a new command line window to test the installation. On the command line, type `java --version` (two minus signs, like the picture):

```
Administrator: cmd                                             -   □   ×

C:\WINDOWS\system32>java --version
openjdk 11.0.15 2022-04-19 LTS
OpenJDK Runtime Environment Microsoft-32930 (build 11.0.15+10-LTS)
OpenJDK 64-Bit Server VM Microsoft-32930 (build 11.0.15+10-LTS, mixed mode)

C:\WINDOWS\system32>
```

Figure A.33 – Testing the Java version

To test the install on a Mac or Linux server, perform the following step:

1. The only difference in testing a Java installation on a Mac or Linux server is the use of the terminal instead of the command line window. Type `java --version` (two minus signs, like the picture) on a new terminal (not the one used for the install):

```
> java --version
openjdk 11.0.16 2022-07-19
OpenJDK Runtime Environment Homebrew (build 11.0.16+0)
OpenJDK 64-Bit Server VM Homebrew (build 11.0.16+0, mixed mode)
~
```

Figure A.34 – Testing the Java version on a Mac

You have now checked whether the correct version is installed.

Summary

The four initial sections in this appendix explained the process of installing the SDK on Windows and Mac. Windows installer or Mac Homebrew were used. The manual process was also described for both types of machines. For Linux users, the process used to install manually on a Mac can be used.

We also explained testing the installation of the Open JDK 11 on those systems.

Appendix B
Custom App Integration with Azure AD

Azure AD has a gallery with thousands of pre-integrated applications to make it easy to provide secure authentication and authorization solutions so that customers, partners, and employees can access the applications they need.

For custom applications, developers can use the Microsoft identity platform (`https://docs.microsoft.com/en-us/azure/active-directory/develop/`) to implement authentication and authorization. Applications integrated with the Microsoft identity platform natively take advantage of the same features provided by Azure AD for identity and security. Applications must register with Azure **Active Directory** (**AD**) (`https://docs.microsoft.com/en-us/azure/active-directory/develop/quickstart-register-app`) and are managed just like any other app.

Acme has several Java-based web applications that it wants to allow access to through Azure AD.

With Azure AD, developers can create applications tailored to specific use cases. Here are some of the different types of applications that you can integrate with Azure AD:

- **Single-page apps (SPAs)**: SPAs are web applications that load a single HTML page and dynamically update that page as the user interacts with the app. Azure AD provides an authentication library for JavaScript, simplifying integration with the Microsoft identity platform.

- **Web app that signs in users**: These applications manage user access using Azure AD. The Microsoft identity platform simplifies signing in users and requesting the necessary tokens.

- **Web app that calls web APIs**: This involves apps that need to access web APIs on behalf of the signed-in user. The Microsoft identity platform provides libraries and guides to handle these scenarios.

- **Protected web API**: Azure AD protects your web APIs using OAuth 2.0 protocols. These APIs can be called by other applications on behalf of a user or using application permissions.

- **Web API that calls web APIs**: This web API needs to call another web API. The Microsoft identity platform enables secure and managed access between these services.

- **Desktop app**: Desktop applications can use Azure AD to authenticate users, acquire tokens to call APIs protected by the Microsoft identity platform, or even access downstream APIs.

- **Daemon app**: These are background services or daemons that run behind the scenes. They may need to call web APIs and can do so using application permissions and application identities without user intervention.

- **Mobile app**: Azure AD simplifies integrating identity management into mobile applications, providing native **software development kits (SDKs)** for both Android and iOS.

This appendix describes adding web applications responsible for signing in users to Azure AD. These applications are not part of the pre-integrated applications found in the Azure AD gallery, nor are they developed using the Microsoft identity platform. Instead, they are custom Java-based web applications developed by Acme.

First, it's essential to understand that Azure AD provides robust, enterprise-grade identity and access management, even for applications not developed using Microsoft's tools or platforms. By integrating these Java-based web applications with Azure AD, Acme can provide its users with a **single sign-on (SSO)** experience and take advantage of Azure AD's security features.

We will cover the following topics in this appendix:

- Enabling SSO for custom web applications

Technical requirements

To build and deploy the application described here, the following is required:

- **Java Development Kit (JDK)** (refer to *Appendix A* for installation instructions to install Java JDK/SDK + **Java Runtime Environment (JRE)**) version 8 and above

- Maven

- **Integrated Development Environment (IDE)** such as **Visual Studio Code (VSCode)** or IntelliJ – optional

- Git (or download the source code from the code repository as a ZIP file)

Enabling SSO for custom web applications

Let's look at enabling SSO for non-gallery custom applications.

Add a non-gallery enterprise application

To add a non-gallery enterprise application, perform the following steps:

1. Go to the Azure AD admin center (`https://aad.portal.azure.com/`) and sign in using one of the following roles: Global Administrator, Cloud Application Administrator, or Application Administrator.

2. Select **Enterprise applications**:

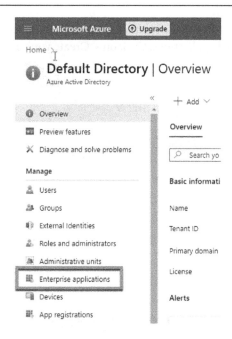

Figure B.1 – Azure AD admin center

3.   The **All applications** pane opens and displays a list of the applications in your Azure AD tenant. In the **Enterprise applications** pane, select + **New application**:

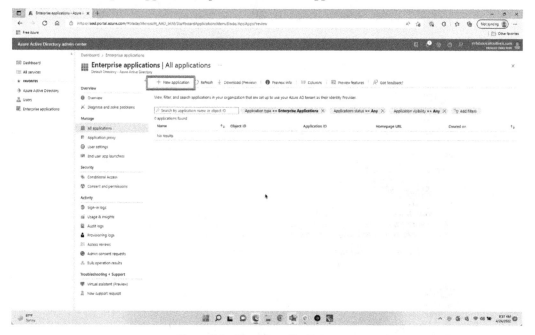

Figure B.2 – The All applications pane

4. The **Browse Azure AD Gallery** pane opens and displays tiles for cloud platforms, on-premises applications, and featured applications. Click on **+ Create your own application**:

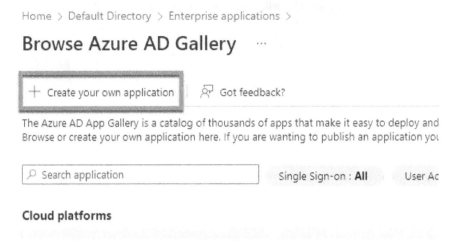

Figure B.3 – Create your own application

5. Enter a name that you want to use to recognize the instance of the application. For example, `SAML Springtest`. Select **Integrate any other application you don't find in the gallery (Non-gallery)**. Click **Create**:

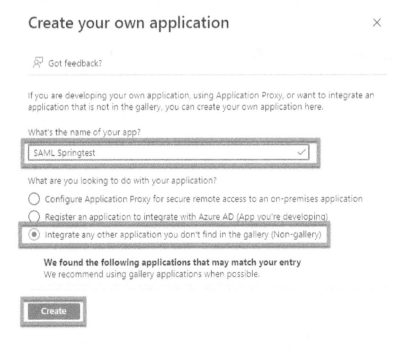

Figure B.4 – The SAML Springtest application

We have added a non-gallery enterprise application.

To simplify the instructions, I have created a Spring web app that supports sign-in by Azure AD account. The web app will redirect you to the Azure AD sign-in page when it's been accessed anonymously. You can download the Springtest application from GitHub (`https://github.com/marcofanti/saml-sso-and-slo-demo-idp-azure-sp-springboot`). You can also create your application following instructions from `https://learn.microsoft.com/en-us/azure/developer/java/spring-framework/configure-spring-boot-starter-java-app-with-azure-active-directory`.

Assign a user account to the SAML Springtest application

As discussed in *Chapter 3*, Acme does not have an identity and access management system that automatically assigns users to roles (and consequently to applications). We will assign one or more users manually to the new application.

To add a user to an application, perform the following steps:

1. Select **Enterprise applications**, and then search for and select **SAML Springtest**.
2. In the left pane, select **Users and groups**.
3. Select + **Add user/group**:

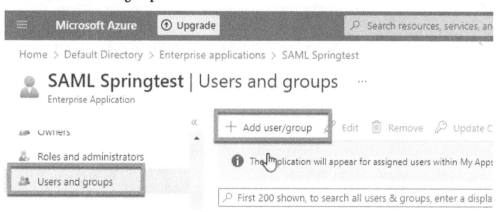

Figure B.5 – Adding users to an application

4. On the **Add Assignment** pane, select **None Selected** under **Users and groups**.
5. Search for and select the user that you want to assign to the application.
6. Click **Select**.

7.   On the **Add Assignment** pane, select **Assign** at the bottom of the pane:

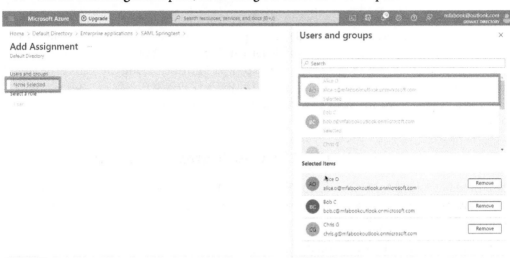

Figure B.6 – Selecting users for an application

We have added a user account to the SAML Springtest application.

Enabling SSO for Acme's Azure AD SAML Toolkit application

Users can sign in to an application using their Azure AD credentials only after enabling SSO access to those applications.

To allow SSO access to an application:

1.   Select **Enterprise applications**, then search for and select **SAML Springtest**.

2.   In the left pane, select **Single sign-on**.

3.   Select **SAML**:

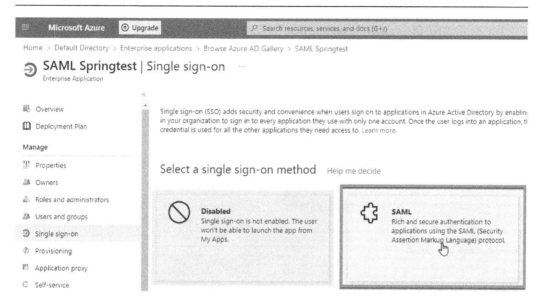

Figure B.7 – Acme's Azure AD SAML Toolkit Single sign-on pane

To begin the configuration of SSO in Azure AD, you must add sign-in and reply URL values and then download a certificate:

1. Select **Edit** in the **Basic SAML Configuration** section on the **Set up Single Sign-On with SAML** pane:

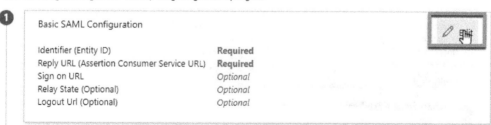

Figure B.8 – The Set up Single Sign-On with SAML pane

2. For **Identifier (Entity ID)**, enter com:acme:springsamltest (or some other unique value for this application).

3. For **Reply URL (Assertion Consumer Service URL)**, enter `https://localhost:8443/SAML/SSO`.

4. Then, select **Save**:

Figure B.9 – The Basic SAML Configuration section

5. In the **SAML Signing Certificate** section, select **Certificate (Raw)** to download the SAML signing certificate and save it to be used later:

Figure B.10 – Saving the SAML certificate

We have enabled SSO for the Azure AD SAML Toolkit application.

Configure SSO in the SAML Springtest app

Using SSO in the application requires you to configure and build the Java application with the appropriate values:

1. To start, clone (or download a ZIP file) the following repository `https://github.com/marcofanti/saml-sso-and-slo-demo-idp-azure-sp-springboot`:

Figure B.11 – Clone the repository

2. Next, rename the certificate file (`.cer`) downloaded in *step 5* from the previous section (if required), and copy it to the `src\main\resources\saml\` directory:

Figure B.12 – Raw certificate file – signature.cer

3. Create the keystores required by the application executing `build.cmd` (or `build.sh`). You can also follow the instructions in the `Readme.md` file provided in the repository:

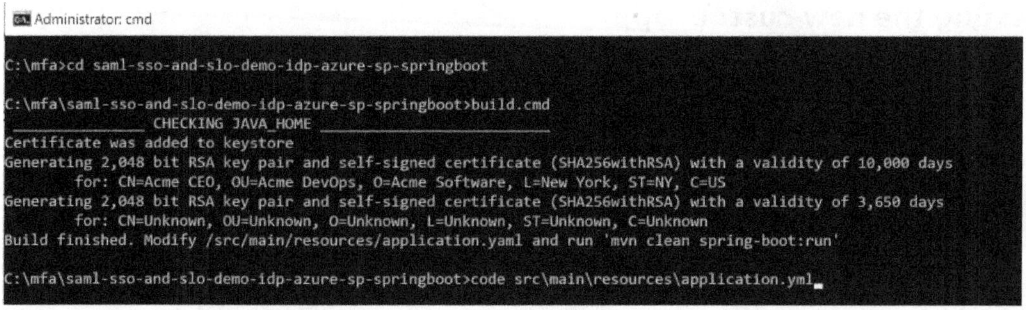

Figure B.13 – Create the keystores required by the SAML Springtest application

4. Copy `entity.id` and `metadata.URL` (also from *step 5* in the previous section) to the `src\main\resources\application.yml` file:

```
18
19   service.provider.entity.id: com:acme:springsamltest
20
21   idp.metadata.url: https://login.microsoftonline.com/9c81c6e0-c312-4a1a-a685-8cffc3fe78d0/federatio
22
23   # you can update credentials if you want, I recommend you to keep as it is for demo purpose
24   saml:
```

Figure B.14 – Copy configuration values to build the new application

5. Execute `mvn clean springboot:run`:

```
Administrator: cmd

C:\mfa>cd saml-sso-and-slo-demo-idp-azure-sp-springboot

C:\mfa\saml-sso-and-slo-demo-idp-azure-sp-springboot>build.cmd
_____ CHECKING JAVA_HOME _____
Certificate was added to keystore
Generating 2,048 bit RSA key pair and self-signed certificate (SHA256withRSA) with a validity of 10,000 days
        for: CN=Acme CEO, OU=Acme DevOps, O=Acme Software, L=New York, ST=NY, C=US
Generating 2,048 bit RSA key pair and self-signed certificate (SHA256withRSA) with a validity of 3,650 days
        for: CN=Unknown, OU=Unknown, O=Unknown, L=Unknown, ST=Unknown, C=Unknown
Build finished. Modify /src/main/resources/application.yaml and run 'mvn clean spring-boot:run'

C:\mfa\saml-sso-and-slo-demo-idp-azure-sp-springboot>code src\main\resources\application.yml

C:\mfa\saml-sso-and-slo-demo-idp-azure-sp-springboot>mvn clean spring-boot:run
```

Figure B.15 – Run the application

We have configured SSO in the SAML Springtest application.

Testing the new custom app

We are now ready to test signing on to the new app. To test SSO, follow these steps:

1. Open a new browser window in InPrivate or incognito mode and browse to the `https://localhost:8443/` URL. Click on **Advanced** and accept the certificate, if required:

Figure B.16 – Test the new application

2. Sign in to the SAML Springtest app using the Azure AD credentials of the user account that you assigned to the application:

Figure B.17 – User and password prompts

If everything was set up correctly, you should see the home page for the SAML Springtest app:

Figure B.18 – Successful login to Acme's application

We have tested the new custom app.

Testing SSO with Acme's Azure AD SAML Toolkit application and the Springtest app

Now let's try SSO. As the name implies, we will log in to one application, and when we try going to a second application, we won't be prompted to sign in again.

Let's start by logging in first to the Acme SAML Toolkit test app (which we configure with **multi-factor authentication (MFA)**):

1. Open a new browser window in InPrivate or incognito mode and browse to the `https://samltoolkit.azurewebsites.net/SAML/Login/9999` URL (replace `9999` with the values that correspond to your own instance).

2. Click **Log in**:

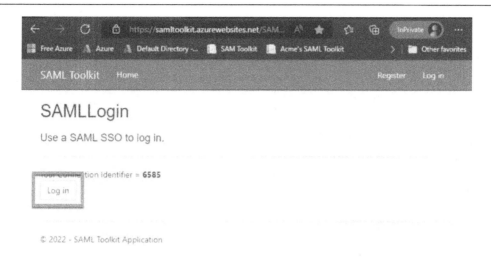

Figure B.19 – Acme's SAML Toolkit log in page

3. Sign in to the Acme Toolkit app using the Azure AD credentials of the user account that you
    assigned to the application (this should be an account also assigned to the Springtest app):

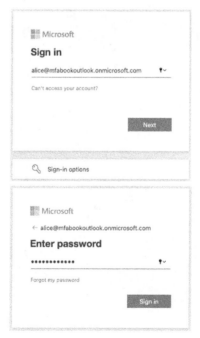

Figure B.20 – User and password prompts

4. Wait for the MFA prompt to appear:

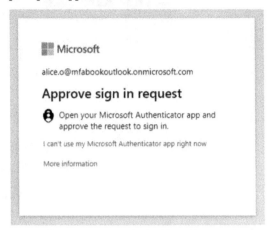

Figure B.21 – MFA prompt

5. Approve the sign-in request on the Authenticator app:

Figure B.22 – The Authenticator app

6. We are in Acme's Azure Toolkit app:

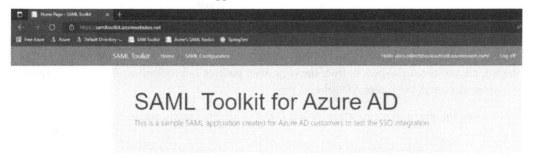

Figure B.23 – The SAML Toolkit app main page

7. Now go to https://localhost:8443/:

Figure B.24 – The Springtest app main page

We have tested SSO with the Azure AD SAML Toolkit application and the Springtest app.

Summary

The process described in this appendix explains how companies can deploy their own custom applications to Azure AD SSO. That, in turn, allows those custom applications to share the features provided by Azure AD for identity and security (including authentication and authorization). Access to those applications can also benefit from the same MFA policies and controls available for enterprise applications already in the Azure AD gallery.

We also tested signing in to the new app as well as SSO with other applications.

Appendix C

Installing Apache Tomcat Software

The **Apache Tomcat** software is an open source web server. Version 10 of the software implements parts of the Jakarta EE platform (`https://projects.eclipse.org/projects/ee4j.jakartaee-platform`), while version 9 and earlier implement parts of the Java EE specification.

In the following sections, we are going to describe the process of installing Apache Tomcat version 9. The manual installation is similar on Windows, Mac, and Linux.

Installing Apache Tomcat

Apache Tomcat 9, just like Java, has an installer for Windows. The manual installation distributions are the same for other systems.

Installing Apache Tomcat on Windows using the installer

Using the installer allows for the configuration of multiple copies of Apache Tomcat 9 on the same Windows machine. It also allows for the configuration of the ports each server is listening to during the installation. This is the preferred method of installation for this book if using Windows. You may need to be an Administrator to perform this installation. Otherwise, follow the manual process in the next section to install Tomcat:

1. Go to the download page for Apache Tomcat version 9 (`https://tomcat.apache.org/download-90.cgi`) and select the **32-bit/64-bit Windows Service Installer** version:

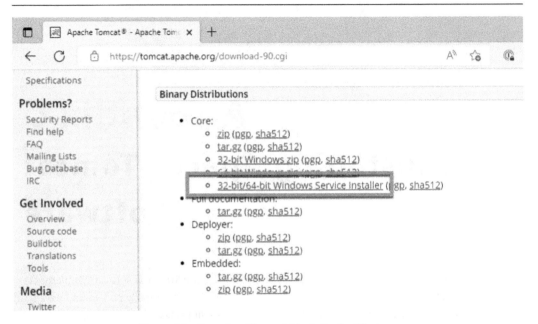

Figure C.1 – Download Tomcat 9 installer for Windows

After the installer is downloaded, the next step is installing and configuring the Apache Tomcat software.

2. Click on **Open file** to start the installer:

Figure C.2 – Starting the installer

3. On the **User Account Control** page, click **Yes** to allow the installer to proceed. This is required because the installer will add Tomcat as a service in Windows:

Figure C.3 – The User Account Control page

4. On the **Welcome** page, confirm the version of the install, and click **Next**:

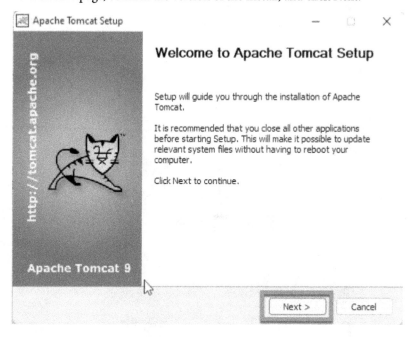

Figure C.4 – Welcome page

5. Accept the license agreement by clicking **I Agree**:

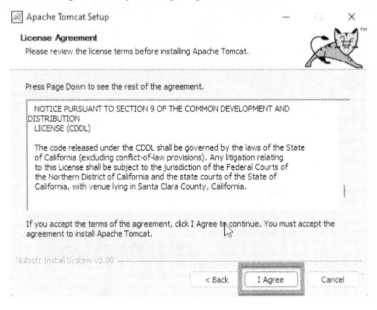

Figure C.5 – License agreement page

6. Make sure that Tomcat is selected and click **Next**:

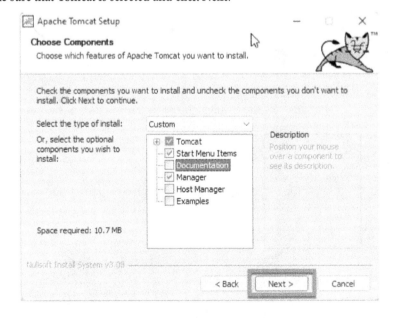

Figure C.6 – The Choose Components page

7.   Choose an unused port under **HTTP/1.1 Connector Port**. You can also create an optional administrator user to manage the Tomcat server. Click **Next**:

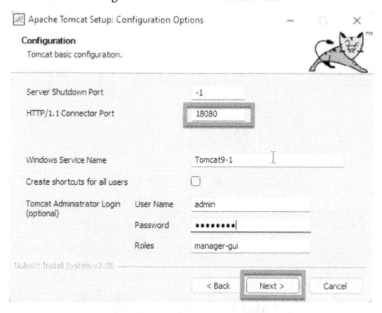

Figure C.7 – Configuration page

8.   Select the path for the Java Virtual Machine. The path for Apache Tomcat 9 should already be selected if the installation from *Appendix A* was completed. Click **Next**:

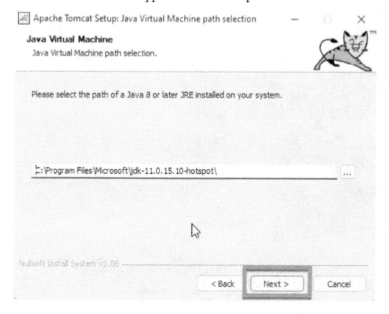

Figure C.8 – Java Virtual Machine selection page

9.   Select the path for the Tomcat installation location. Click **Install**:

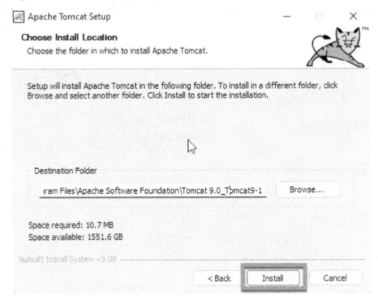

Figure C.9 – Install location page

10. Select **Run Apache Tomcat** and click **Finish**:

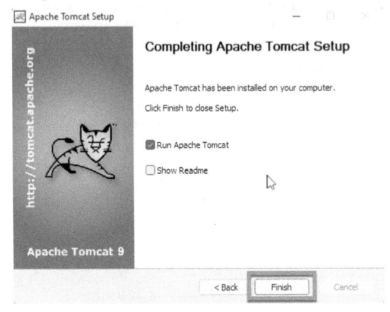

Figure C.10 – Finish installation

This completes the installation. To test whether Tomcat was installed correctly, open a browser with the localhost URL and the port selected during installation. In this example, it is `http://localhost:18080`:

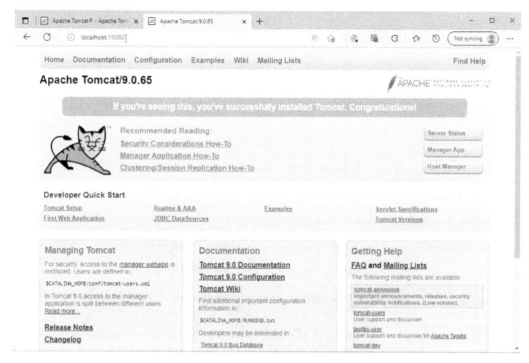

Figure C.11 – Apache Tomcat home page

In some scenarios, you may need more than one web server running on the same machine.

To install a second (or third) instance of Tomcat, repeat *steps 1-4* from the previous section:

1. Make sure **Tomcat** and **Examples** are selected and click **Next**:

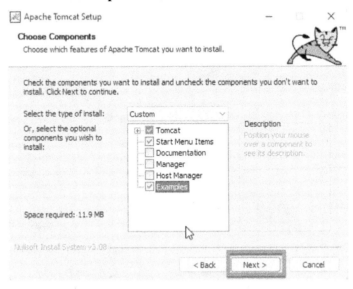

Figure C.12 – The Choose Components page

2. Choose an unused port under **HTTP/1.1 Connector Port** (different from the previous installations). Also, change the name of the service to something different from the previous installations. Click **Next**:

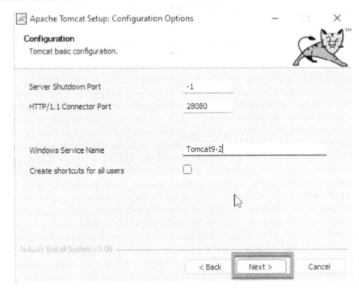

Figure C.13 – Configuration page

3.   Select the path for the Tomcat installation location. Click **Install**:

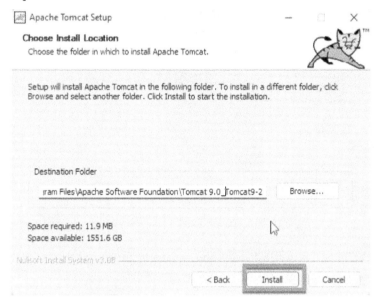

Figure C.14 – Choose Install Location page

4.   Select **Run Apache Tomcat** and click **Finish**:

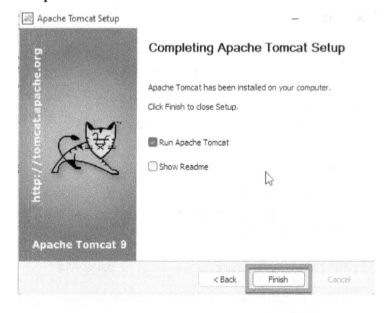

Figure C.15 – Finish installation

This completes the installation. As we selected to install examples, we will test whether the Tomcat examples were installed correctly, opening a browser with the example URL and port selected during the installation. In this example, the URL is `http://localhost:28080/examples`:

Apache Tomcat Examples

- Servlets examples
- JSP Examples
- WebSocket Examples

Figure C.16 – Apache Tomcat Examples page

5. After Apache Tomcat is installed, delete the installation file if no more instances are going to be installed (click on the trash can):

Figure C.17 – Deleting the installer file

This completes the installation. The next section describes how to install Apache Tomcat 9 on a Mac. You can skip the section if installing Java on a Mac is not required.

Installing Apache Tomcat 9 on a Mac

There are multiple ways to install Apache Tomcat 9 on a Mac. One of the easiest ways is to use Homebrew. **Homebrew** is an open source software package management system that simplifies software for installation for Mac users.

Using Homebrew to install Apache Tomcat 9 on a Mac

Apache Tomcat 9 is available as a binary package and is supported for both Intel and Apple Silicon Macs.

To check the Homebrew formulae and install the JDK, perform the following steps:

1. Validate the version and supported platforms for Apache Tomcat 9 on the Homebrew page: `https://formulae.brew.sh/formula/tomcat@9`. Copy the `install` command from that page as well:

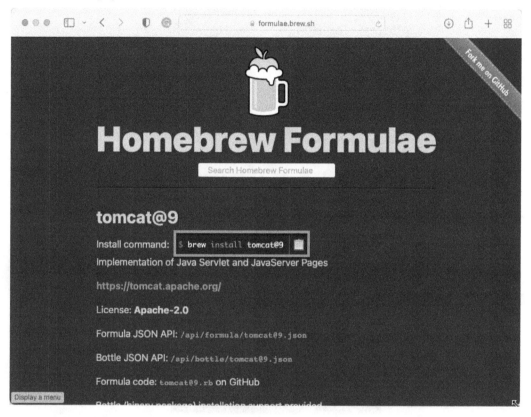

Figure C.18 – tomcat@9 homebrew formulae

2. Run `brew install tomcat@9` on a Mac terminal:

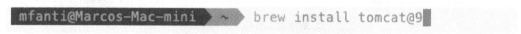

Figure C.19 – Installing Apache Tomcat 9 using homebrew

3.   Check whether the installation completed successfully. Copy the command to run Tomcat. Also, copy the path up to `tomcat@9`. That's `$CATALINA_HOME` for this installation:

```
3db260d88a6398fbf8d23b277--tomcat@9-9.0.65.bottle_manifest.json
==> Downloading https://ghcr.io/v2/homebrew/core/tomcat/9/blobs/sha256:491d0df379ee62963762be3d1b2b46df3e2e
Already downloaded: /Users/mfanti/Library/Caches/Homebrew/downloads/2158d20d7563b34f46ba18c5150c0aab508d736
2f5ccf2ce20d4e79e52a7c7b1--tomcat@9I-9.0.65.all.bottle.tar.gz
==> Pouring tomcat@9--9.0.65.all.bottle.tar.gz
==> Caveats
Configuration files: /opt/homebrew/etc/tomcat@9

tomcat@9 is keg-only, which means it was not symlinked into /opt/homebrew,
because this is an alternate version of another formula.

If you need to have tomcat@9 first in your PATH, run:
  echo 'export PATH="/opt/homebrew/opt/tomcat@9/bin:$PATH"' >> ~/.zshrc


To restart tomcat@9 after an upgrade:
  brew services restart tomcat@9
                                          ou can just run:
  /opt/homebrew/opt/tomcat@9/bin/catalina run

    /opt/homebrew/Cellar/tomcat@9/9.0.65: 628 files, 15.3MB
==> Running `brew cleanup tomcat@9`...
Disable this behaviour by setting HOMEBREW_NO_INSTALL_CLEANUP.
Hide these hints with HOMEBREW_NO_ENV_HINTS (see `man brew`).
mfanti@Marcos-Mac-mini                                        1340   13:33:14
```

Figure C.20 – Homebrew post-install commands

4.   Use your favorite editor to edit the `$CATALINA_HOME/libexec/conf/server.xml` file:

```
mfanti@Marcos-Mac-mini      code /opt/homebrew/opt/tomcat@9/libexec/conf/server.xml
```

Figure C.21 – Editing server.xml

5.   Look for `8005` (the `SHUTDOWN` port) and change it to `18005`:

```
<Server port="8005" shutdown="SHUTDOWN">
```

Figure C.22 – SHUTDOWN port

6.   Look for the connector ports `8080` and `8443` and change them to `18080` and `18443`, respectively:

```
<Connector port="8080" protocol="HTTP/1.1"
           connectionTimeout="20000"
           redirectPort="8443" />
```

Figure C.23 – Connector ports

7.   Use the command copied in *step 3* (`$CATALINA_HOME/bin/catalina run`) to start Tomcat:

```
~  /opt/homebrew/opt/tomcat@9/bin/catalina run
```

Figure C.24 – Starting Tomcat

This completes the installation. To test whether Tomcat was installed correctly, open a browser with the localhost URL and port selected during the installation. In this example, it is `http://localhost:18080`:

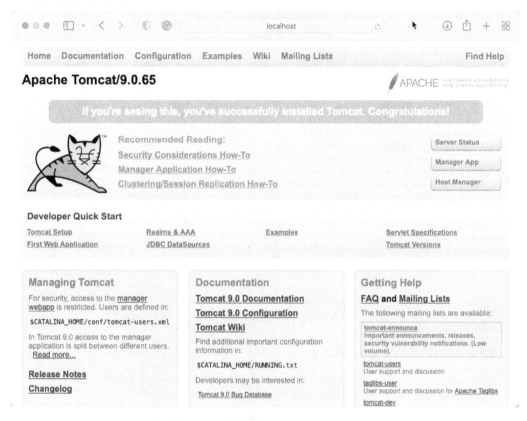

Figure C.25 – Apache Tomcat home page

This completes the installation using Homebrew. The next subsection describes how to install Apache Tomcat 9 manually on a Mac. You can also use the steps in the next subsection to install a second instance of Tomcat 9 on the same server.

Installing Apache Tomcat 9 manually on a Mac (or a Linux server)

Here are the steps to follow for a manual installation on a Mac (or a Linux server):

1. Go to the download page for Apache Tomcat version 9 (`https://tomcat.apache.org/download-90.cgi`) and select Apache Tomcat 9, the **Core Distribution** and the **zip** version:

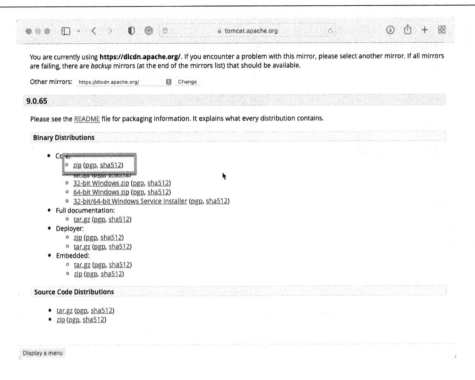

Figure C.26 – Downloading the Apache Tomcat 9 zip file for Mac

2. Unzip the file in the /tmp directory:

```
unzip ~/Downloads/apache-tomcat-9*.zip -d /tmp
```

Figure C.27 – Unzipping Apache Tomcat 9 in /tmp

3. Move the extracted folder into the location owned by the user that is going to run the server. That's $CATALINA_HOME for this installation:

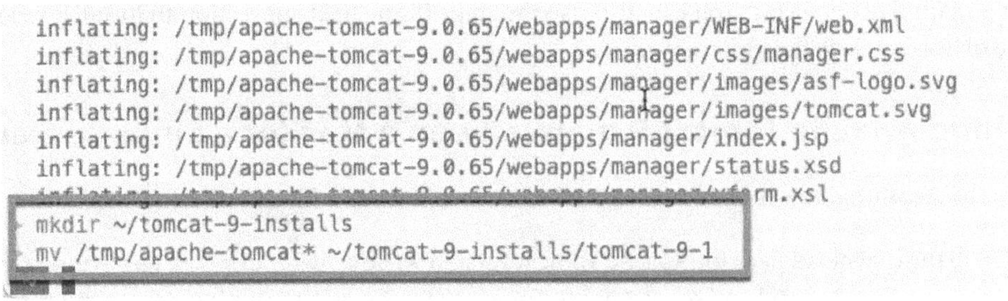

```
inflating: /tmp/apache-tomcat-9.0.65/webapps/manager/WEB-INF/web.xml
inflating: /tmp/apache-tomcat-9.0.65/webapps/manager/css/manager.css
inflating: /tmp/apache-tomcat-9.0.65/webapps/manager/images/asf-logo.svg
inflating: /tmp/apache-tomcat-9.0.65/webapps/manager/images/tomcat.svg
inflating: /tmp/apache-tomcat-9.0.65/webapps/manager/index.jsp
inflating: /tmp/apache-tomcat-9.0.65/webapps/manager/status.xsd
inflating: /tmp/apache-tomcat-9.0.65/webapps/manager/xform.xsl
mkdir ~/tomcat-9-installs
mv /tmp/apache-tomcat* ~/tomcat-9-installs/tomcat-9-1
```

Figure C.28 – Moving Tomcat to a new $CATALINA_HOME

4. Use your favorite editor to edit the $CATALINA_HOME/conf/server.xml file:

```
code ~/tomcat-9-installs/tomcat-9-1/conf/server.xml
```

Figure C.29 – Editing server.xml

5. Look for 8005 (the SHUTDOWN port) and change it to 28005. If that port is already in use, choose another unused port:

```
<Server port="8005" shutdown="SHUTDOWN">
```

Figure C.30 – SHUTDOWN port

6. Look for the connector ports 8080 and 8443 and change them to 28080 and 28443, respectively. If those ports are already in use, choose other unused ports instead:

```
<Connector port="8080" protocol="HTTP/1.1"
           connectionTimeout="20000"
           redirectPort="8443" />
```

Figure C.31 – Connector ports

7. Change the mode of the catalina.sh file in the bin directory to executable ($CATALINA_HOME/bin/catalina.sh):

```
> chmod +x ~/tomcat-9-installs/tomcat-9-1/bin/catalina.sh
```

Figure C.32 – Start Tomcat

8. Use catalina.sh ($CATALINA_HOME/bin/catalina.sh run) to start Tomcat:

```
~/tomcat-9-installs/tomcat-9-1/bin/catalina.sh run
```

Figure C.33 – Starting Tomcat

9. Use catalina.sh ($CATALINA_HOME/bin/catalina.sh run) to start Tomcat:

```
> ~/tomcat-9-installs/tomcat-9-1/bin/catalina.sh run
Using CATALINA_BASE:   /Users/mfanti/tomcat-9-installs/tomcat-9-1
Using CATALINA_HOME:   /Users/mfanti/tomcat-9-installs/tomcat-9-1
Using CATALINA_TMPDIR: /Users/mfanti/tomcat-9-installs/tomcat-9-1/temp
Using JRE_HOME:        /Users/mfanti/.sdkman/candidates/java/current
Using CLASSPATH:       /Users/mfanti/tomcat-9-installs/tomcat-9-1/bin/bootstrap.jar:/Users/mfanti/tom
cat-9-installs/tomcat-9-1/bin/tomcat-juli.jar
Using CATALINA_OPTS:
```

Figure C.34 – Confirming the correct $CATALINA_HOME

To test whether the Tomcat example was installed correctly, we can open a browser with the **examples** URL and port selected during the installation. In this example, it is `http://localhost:28080/examples`:

Apache Tomcat Examples

- Servlets examples
- JSP Examples
- WebSocket Examples

Figure C.35 – Apache Tomcat Examples page

With that, we have installed Apache Tomcat 9 on a Mac.

Summary

This appendix explained the process of installing the Apache Tomcat version 9 web server. The Windows installer or Homebrew on Mac was used. The manual process was also described for Macs. For Windows and Linux users, the process used to install manually on a Mac can also be followed.

We also explained how to test the installation of multiple instances of the web servers on those systems.

Index

packtpub.com

Subscribe to our online digital library for full access to over 7,000 books and videos, as well as industry leading tools to help you plan your personal development and advance your career. For more information, please visit our website.

Why subscribe?

- Spend less time learning and more time coding with practical eBooks and Videos from over 4,000 industry professionals

- Improve your learning with Skill Plans built especially for you

- Get a free eBook or video every month

- Fully searchable for easy access to vital information

- Copy and paste, print, and bookmark content

Did you know that Packt offers eBook versions of every book published, with PDF and ePub files available? You can upgrade to the eBook version at packtpub.com and as a print book customer, you are entitled to a discount on the eBook copy. Get in touch with us at customercare@packtpub.com for more details.

At www.packtpub.com, you can also read a collection of free technical articles, sign up for a range of free newsletters, and receive exclusive discounts and offers on Packt books and eBooks.

Other Books You May Enjoy

If you enjoyed this book, you may be interested in these other books by Packt:

Keycloak - Identity and Access Management for Modern Applications

Stian Thorgersen, Pedro Igor Silva

ISBN: 978-1-80056-249-3

- Understand how to install, configure, and manage Keycloak
- Secure your new and existing applications with Keycloak
- Gain a basic understanding of OAuth 2.0 and OpenID Connect
- Understand how to configure Keycloak to make it ready for production use
- Discover how to leverage additional features and how to customize Keycloak to fit your needs
- Get to grips with securing Keycloak servers and protecting applications

Cloud Auditing Best Practices

Shinesa Cambric, Michael Ratemo

ISBN: 978-1-80324-377-1

- Understand the cloud shared responsibility and role of an IT auditor
- Explore change management and integrate it with DevSecOps processes
- Understand the value of performing cloud control assessments
- Learn tips and tricks to perform an advanced and effective auditing program
- Enhance visibility by monitoring and assessing cloud environments
- Examine IAM, network, infrastructure, and logging controls
- Use policy and compliance automation with tools such as Terraform

Packt is searching for authors like you

If you're interested in becoming an author for Packt, please visit `authors.packtpub.com` and apply today. We have worked with thousands of developers and tech professionals, just like you, to help them share their insight with the global tech community. You can make a general application, apply for a specific hot topic that we are recruiting an author for, or submit your own idea.

Share Your Thoughts

Now you've finished *Implementing Multifactor Authentication*, we'd love to hear your thoughts! Scan the QR code below to go straight to the Amazon review page for this book and share your feedback or leave a review on the site that you purchased it from.

https://packt.link/r/1803246960

Your review is important to us and the tech community and will help us make sure we're delivering excellent quality content.

Download a free PDF copy of this book

Thanks for purchasing this book!

Do you like to read on the go but are unable to carry your print books everywhere?

Is your eBook purchase not compatible with the device of your choice?

Don't worry, now with every Packt book you get a DRM-free PDF version of that book at no cost.

Read anywhere, any place, on any device. Search, copy, and paste code from your favorite technical books directly into your application.

The perks don't stop there, you can get exclusive access to discounts, newsletters, and great free content in your inbox daily

Follow these simple steps to get the benefits:

1. Scan the QR code or visit the link below

https://packt.link/free-ebook/9781803246963

2. Submit your proof of purchase
3. That's it! We'll send your free PDF and other benefits to your email directly

www.ingramcontent.com/pod-product-compliance
Lightning Source LLC
Chambersburg PA
CBHW081451050326
40690CB00015B/2762